Springer Texts in Statistics

Advisors:
George Casella   Stephen Fienberg   Ingram Olkin

Springer
New York
Berlin
Heidelberg
Barcelona
Hong Kong
London
Milan
Paris
Singapore
Tokyo

Springer Texts in Statistics

(continued after index)

V.G. Kulkarni

Modeling, Analysis, Design, and Control of Stochastic Systems

With 23 Illustrations

 Springer

V.G. Kulkarni
Department of Operations Research
University of North Carolina
Chapel Hill, NC 27599
USA

Library of Congress Cataloging-in-Publication Data
Kulkarni, Vidyadhar G.
   Modeling, analysis, design, and control of stochastic systems /
V.G. Kulkarni.
      p.   cm. — (Springer texts in statistics)
   Includes bibliographical references and index.
   ISBN 0-387-98725-8 (hardcover : alk. paper)
   1. Stochastic processes.   I. Title.   II. Series.
QA274.K84   1999
519.2′3—dc21   98-52791

Printed on acid-free paper.

Production managed by Jenny Wolkowicki; manufacturing supervised by Nancy Wu.
Photocomposed pages prepared from the author's Plain TeX files.
Printed and bound by R.R. Donnelley and Sons, Harrisonburg, VA.
Printed in the United States of America.

9 8 7 6 5 4 3 2 (Corrected second printing, 2000)

ISBN 0-387-98725-8   SPIN 10782824

Springer-Verlag   New York Berlin Heidelberg
A member of BertelsmannSpringer Science+Business Media GmbH

About This Book

Who Is This Book for?

This book is meant to be used as a textbook in a junior or senior level undergraduate course in stochastic models. Students are expected to be undergraduate students in engineering, operations research, computer science, mathematics, statistics, business administration, public policy, or any other discipline with a mathematical core.

Students are expected to be familiar with elementary matrix operations (additions, multiplications, solving systems of linear equations; but not eigenvalues, eigenvectors), first-year calculus (derivatives and integrals of simple functions; but not differential equations), and probability (which is reviewed in Chapters 1 to 4 of this book).

What Is the Philosophy of This Book?

As the title suggests, this book addresses four aspects of using stochastic methodology to study real systems.

(1) **Modeling.** The first step is to understand how a real system operates, and what is the purpose of studying it. This enables us to make assumptions to create a model that is simple yet sufficiently true to the real system so that the answers provided by the model will have some credibility. In this book this step is emphasized repeatedly with the use of a large number of real life modeling examples.

(2) **Analysis.** The second step is to do a careful analysis of the model and compute the answers. To facilitate this step the book develops special classes of stochastic processes in Chapters 5, 6, and 7: discrete-time Markov chains, continuous time Markov chains, renewal processes, cumulative processes, semi-Markov processes, etc. For each of these classes, we develop tools to compute the transient distributions, limiting distributions,

cost evaluations, first passage times, etc. These tools generally involve matrix computations, and can be done easily in any matrix oriented language, e.g., MATLAB. Chapter 8 applies these tools to queueing systems.

(3) **Design.** In practice, a system is described by a small number of parameters, and we are interested in setting the values of these parameters so as to optimize the performance of the system. This is called "designing" a system. The performance of the system can be computed as a function of the system parameters using the tools developed here. Then the appropriate parameter values can be determined to minimize or maximize this function. This is illustrated by several examples in Chapter 9.

(4) **Control.** In some applications, the system can be controlled dynamically. Thus instead of finding optimal parameters as in the design aspect, the aim here is to find an optimal operating policy. Chapter 10 shows how this can be done using linear programming.

How Is This Book Intended to Be Used?

Typically, the book will be used in a one-semester course in stochastic models. Students taking this course will be expected to have a background in probability. Hence, Chapters 1 through 4 should be used to revise the material. Chapters 5, 6, and 7 should be covered completely. Chapters 8, 9, and 10 should be covered as time permits. Furthermore, Chapter 10 requires the availability of a linear program solver.

The book may also be used in two-quarter sequence of courses in probability and stochastic processes. In this case, we would cover Chapters 1 through 5 in the first quarter, and Chapters 6 through 10 in the second quarter.

There are many running examples in this book. Hence the instructor should try to use them in that spirit. Similarly, there are many running problems in the problem sections. The instructor may wish to use a running series of problems for homeworks.

What Is So Different About This Book?

This book requires a new mindset: a numerical answer to a problem is as valid as an algebraic answer to a problem! (More and more, students already have it; it is the teachers who are lagging!) Since computational power is now conveniently and cheaply available, the emphasis in this book is on using the computer to obtain numerical answers, rather than restricting ourselves to analytically tractable examples. This emphasis is not so obvious in the chapters on probability, but is very clear in the remaining chapters.

There are several consequences of this new mindset: the discussion of the transient analysis of stochastic processes is no longer minimized. Indeed, transient analysis is just as easy as the limiting analysis when done on a computer. Second, the problems at the end of the chapters are designed to be fairly easy, but may require the use of computers to do numerical experimentations.

Software to Accompany This Book

A software package called MAXIM is available for use with this textbook. It is a collection of over 80 programs written in MATLAB. These programs can be used directly as function files from MATLAB. The user's manual describing these programs is in a file called "readme". These programs can be accessed via a Graphical User Interface provided in a software package called MAXIMGUI. The GUI is designed to run on PCs with Windows 95 or above. Since the software is an evolving organism, I have decided not to include any information about it in this book, for fear that it will very soon become outdated. PC users can download the zip files containing the MAXIM and the MAXIMGUI software from ftp://ftp.mathworks.com/pub/books/kulkarni/MAXIM.zip or ftp://ftp.mathworks.com/pub/books/kulkarni/MAXIMGUI.zip. UNIX users can download the tar files containing the MAXIM and the MAXIMGUI software from ftp://ftp.mathworks.com/pub/books/kulkarni/MAXIM.tar or ftp://ftp.mathworks.com/pub/books/kulkarni/MAXIMGUI.tar. The user will need to have MATLAB installed on his/her machine in order to use the software. More up-to-date information is also available on the author's home page at www.unc.edu/~vkulkarn.

University of North Carolina
Chapel Hill, NC

V.G. Kulkarni

March, 1999.

Acknowledgments

This book is the end result of teaching an undergraduate course on stochastic models over the last several years. This means that several batches of undergraduate students have helped in the evolution of this material. I thank them all (they are too numerous to mention by name) for their willingness to be subjected to a book under preparation, and for their help in improving it.

I wish to thank Shau-Shiang Ja and Wendy Sense for their invaluable help in developing the graphical user interface for the MAXIM software package. I also wish to thank Tugrul Sanli for his immense help in developing the solutions manual.

Last, but not the least, I thank my wife Radhika, and my three sons, Milind, Ashwin, and Arvind, for their patience and understanding while I was working on this book.

Contents

1

Probability

1.1. Probability Model

Random phenomena are common in natural as well as in man-made systems. Weather is a natural random phenomenon, while the stock market is a man-made random phenomenon. Study of such random phenomena is vital if we are to understand their behavior, and predict (and possibly control) their future. The subject of probability provides us with a mathematical tool to do this.

The basic step in studying a random phenomenon (sometimes called a random experiment) is to build a probability model to describe it. Such a probability model has three basic components:

(1) sample space;

(2) events; and

(3) probability of events.

We discuss them one by one and explain with the help of examples.

1.2. Sample Space

Definition 1.1 (Sample Space, Ω). A sample space is the set of all possible outcomes of a random phenomenon.

A sample space is generally denoted by the Greek letter Ω (omega). Elements of Ω, generally denoted by ω, are called the *outcomes* or *sample points*.

Example 1.1 (Coin Tossing). Consider a random experiment of tossing a coin. Assuming that the coin will never stand on its edge, there are two possible outcomes (sample points): Head (H) and Tail (T). Hence the sample space for this random experiment is

$$\Omega = \{H, T\}.$$ □

Example 1.2 (Clinical Trial). Consider a clinical experiment for testing a new experimental drug for stomach ulcers. The outcome of the experiment is recorded as a Success (S) or a Failure (F). Note that these terms will have to be defined precisely so as to avoid ambiguity. Thus the sample points are S and F and the sample space is

$$\Omega = \{S, F\}.$$ □

Example 1.3 (Two Clinical Trials). Suppose the above experimental drug is administered to two patients. Then the outcome for the two trials together is a pair (x, y), where x is the outcome of the trial for the first patient, and y is the outcome for the second patient. Thus the sample space is as given below:

$$\Omega = \{(S, S), (S, F), (F, S), (F, F)\}.$$ □

Example 1.4 (Die Tossing). Suppose we toss a six-sided die and observe its outcome. The sample space for this experiment is given by

$$\Omega = \{1, 2, 3, 4, 5, 6\}.$$ □

Example 1.5 (Telephony). Suppose we are observing the utilization of a particular 800 service. We do this by keeping track of the number of calls received during every hour. Suppose we are interested in the number of calls received during 1:00 p.m. to 2:00 p.m. on Monday. This number can be any nonnegative integer. Hence the sample space is given by

$$\Omega = \{0, 1, 2, 3, \ldots\}.$$ □

Example 1.6 (Coin Tossing). Suppose we repeatedly toss a coin until a head is observed. The sample space can be written as

$$\Omega = \{H, TH, TTH, TTTH, \ldots\}.$$

Note that we can also write the sample space as

$$\Omega = \{1, 2, 3, \ldots\}.$$

The number k in the last sample space represents the outcome in which a string of $k - 1$ tails was followed by a head. □

Example 1.7 (Rainfall). Suppose we are studying the amount of rainfall at a particular site. In particular we measure the amount of rainfall (in inches) recorded in a given calendar year. The sample space for this random phenomenon is given by

$$\Omega = [0, \infty).$$ □

Example 1.8 (Temperature). Suppose we are also studying the temperature variation at the site in the above example. We do this by keeping track of the maximum and minimum temperatures recorded during the calendar year. A typical outcome of this random phenomenon is a pair of numbers (x, y), where x is the maximum and y is the minimum temperature. Hence the sample space is given by

$$\Omega = \{(x, y) : -\infty < y \le x < \infty\}.$$

The above examples show that sample spaces can be very complicated sets depending on the nature of the random phenomenon we are interested in.

1.3. Events

The second component of the probability model of a random phenomena is a collection of events. In studying a random phenomenon we are interested in the occurrence or non-occurrence of "events." It is important to distinguish between the sample points and the events. A formal definition of an event is given below:

Definition 1.2 (Event). An event is a subset of the sample space.

Since Ω is a subset of Ω, we see that Ω is an event. It is called the *universal event*. Similarly the null set $\emptyset$ is a subset of Ω. Hence it is also an event, called the *null event*. Thus the universal event always takes place, while the null event never takes place.

Example 1.9 (Clinical Trial). For Example 1.2, there are at most four events as listed below:

$$\{\emptyset, \{S\}, \{F\}, \Omega\}.$$

Example 1.10 (Finite Ω). Consider a random phenomenon with a finite sample space

$$\Omega = \{1, 2, 3, \ldots, N\}.$$

There are 2^N possible events, including Ω and $\emptyset$, for this sample space.

Example 1.11 (Die Tossing). Consider Example 1.4. The event $E =$ "get an even number on the die" is given by the subset

$$E = \{2, 4, 6\}.$$

Example 1.12 (Two Clinical Trials). Consider Example 1.3. Let E_i be the event that we observe i successes in the two trials ($0 \le i \le 2$). Then $E_0 = \{(F, F)\}$, $E_1 = \{(S, F), (F, S)\}$, and $E_2 = \{(S, S)\}$.

Since events are by definition subsets of Ω, do the usual set operations have any meaning in event terminology? Indeed! This correspondence is given in Table 1.1 (here $E, E_1, E_2, \ldots$ are events).

TABLE 1.1. Correspondence between event and set-theoretic terminology.

Event Description	Set-Theoretic Notation
E_1 or E_2	$E_1 \cup E_2$
E_1 and E_2	$E_1 \cap E_2$ or $E_1 E_2$
Not E	E^c or $\bar{E}$
At least one of $E_1, E_2, \ldots$	$\bigcup_{n=1}^{\infty} E_n$
All of $E_1, E_2, \ldots$	$\bigcap_{n=1}^{\infty} E_n$

Events $E_1, E_2, \ldots$ are said to be *exhaustive* if at least one of the events always takes place. Mathematically, this is equivalent to saying that

$$\bigcup_{n=1}^{\infty} E_n = \Omega.$$

Similarly, events $E_1, E_2, \ldots$ are said to be *mutually exclusive* or *disjoint* if at most one of the events can take place. This is equivalent to saying

$$E_i \cap E_j = \emptyset, \text{ if } i \neq j.$$

Thus when $E_1, E_2, \ldots$ are mutually exclusive and exhaustive they define a *partition* of Ω, i.e., each sample point $\omega \in \Omega$ belongs to one and only one E_n.

Example 1.13 (Die Tossing). Consider Example 1.4. Let E_1 be the event that we get an even number, and let E_2 be the event that we get an odd number on the die. Then $E_1 = \{2, 4, 6\}$ and $E_2 = \{1, 3, 5\}$ are mutually exclusive and exhaustive. Let E_3 denote that we get a number larger than 3 on the die. Then $E_3 = \{4, 5, 6\}$. The event "E_1 or E_3" is given by $\{2, 4, 5, 6\}$. The event "E_2 and E_3" is given by $\{5\}$.   □

1.4.   Probability of Events

The third component of a probability model is the probability function. Intuitively, the probability function associates a numerical value to each event that describes the likelihood of observing the event. A formal definition is given below.

Definition 1.3 (Probability of Events, P). Probability of an event E, written $P(E)$, is a number representing the likelihood of occurrence of the event E.

Of course, the P function cannot be arbitrary. It must satisfy certain requirements if it is to provide a consistent description of the random phenomenon. These requirements are listed below:

Axioms of Probability.

(1) $0 \le P(E) \le 1$.

(2) $P(\Omega) = 1$.

(3) If $E_1, E_2, \ldots$ are disjoint then

$$P\left(\bigcup_{n=1}^{\infty} E_n\right) = \sum_{n=1}^{\infty} P(E_n).$$

Intuitively, an event that occurs with certainty is assigned probability 1, while an event that never occurs is assigned probability 0. Hence the probabilities of all events must be bounded below by 0 and above by 1. This is Axiom 1. Since Ω is the set of all possible outcomes, one of them must occur by definition. Hence the event Ω must be assigned probability 1. This is Axiom 2. To explain Axiom 3, consider two disjoint events E_1 and E_2. Since $E_1 \cap E_2 = \emptyset$, the number of cases in which event $E_1 \cup E_2$ can occur is the sum of the number of cases in which E_1 can occur and those in which E_2 can occur. This intuitively suggests that the probability of $E_1 \cup E_2$ should be the sum of the probability of E_1 and that of E_2. Axiom 3 states that this must hold for even a countable number of disjoint events. (Sometimes this is called the *axiom of countable additivity*.) A simple consequence of Axiom 3 is that

$$P(\emptyset) = 0.$$

Example 1.14 (Coin Tossing). Consider Example 1.1. Suppose the coin is "fair." Intuitively, this means that the probability of getting H or T is the same. Thus the probability function reflecting the above belief is

$$P(\emptyset) = 0, \qquad P(\{H\}) = P(\{T\}) = 0.5, \qquad P(\{H, T\}) = 1.$$

Note that this satisfies all the axioms of probability.   □

Example 1.15 (Equally Likely Outcomes). Consider a probability model with finite sample space $\Omega = \{1, 2, \ldots, N\}$ as in Example 1.10. Suppose each of these N outcomes is equally likely. Let $E_i = \{i\}$, for $1 \le i \le N$. Thus we have

$$P(E_i) = \frac{1}{N}.$$

Let $m(E)$ be the number of outcomes that result in an event $E \subseteq \Omega$. Then the axioms of probability imply that we must have

$$P(E) = \frac{m(E)}{N}.$$

Thus in probability models with equally likely outcomes we can compute the probability of an event as the ratio of the number of favorable outcomes to the total number of outcomes.   □

Example 1.16 (Random Pairing). Suppose n couples are at a party. The host wants to play a game in which n men are to be randomly paired with n women. What is the probability that the ith woman gets paired with her husband?

The sample space Ω for this problem can be taken to be the set of all permutations of the integers $\{1, 2, \ldots, n\}$. For example, a permutation $(3, 1, 2, \ldots)$ represents the pairing of man number 3 with woman number 1, man number 1 with woman number 2, etc. Let E_i be the event that the ith woman gets paired with her husband. Then E_i is the set of all permutations with integer i in the ith position. There are $(n-1)!$ such permutations. Since all permutations are equally likely (the pairing is assumed to be random), we have

$$P(E_i) = \frac{m(E_i)}{m(\Omega)} = \frac{(n-1)!}{n!} = \frac{1}{n}.$$

What is the probability that women i and j get paired with their respective husbands ($i \neq j$)? Here we are interested in the event $E_i E_j$, which consists of all permutations with number i in position i and number j in position j. There are $(n-2)!$ such permutations. Hence we get

$$P(E_i E_j) = \frac{m(E_i E_j)}{m(\Omega)} = \frac{(n-2)!}{n!} = \frac{1}{n(n-1)}. \qquad \square$$

Next we study several important consequences of the axioms of probability.

Theorem 1.1.

$$P(E^c) = 1 - P(E). \qquad (1.1)$$

Proof. E and E^c are mutually exclusive and exhaustive events. Hence

$$1 = P(\Omega) = P(E \cup E^c) = P(E) + P(E^c).$$

Equation (1.1) follows from this. $\qquad \square$

Example 1.17 (Star Trek). There are exactly 100 episodes of the original "Star Trek" series. A TV station airs a randomly picked episode every Sunday night at 9:00 p.m. A trekkie student watches them religiously every week. What is the probability that the student watches at least one episode more than once during a 16-week semester?

Let Ω be the set of all possible sequences of 16 episodes from the 100, allowing repetitions. Let E be the event of interest. Then E^c is the event that each of the 16 episodes are distinct. There are $100!/(100-16)! = (100)(99)(98)\cdots(85)$ sequences of 16 distinct episodes. Hence

$$\begin{aligned}
P(E) &= 1 - P(E^c) \\
&= 1 - \frac{m(E^c)}{m(\Omega)} \\
&= 1 - \frac{100!/(100-16)!}{100^{16}} \\
&= 1 - .2816 = .7184.
\end{aligned}$$

Note that computing $m(E^c)$ is easy, while computing $m(E)$ is difficult in this problem. Hence Theorem 1.1 is useful here. $\qquad\square$

Theorem 1.2. *Let E and F be two events, not necessarily disjoint. Then*

$$P(E \cup F) = P(E) + P(F) - P(EF). \qquad (1.2)$$

Proof. The events EF^c, E^cF, and EF are mutually exclusive and

$$E = EF^c \cup EF,$$

$$F = E^cF \cup EF,$$

and

$$E \cup F = EF^c \cup E^cF \cup EF.$$

Hence from Axiom 3, we get

$$P(E) = P(EF^c) + P(EF),$$

$$P(F) = P(E^cF) + P(EF),$$

and

$$P(E \cup F) = P(EF^c) + P(E^cF) + P(EF).$$

The theorem follows from these equations. $\qquad\square$

We illustrate the usefulness of the above theorem by means of two examples.

Example 1.18 (Die Tossing). Consider the events E_2 and E_3 of Example 1.13. Assume the die is fair, i.e., all outcomes are equally likely. Then

$$P(E_2 \cup E_3) = P(E_2) + P(E_3) - P(E_2E_3)$$
$$= \tfrac{3}{6} + \tfrac{3}{6} - \tfrac{1}{6} = \tfrac{5}{6}. \qquad\square$$

Example 1.19 (Random Pairing). Consider the events E_i and E_j of Example 1.16. We can compute the probability that either woman i or woman j gets paired with her own husband as

$$P(E_i \cup E_j) = P(E_i) + P(E_j) - P(E_iE_j)$$
$$= \frac{1}{n} + \frac{1}{n} - \frac{1}{n(n-1)} = \frac{2n-3}{n(n-1)}.$$

Note that it would be difficult to compute this probability by directly computing the number of permutations in which either woman i or woman j gets paired with her own husband. $\qquad\square$

The next theorem gives the extension of Theorem 1.2 to the union of n events.

Theorem 1.3 (Inclusion–Exclusion Principle). *Let $E_i, 1 \leq i \leq n$, be n events, not necessarily disjoint. Then*

$$P\left(\bigcup_{i=1}^{n} E_i\right) = \sum_{i=1}^{n} P(E_i)$$
$$- \sum_{i<j} P(E_i E_j)$$
$$+ \sum_{i<j<k} P(E_i E_j E_k)$$
$$\vdots$$
$$+ (-1)^{n+1} P\left(\bigcap_{i=1}^{n} E_i\right). \tag{1.3}$$

Proof. The proof follows by induction on n. We illustrate with the case of $n = 3$. Think of $E_2 \cup E_3$ as one event and apply Theorem 1.2 to get

$$
\begin{aligned}
P(E_1 \cup E_2 \cup E_3) &= P(E_1) + P(E_2 \cup E_3) - P(E_1 \cap (E_2 \cup E_3)) \\
&= P(E_1) + P(E_2) + P(E_3) - P(E_2 E_3) - P(E_1 E_2 \cup E_1 E_3) \\
&= P(E_1) + P(E_2) + P(E_3) - P(E_2 E_3) \\
&\quad - (P(E_1 E_2) + P(E_1 E_3) - P(E_1 E_2 E_1 E_3)) \\
&= P(E_1) + P(E_2) + P(E_3) - P(E_2 E_3) - P(E_1 E_2) \\
&\quad - P(E_1 E_3) + P(E_1 E_2 E_3). \qquad \square
\end{aligned}
$$

Example 1.20 (Random Pairing). Consider Example 1.16. We want to compute the probability that at least one woman gets paired with her own husband. This probability is given by $P(\cup_{i=1}^{n} E_i)$, which can be evaluated by using (1.3). From Example 1.16 we see that

$$P(E_i) = \frac{1}{n} \quad \text{for} \quad 1 \leq i \leq n.$$

There are n terms like this in the first sum on the right-hand side of (1.3). Similarly, from Example 1.16 we see that

$$P(E_i E_j) = \frac{1}{n(n-1)} \quad \text{for} \quad 1 \leq i < j \leq n,$$

and there are $n(n-1)/2!$ terms like this in the second sum. By a similar argument we get

$$P(E_i E_j E_k) = \frac{1}{n(n-1)(n-2)} \quad \text{for} \quad 1 \leq i < j < k \leq n,$$

and there are $n(n-1)(n-2)/3!$ terms like this in the third sum. Proceeding this way we get the last term as

$$P\left(\bigcap_{i=1}^{n} E_i\right) = \frac{1}{n!}.$$

Substituting in (1.3) we get

$$P\left(\bigcup_{i=1}^{n} E_i\right) = \sum_{i=1}^{n} \frac{1}{n}$$

$$- \sum_{i<j} \frac{1}{n(n-1)}$$

$$+ \sum_{i<j<k} \frac{1}{n(n-1)(n-2)}$$

$$\vdots$$

$$+ (-1)^{n+1} \frac{1}{n!}$$

$$= 1 - \frac{1}{2!} + \frac{1}{3!} - \cdots + (-1)^{n+1} \frac{1}{n!}.$$

As $n \to \infty$ the above probability converges to

$$1 - \frac{1}{2!} + \frac{1}{3!} - \cdots = 1 - e^{-1}.$$

This is somewhat surprising, since we would have naively guessed that the probability of at least one correct pairing should approach 1 as the number of couples goes to infinity. □

1.5. Conditional Probability

In the previous section we saw how to set up a probability model of a random phenomenon, and how it helps us compute probabilities of complicated events of interest. In practice, we frequently face a situation where we have partial knowledge about the phenomenon, and are interested in computing the probabilities of events assuming this knowledge. We need the concept of conditional probabilities to handle such a situation. It is formally defined below:

Definition 1.4 (Conditional Probability). The conditional probability of an event E, given that an event F has occurred, is denoted by $P(E|F)$ and is given by

$$P(E|F) = \frac{P(EF)}{P(F)} \tag{1.4}$$

assuming $P(F) > 0$. □

Three remarks are in order about the above definition. First, if $P(F) = 0$, then $P(EF)$ is also 0 and the right-hand side of (1.4) reduces to 0/0, an undefined quantity. Practically speaking, if $P(F) = 0$, the event F will never take place, and we won't need to talk about $P(E|F)$. Second, $E|F$ does not imply that event E takes place after event F. There is no chronological order implied. The probability of E, given that event F is going to take place in the future, is also given by (1.4). Third, we can write (1.4) as

$$P(EF) = P(E|F)P(F).\tag{1.5}$$

This equation is valid even if $P(F) = 0$.

Example 1.21 (Die Tossing). Consider the events E_2 and E_3 of Example 1.13. We have

$$P(E_2|E_3) = \frac{P(E_2E_3)}{P(E_3)} = \frac{P(\{5\})}{P(\{4, 5, 6\})} = \frac{1/6}{3/6} = \frac{1}{3}. \qquad \square$$

Example 1.22 (Random Pairing). Consider Example 1.16. Given that woman 2 has been paired with her own husband, what is the probability that woman 1 is also paired with her own husband?

We want $P(E_1|E_2)$. Using the results of Example 1.16 we get

$$P(E_1|E_2) = \frac{P(E_1E_2)}{P(E_2)} = \frac{1/(n(n-1))}{1/n} = \frac{1}{n-1}. \qquad \square$$

An immediate extension of (1.5) is

$$P(E_1E_2 \ldots E_n) = P(E_n|E_1E_2 \ldots E_{n-1})P(E_{n-1}|E_1 \ldots E_{n-2}) \ldots P(E_2|E_1)P(E_1).\tag{1.6}$$

This can be verified by direct substitution using (1.5).

Example 1.23 (Urn Models). Consider an urn that contains r red balls and b blue balls. An experiment involves picking a ball at random from the urn, and replacing it with a balls of the same color. Suppose this experiment is repeated three times. Compute the probability that a red ball is drawn in each of these experiments.

Let E_i be the event that the ith experiment yields a red ball, $1 \le i \le 3$. Then we want to compute $P(E_1E_2E_3)$. We have

$$P(E_1) = \frac{r}{r+b},$$

$$P(E_2|E_1) = \frac{r+a}{r+a+b},$$

$$P(E_3|E_1E_2) = \frac{r+2a}{r+2a+b}.$$

Substituting in (1.6) we get

$$P(E_1E_2E_3) = \frac{r(r+a)(r+2a)}{(r+b)(r+a+b)(r+2a+b)}. \qquad \square$$

It is important to note that $P(\cdot|F)$ satisfies the axioms of probability, i.e.,

(1) $0 \le P(E|F) \le 1$, for all events E.

(2) $P(\Omega|F) = 1$.

(3) If $E_1, E_2, \ldots$ are disjoint, then

$$P\left(\bigcup_{n=1}^{\infty} E_n | F\right) = \sum_{n=1}^{\infty} P(E_n | F).$$

Example 1.24 (Playing Cards). A standard deck of playing cards contains 52 cards divided evenly among four suits: spades, clubs, diamonds, and hearts. Spades and clubs are black, diamonds and hearts are red. Suppose two cards are dealt at random from this deck. Given that both are black, what is the probability that at least one is a spade?

Let E be the event that both cards are black, and let E_i be the event that the ith card is a spade. We want to compute $P(E_1 \cup E_2 | E)$. We have

$$P(E) = \frac{26}{52} \cdot \frac{25}{51},$$

$$P(E_1 E) = \frac{13}{52} \cdot \frac{25}{51} = P(E_2 E),$$

$$P(E_1 E_2 E) = \frac{13}{52} \cdot \frac{12}{51}.$$

Then

$$\begin{aligned}
P(E_1 \cup E_2 | E) &= P(E_1 | E) + P(E_2 | E) - P(E_1 E_2 | E) \\
&= \frac{P(E_1 E)}{P(E)} + \frac{P(E_2 E)}{P(E)} - \frac{P(E_1 E_2 E)}{P(E)} \\
&= \frac{13 \cdot 25}{26 \cdot 25} + \frac{13 \cdot 25}{26 \cdot 25} - \frac{13 \cdot 12}{26 \cdot 25} \\
&= \frac{13 \cdot 38}{26 \cdot 25} = .76.
\end{aligned}$$

$\square$

1.6.   Law of Total Probability

The concept of conditional probability introduced in the previous section yields a very useful tool for computing probabilities of complicated events. It is given in the following theorem:

Theorem 1.4 (Law of Total Probability). *Let $E_1, E_2, E_3, \ldots$ be a set of mutually exclusive and exhaustive events. Then for an event E*

$$P(E) = \sum_{n=1}^{\infty} P(E | E_n) P(E_n). \tag{1.7}$$

Proof. Since $E_n, n \geq 1$, are mutually exclusive, $E \cap E_n, n \geq 1$, are disjoint, and since they are exhaustive

$$E = \bigcup_{n=1}^{\infty} E E_n.$$

Hence, from the axiom of countable additivity, and (1.5) we get

$$P(E) = P\left(\bigcup_{n=1}^{\infty} E E_n\right)$$

$$= \sum_{n=1}^{\infty} P(E E_n)$$

$$= \sum_{n=1}^{\infty} P(E|E_n)P(E_n)$$

which proves the theorem.   □

The above theorem is useful when the probabilities $P(E_n)$ and conditional probabilities $P(E|E_n)$ are easy to compute. As a special case, using $E_1 = F$ and $E_2 = F^c$ as two mutually exclusive and exhaustive events, we get

$$P(E) = P(E|F)P(F) + P(E|F^c)P(F^c). \tag{1.8}$$

We illustrate with two examples.

Example 1.25 (Urn Models). Consider Example 1.23. What is the probability that a red ball is drawn on the second experiment?

Let E_i be as in Example 1.23. We want $P(E_2)$. We have

$$P(E_1) = \frac{r}{r+b}, \qquad P(E_1^c) = \frac{b}{r+b},$$

and

$$P(E_2|E_1) = \frac{r+a}{r+a+b}, \qquad P(E_2|E_1^c) = \frac{r}{r+a+b}.$$

Using (1.8) we get

$$P(E_2) = P(E_2|E_1)P(E_1) + P(E_2|E_1^c)P(E_1^c)$$

$$= \frac{r+a}{r+a+b} \cdot \frac{r}{r+b} + \frac{r}{r+a+b} \cdot \frac{b}{r+b}$$

$$= \frac{r}{r+b}. \tag{1.9}$$

Can we compute $P(E_3)$ the same way? Conditioning on E_1, we get

$$P(E_3) = P(E_3|E_1)P(E_1) + P(E_3|E_1^c)P(E_1^c). \tag{1.10}$$

Now, $P(E_3|E_1)$ is the same as the probability of getting a red ball on the *second* experiment using an urn with $r + a$ red balls and b blue balls. Similarly, $P(E_3|E_1^c)$ is the same as the probability of getting a red ball on the *second* experiment using an urn with r red balls and $b + a$ blue balls. Hence from the preceding computations we get

$$P(E_3|E_1) = \frac{r+a}{r+a+b}, \qquad P(E_3|E_1^c) = \frac{r}{r+a+b}.$$

Substituting in (1.10) we get

$$P(E_3) = \frac{r+a}{r+a+b} \cdot \frac{r}{r+b} + \frac{r}{r+a+b} \cdot \frac{b}{r+b} = \frac{r}{r+b}.$$

Proceeding in this fashion we can show that

$$P(E_n) = \frac{r}{r+b}$$

for all $n \geq 1$!   □

Example 1.26 (Random Coin Tossing). An urn contains n coins. When the ith coin is tossed, it turns up heads with probability p_i, $1 \leq i \leq n$. Suppose a coin is picked at random from the urn and tossed. What is the probability of getting a head?

Let E_i be the event that the ith coin is picked, $1 \leq i \leq n$, and let E be the event that it lands heads up. We have

$$P(E_i) = \frac{1}{n}, \qquad P(E|E_i) = p_i, \qquad 1 \leq i \leq n.$$

Then, using (1.7), we get

$$P(E) = \sum_{i=1}^{n} p_i \left(\frac{1}{n}\right) = \frac{1}{n} \sum_{i=1}^{n} p_i.$$

□

1.7. Bayes' Rule

Let $E_1, E_2, \ldots$ be a set of mutually exclusive and exhaustive events. Suppose we know $P(E_i)$, $i \geq 1$. Let E be another event, and suppose we also know $P(E|E_i)$, $i \geq 1$. The event E is observable, while the events E_i are not. We want to compute $P(E_i|E)$. In other words, we want to revise our estimate of $P(E_i)$ given that event E has occurred. The result in the following theorem, called Bayes' rule, helps us do precisely this:

Theorem 1.5 (Bayes' Theorem). *Let $E_1, E_2, \ldots$ be a set of mutually exclusive and exhaustive events, and let E be another event. Then*

$$P(E_i|E) = \frac{P(E|E_i)P(E_i)}{\sum_{n=1}^{\infty} P(E|E_n)P(E_n)}. \tag{1.11}$$

Proof. From the law of total probability we get

$$P(E) = \sum_{n=1}^{\infty} P(E|E_n)P(E_n),$$

and by (1.5) we get

$$P(E_i E) = P(E|E_i)P(E_i).$$

Substituting in

$$P(E_i|E) = \frac{P(E_i E)}{P(E)}$$

we get (1.11).   □

We illustrate with several examples below.

Example 1.27 (Random Coin Tossing). Consider Example 1.26. Suppose a coin is picked randomly from the urn, and we do not know which coin it is. What is the probability that it is the ith coin given that it landed heads?

Using the results of Example 1.26 and (1.11) we get

$$P(E_i|E) = \frac{P(E_i E)}{P(E)} = \frac{p_i}{\sum_{r=1}^{n} p_r}.$$ □

Example 1.28 (AIDS). A new AIDS blood test has been proposed to test potential clients by an insurance company. Clinical trials have been done for the test using a controlled environment and they have produced the following estimates: if a person has AIDS the test is 99% successful in identifying it; if a person does not have AIDS the test says so 99.5% of the time. The proponents of the test are claiming that these numbers show that the test is highly effective. Is the test reliable enough for the insurance company to deny AIDS coverage to those clients who test positive by citing the "pre-existing conditions" clause?

In effect, we are asking for the probability that a person actually has AIDS given that the test says so. Let us say that if this probability is close to 1 (say at least .99) the test is a reliable indicator of the presence of AIDS. Let E_1 be the event that a person has AIDS, while E_2 is the event that he (or she) does not. Let E be the event that the test is positive. We want to compute $P(E_1|E)$. Suppose it is estimated that one million persons from a population of 250 million are infected with AIDS. We have

$$P(E_1) = 1/250, \qquad P(E_2) = 249/250,$$

and

$$P(E|E_1) = .99, \qquad P(E|E_2) = .005.$$

Then using Bayes' rule we get

$$P(E_1|E) = \frac{P(E|E_1)P(E_1)}{P(E|E_1)P(E_1) + P(E|E_2)P(E_2)}$$
$$= \frac{.99(1/250)}{.99(1/250) + .005(249/250)}$$
$$= \frac{.99}{2.24} = .4419.$$

Thus there is only a 44% chance that a person actually has AIDS given that the test is positive! This is far too small to be taken as a reliable indicator.   □

1.8. Independence

The next important concept in probability is that of independence. Intuitively, we say that two events are independent if information about one does not affect the likelihood of the occurrence of the other. This is made precise in the definition below:

Definition 1.5 (Independent Events). Events E and F are said to be independent of each other if

$$P(EF) = P(E)P(F). \tag{1.12}$$

If events E and F are independent, and $P(F) > 0$, then we have

$$P(E|F) = \frac{P(EF)}{P(F)} = \frac{P(E)P(F)}{P(F)} = P(E).$$

Thus, the occurrence of event F does not alter the probability of the occurrence of event E. Similarly, we see that the independence of E and F implies that

$$P(F|E) = P(F).$$

Thus the definition of independence is consistent with our intuition.

Example 1.29 (Two Dice). A six-sided die is tossed twice in such a way that all the outcomes are equally likely. Let E_i be the event that the ith throw shows an even number. Are E_1 and E_2 independent?

The sample space for this experiment is

$$\Omega = \{(i, j) : 1 \le i, j \le 6\}.$$

Here i is the outcome on the first throw, and j is the outcome on the second throw. We have

$$E_1 = \{(i, j) : i = 2, 4, 6; \ 1 \le j \le 6\},$$
$$E_2 = \{(i, j) : 1 \le i \le 6; \ j = 2, 4, 6\},$$
$$E_1 E_2 = \{(i, j) : i = 2, 4, 6; \ j = 2, 4, 6\}.$$

Since all 36 outcomes in Ω are equally likely, we have

$$P(E_1) = \frac{m(E_1)}{m(\Omega)} = \frac{18}{36} = .5,$$

$$P(E_2) = \frac{m(E_2)}{m(\Omega)} = \frac{18}{36} = .5,$$

and

$$P(E_1 E_2) = \frac{m(E_1 E_2)}{m(\Omega)} = \frac{9}{36} = .25.$$

Thus

$$P(E_1 E_2) = P(E_1)P(E_2)$$

and hence the events E_1 and E_2 are independent. Indeed, under the assumption of "equally like outcomes," if E_i is any event based on the outcome of the ith throw, we can show that E_1 and E_2 are independent. We say that the two tosses are independent. □

Example 1.30 (Random Pairing). Consider Example 1.16. Are events E_i and E_j independent, assuming $i \neq j$?

From Example 1.16 we see that

$$P(E_i) = P(E_j) = \frac{1}{n},$$

while

$$P(E_i E_j) = \frac{1}{n(n-1)} \neq P(E_i)P(E_j).$$

Hence E_i and E_j are not independent. □

Example 1.31 (Two Clinical Trials). Consider Example 1.3. Suppose

$$P(\{(S, S)\}) = .64, \quad P(\{(S, F)\}) = P(\{(F, S)\}) = .16, \quad P(\{(F, F)\}) = .04.$$

Are the two clinical trials independent?

First of all, we haven't defined what we mean by "independent trials." We shall say the trials are independent if all events on trial 1 are independent of all events on trial 2. In Example 1.9 we had enumerated all events for a single clinical trial. Using that, we see that we need to check for the independence of four pairs of events. We describe one such check below. Let E_i be the event that a success is obtained on trial i, $i = 1, 2$. Then

$$P(E_1) = P(\{(S, S), (S, F)\}) = .8 = P(E_2).$$

Similarly,

$$P(E_1 E_2) = P(\{(S, S)\}) = .64 = P(E_1)P(E_2).$$

Hence E_1 and E_2 are independent. In a similar way we can show that all events on trial 1 are independent of all events on trial 2. Hence the two trials are independent. □

Next we discuss the independence of three or more events.

Definition 1.6 (Mutual Independence). Events $E_1, E_2, \ldots, E_n$ are said to be mutually independent if for any subset $S \subseteq \{1, 2, \ldots, n\}$

$$P\left(\bigcap_{i \in S} E_i\right) = \prod_{i \in S} P(E_i).$$

Thus mutual independence of the three events E_1, E_2, E_3 implies the following equalities:

$$P(E_1 E_2) = P(E_1)P(E_2),$$
$$P(E_2 E_3) = P(E_2)P(E_3),$$
$$P(E_1 E_3) = P(E_1)P(E_3),$$
$$P(E_1 E_2 E_3) = P(E_1)P(E_2)P(E_3).$$

In general, the mutual independence of n events implies $2^n - (n + 1)$ equalities of the type shown above. In this book we use the term "independence" to mean "mutual independence."

Example 1.32 (Series System). Consider a system consisting of n components. Each component can be in one of two states: up or down. The system as a whole is up if all the components are up, and is down otherwise. Such a system is called a series system. Suppose the n components are independent, and that the probability that the ith component is up is p_i. What is the probability that the system is up?

Let E_i be the event that the ith component is up, and let E be the event that the system is up. We are given that $E_i, 1 \leq i \leq n$, are mutually independent. The series nature of the system implies

$$E = \bigcap_{i=1}^{n} E_i.$$

Hence the probability that the system is up is given by

$$P(E) = P\left(\bigcap_{i=1}^{n} E_i\right) = \prod_{i=1}^{n} P(E_i) = \prod_{i=1}^{n} p_i.$$

Notice that

$$P(E) \leq \min_{1 \leq i \leq n} \{p_i\},$$

implying that the series system is weaker than its weakest link!   □

Example 1.33 (Parallel System). Consider a system consisting of n components as in Example 1.32. The system as a whole is up if at least one of the n components is up, and is down otherwise. Such a system is called a parallel system. What is the probability that the system is up?

Let the events E_i and E be as in Example 1.32. The parallel nature of the system implies

$$E^c = \bigcap_{i=1}^{n} E_i^c.$$

Hence the probability that the system is up is given by

$$P(E) = 1 - P(E^c) = 1 - P\left(\bigcap_{i=1}^{n} E_i^c\right) = 1 - \prod_{i=1}^{n} P(E_i^c) = 1 - \prod_{i=1}^{n} (1 - p_i).$$

Notice that

$$P(E) \geq \max_{1 \leq i \leq n} \{p_i\},$$

implying that the parallel system is stronger than its strongest link!   □

In Examples 1.29 and 1.31 we alluded to independent die tossing and clinical trials. In probability models we frequently use this concept of independent trials (or experiments). Hence we formally define the concept below:

Definition 1.7 (Independent Trials). If events based on the outcomes of different trials are mutually independent, the trials are called independent.

Example 1.34 (Independent Coin Tossing). Suppose a coin lands heads up with probability p and tails with probability $1 - p$. What is the probability of getting exactly k heads in n independent tosses of the coin? ($0 \leq k \leq n$.)

The sample space can be taken to be the set of all sequences of H and T of length n. Independent tossing implies that any event on a given toss is independent of any other event on any other toss. Now let E_k^n be the event that exactly k heads are obtained in n tosses. It consists of all the sequences of length n containing exactly k H's and $n - k$ T's. There is a total of

$$\binom{n}{k} = \frac{n!}{k!\,(n-k)!}$$

sequences of this type. Using the independence of tosses we see that the probability of observing any one of these sequences is $p^k(1 - p)^{n-k}$. Hence we get

$$P(E_k^n) = \binom{n}{k} p^k (1 - p)^{n-k}.$$

Note in particular that

$$P(E_n^n) = P(\text{all } n \text{ heads}) = p^n$$

and

$$P(E_0^n) = P(\text{all } n \text{ tails}) = (1 - p)^n$$

as expected.   □

Example 1.35 (Bayesian Updating). Consider Example 1.26. Suppose the randomly chosen coin is tossed N times independently, and observed to produce N heads. What is the probability that it is the ith coin?

Let E_i be as in Example 1.26 and let E be the event that we get N heads in N independent trials. We want to compute $P(E_i|E)$. From Example 1.34 we get

$$P(E|E_i) = p_i^N, \qquad 1 \le i \le n.$$

Using the results of Example 1.27 we get

$$P(E_i|E) = \frac{p_i^N}{\sum_{r=1}^n p_r^N}. \qquad \qquad \Box$$

Example 1.36 (Probability of First Event). Suppose independent trials of a random experiment are performed repeatedly. Each trial results in one and only one of three outcomes, $1, 2, 3$. Assume that the probability of getting outcome i on any given trial is p_i, with $p_1 + p_2 + p_3 = 1$. What is the probability that outcome 1 is observed before outcome 2?

Let E_i^n be the event that the nth trial results in outcome i. Let F_n be the event that outcome 1 is observed for the first time on the nth trial and outcome 2 does not occur in the first n trials. Then F_n, $n \ge 1$, are mutually exclusive and the event "1 occurs before 2" is given by

$$F = \bigcup_{n=1}^\infty F_n.$$

Now,

$$F_n = \left(\bigcap_{k=1}^{n-1} E_3^k \right) \cap E_1^n.$$

Using the independence of trials, we get

$$P(F_n) = p_3^{(n-1)} p_1.$$

Hence

$$P(F) = \sum_{n=1}^\infty P(F_n) = \sum_{n=1}^\infty p_3^{(n-1)} p_1 = \frac{p_1}{1 - p_3} = \frac{p_1}{p_1 + p_2}.$$

In other words the probability of observing outcome i before outcome j in a sequence of independent trials is proportional to the probability of observing i on one trial. $\qquad \Box$

1.9. Problems

CONCEPTUAL PROBLEMS

1.1. Let E, F, and G be three events. Give the set-theoretic notation for the following compound events:

    (1) E and F but not G;

    (2) E or F but not both;

    (3)   none of E or F or G;

    (4)   exactly two of the three events.

1.2. Give the event description of the following sets:

    (1) $(E \cap F) \cup G$;

    (2) $(EFG)^c$;

    (3) $E \cap (F \cup G)$;

    (4) $E \cup FG$.

1.3. Prove DeMorgan's Laws:

    (1) $\left(\bigcup_{i=1}^{n} E_i \right)^c = \bigcap_{i=1}^{n} E_i^c$;

    (2) $\left(\bigcap_{i=1}^{n} E_i \right)^c = \bigcup_{i=1}^{n} E_i^c$.

1.4. Show that $P(\emptyset) = 0$.

1.5. Show that $P(E) \leq P(F)$ if $E \subseteq F$.

1.6. Show that

$$P\left(\bigcup_{i=1}^{n} E_i \right) \leq \sum_{i=1}^{n} P(E_i).$$

1.7. Let E_n, $n \geq 1$, be a set of mutually exclusive events. Suppose $P(F) > 0$. Show that

$$P\left(\bigcup_{n=1}^{\infty} E_n | F \right) = \sum_{n=1}^{\infty} P(E_n | F).$$

1.8. Assuming $P(F) > 0$, show that

$$P(E_1 \cup E_2 | F) = P(E_1 | F) + P(E_2 | F) - P(E_1 E_2 | F).$$

1.9. Show that

$$P(EF | G) = P(E | FG)P(F | G).$$

1.10. Show that the universal event and the null event are independent of all other events.

1.11. Suppose event E is independent of itself. Show that E must be either the universal event or the null event.

1.12. Show that if E and F are independent, then so are E^c and F^c.

1.13. *Conditional Independence.* Events E and F are said to be conditionally independent, given G, if

$$P(EF|G) = P(E|G)P(F|G).$$

Consider a probability model with $\Omega = \{1, 2, 3\}$, with equally likely outcomes. Define three events $E = \{1, 2\}$, $F = \{2, 3\}$, $G = \{2\}$. Show that E and F are conditionally independent given G, but that E and F are not independent.

1.14. Definition 1.6 gives the definition of mutual independence. A less stringent form of independence is called *pairwise independence* and is defined as follows. Events $E_1, E_2, \ldots, E_n$ are said to be pairwise independent if

$$P(E_i E_j) = P(E_i)P(E_j)$$

for all pairs $i \neq j$. Consider a probability model with $\Omega = \{1, 2, 3, 4\}$, with equally likely outcomes. Define three events $E = \{1, 2\}$, $F = \{1, 3\}$, $G = \{1, 4\}$. Show that E, F, G are pairwise independent but not mutually independent.

1.15. Let E_i, $1 \leq i \leq n$, be n independent events. Show that

$$P\left(\bigcup_{i=1}^{n} E_i\right) = 1 - \prod_{i=1}^{n}(1 - P(E_i)).$$

COMPUTATIONAL PROBLEMS

1.1. Give the sample space for a random experiment that involves:

(1)   Tossing two coins.

(2)   Measuring the amount of water overflow from a dam during 1 month.

1.2. Give the sample space for a random experiment that involves:

(1)   Testing three light bulbs to see if they work or not.

(2)   Measuring the dollar sales from a store during 1 month.

(3)   Counting the number of cars crossing a road during 1 hour.

1.3. Give the sample space for a random experiment that involves:

(1)   Testing the functional states of two machines, assuming that each machine can be in one of three states: new, good, bad.

(2)   Counting the number of arrivals at a store during 1 business day.

(3)   Noting the price of a given stock at the end of 1 week.

1.4. Consider the random experiment of testing the three light bulbs as in Computational Problem 1.2, part 1. Describe the events E_i = exactly i light bulbs are working, $0 \leq i \leq 3$, as subsets of the sample space.

1.5. Consider the random experiment of counting the number of arrivals at a store in 1 day. Describe the following events as subsets:

(**1**)   Exactly thirteen customers arrived.

(**2**)   More than ten customers arrived.

(**3**)   No more than five customers arrived.

(**4**)   At least twenty customers arrived.

1.6. Consider the random experiment of drawing three cards without replacement from a full deck. Describe the following events as subsets:

(**1**)   All three cards are red.

(**2**)   At least one card is a spade.

(**3**)   None of the three cards is a diamond.

(**4**)   At least one card is red and one card is black.

Note: First decide upon a sample space that can be used in all four events.

1.7. Consider Computational Problem 1.4. Suppose all possible outcomes are equally likely. Compute the probability of E_i, $0 \leq i \leq 3$.

1.8. Consider Computational Problem 1.6. Suppose the drawing without replacement is done in a uniform random fashion. Compute the probabilities of the four events given there.

1.9. A class consists of twelve boys and ten girls. A group of six students is chosen from this group. What is the probability that we get a group consisting of three boys and three girls?

1.10. A coin is tossed three times so that all possible outcomes are equally likely. Let E_i be the event that the ith toss produces a head, $1 \leq i \leq 3$. Compute:

(**1**) $P(E_1 \cup E_2)$;

(**2**) $P(E_1 \cap E_2)$;

(**3**) $P(E_1 \cup E_2 \cup E_3)$.

1.11. Let E_i be as in Computational Problem 1.10. Let F be the event that the number of observed heads is odd. Compute:

(**1**) $P(E_1 \cup F)$;

(2) $P(E_2 \cap F)$;

(3) $P((E_1 \cup E_2) \cap F)$.

1.12. Suppose that three balls are drawn without replacement from an urn containing r red balls and b blue balls. Let E_i be the probability that the ith ball drawn is a red one. Compute:

(1) $P(E_1 \cup E_2)$;

(2) $P(E_2 \cap E_3)$;

(3) $P(E_1 \cup E_2 \cup E_3)$.

1.13. Let E_i be as in Computational Problem 1.12. Let F be the event that the number of observed red balls is odd. Compute:

(1) $P(E_1 \cup F)$;

(2) $P(E_2 \cap F)$;

(3) $P((E_1 \cup E_2) \cap F)$.

1.14. Suppose that there are n students at a party. Assuming that none of them is born on Feb. 29, compute the probability that at least two students have the same birthday. For what values of n is this probability greater than .5?

1.15. Susan has written n New Year's cards to n friends. Unfortunately, she sealed all the envelopes without noting whose card is in which envelope. If she writes the n addresses on the n envelopes in a completely random order, what is the probability that nobody gets the correct card?

1.16. Consider Computational Problem 1.11. Compute:

(1) $P(E_1|F)$;

(2) $P(F|E_1)$.

1.17. Consider Computational Problem 1.12. Compute:

(1) $P(E_2|E_1)$;

(2) $P(E_1|E_2)$.

1.18. Consider Computational Problem 1.13. Compute:

(1) $P(E_2|F)$;

(2) $P(E_1 \cup E_2|F)$.

1.19. Consider Computational Problem 1.15. Let E_i be the event that the ith friend receives the correct card. Compute $P(E_1^c|E_2)$.

1.20. Consider the random experiment of Computational Problem 1.6. Compute the probability that:

(1)   at least one card is black given that the first card is red;

(2)   the first card is a spade given that all three cards are black; and

(3)   none of the three cards is a diamond given that all the cards are red.

1.21. Consider the random experiment of Computational Problem 1.6. Compute the probability that the third card drawn is a black one.

1.22. Based on the experience on similar oil fields, an oil executive has determined that the probability that this oil field contains oil is .6. Before starting the drilling she decides to order a seismological test to see if there is oil in the ground. Unfortunately the test is not entirely accurate. It concludes that there is oil with probability .9 if there is indeed oil in the ground. If there is no oil in the ground, it concludes so with probability .8. Suppose the test results come back positive (i.e., the test says there is oil in the ground). What is the probability that there is indeed oil in the ground given this test result?

1.23. A machine can be in two conditions: good or bad. Every day the machine produces one item which itself can be good or bad. If the machine is in good condition, the produced item is good with probability 1, or else it is good with probability p. Suppose that initially the machine is in either good or bad condition with equal probability. What is the probability that the machine is in good condition given that the item is good?

1.24. Consider Computational Problem 1.23. Suppose the machine changes states as follows. If the machine is in good condition at the beginning of a day, it is in good condition at the beginning of the next day with probability r. If the machine is in bad condition at the beginning of a day, it is in bad condition at the beginning of the next day with probability 1. What is the probability that the machine is in bad condition at the beginning of the third day, given that it is equally likely to be in good or bad condition at the beginning and that the items it produced on days 1 and 2 were both good?

1.25. A box contains three coins. Coin 1 is a two-headed coin, coin 2 is a two-tailed coin, while coin 3 is a fair coin. One coin is drawn at random from this box and tossed. Given that it lands heads, compute the probability that it is a two-headed coin.

1.26. Suppose that a coin is tossed repeatedly in an independent fashion. Suppose the probability that it lands heads on any one toss is .6.

(1)   What is the probability that the first head is observed on the third toss?

(2)   What is the probability that the first head is observed on the kth toss ($k \geq 1$)?

1.27. Suppose that a coin is tossed repeatedly in an independent fashion. Suppose the probability that it lands heads on any one toss is p. What is the probability that we observe two consecutive heads before two consecutive tails?

1.28. Consider a system consisting of three independent components. Let p_i be the probability that the ith component is functional, with $p_1 = .95$, $p_2 = .92$, and $p_3 = .98$. The system needs at least two components to work in order to function correctly. Compute the probability that the system works.

1.29. Five radio towers are erected in a straight line, exactly 8 miles apart. The signal from each tower can travel 16.6 miles. Each tower is functional (capable of receiving and transmitting messages) with probability .98, and they are independent of each other. What is the probability that a message from the first tower will reach the fifth tower? Assume that each tower that receives a message transmits it further.

1.30. An airplane has four jet engines, two on each wing. The probability that an engine performs without any problems for the duration of a flight is .99. All engines are independent of each other. The plane needs at least one functioning engine on each wing to fly properly. What is the probability that the plane flight is successful?

2

Univariate Random Variables

2.1. Random Variables

In Chapter 1 we saw that the first step in the study of a random phenomenon is the construction of a probability model: that is, its sample space, the set of events, and the probability function. In several applications we are not only interested in the probabilities of events, but also in specific numerical values associated with the outcomes. We illustrate with an example:

Example 2.1 (Two-Coin Tossing). Consider a random experiment of tossing two coins. The sample space is given by

$$\Omega = \{(T, T), (T, H), (H, T), (H, H)\}.$$

Suppose we are interested in the number of heads obtained. This means that we associate a numerical value 0 to the outcome (T, T), value 1 to the outcomes (T, H) and (H, T), and value 2 to (H, H). Thus when the outcome (T, T) occurs we observe a value 0, etc.   □

Note that we may assign the same numerical value to more than one outcome, however a given outcome has only one numerical value associated with it. Thus we can say that we are interested in a numerical *function* of the outcomes or of the sample points. Since the outcomes are random, the actual value taken by this function in a given trial of the experiment is also random. A natural term for such functions would be *random functions*. Unfortunately, the standard name for them is *random variables*, which is misleading because there is nothing variable about the function, but we shall stick with it. A formal definition follows:

Definition 2.1 (Random Variable). A random variable is a function from the sample space into the set of real numbers $(-\infty, \infty)$.

It is common practice to use upper-case italic letters like X, Y, W, Z, etc. to represent the random variables. Thus when we say that X is a random variable, we mean that X is a function $X : \Omega \rightarrow (-\infty, \infty)$, and $X(\omega)$ is the value of the function at the sample point $\omega \in \Omega$.

Example 2.2 (Two-Coin Tossing). Consider the experiment of Example 2.1. Let X be the random variable representing the number of heads obtained in two tosses. Then X is the function given below:

$$X((T, T)) = 0; \qquad X((T, H)) = 1;$$
$$X((H, T)) = 1; \qquad X((H, H)) = 2. \qquad \qquad \square$$

Example 2.3 (Two-Dice Tossing). Consider the experiment of tossing two dice and observing the individual outcomes. A typical sample point is (i, j), where i is the number showing on the first die, and j on the second. The sample space is

$$\Omega = \{(i, j) : 1 \leq i, j \leq 6\}.$$

Let X be the random variable representing the sum of the two dice outcomes. Thus X is the following function:

$$X((i, j)) = i + j, \qquad (i, j) \in \Omega.$$

Let Y be the random variable representing the maximum of the two dice outcomes. Then Y is given by

$$Y((i, j)) = \max\{i, j\}, \qquad (i, j) \in \Omega. \qquad \qquad \square$$

The above example shows that we can define more than one random variable on the same sample space. This is natural since we may be interested in several different numerical aspects of the same random phenomenon.

Example 2.4 (Lifetimes). Consider a random experiment to test the lifetime of a light bulb. The experiment consists of noting the time when the light bulb is turned on, and the time when it fails. A typical sample point is (s, f) where s is the start time and f is the failure time of the light bulb. The sample space is

$$\Omega = \{(s, f) : 0 \leq s \leq f < \infty\}.$$

Let X be the random variable representing the lifespan of the light bulb. Then X is given by

$$X((s, f)) = f - s, \qquad (s, f) \in \Omega.$$

Note that the lifespan can be any nonnegative real number. $\qquad \qquad \square$

The next question is: how do we describe the random nature of a random variable? We answer this question in the next section.

2.2. Cumulative Distribution Function

In practice we are interested in the probability that a random variable takes values in a given set. Generally these sets are intervals or some combination of intervals in the real line $(-\infty, \infty)$. Thus we may want to know the probability of $\{X \in (0, 10)\}$. How do we define this probability? First we think of $\{X \in (0, 10)\}$ as a shorthand notation for the set $\{\omega \in \Omega : X(\omega) \in (0, 10)\} \subseteq \Omega$. Hence the natural way is to define

$$P(\{X \in (0, 10)\}) = P(\{\omega \in \Omega : X(\omega) \in (0, 10)\}).$$

This motivates the following general definition:

Definition 2.2 (Probability Function of X). Let E be a subset of the real line. Then

$$P(\{X \in E\}) = P(\{\omega \in \Omega : X(\omega) \in E\}).$$

If $E = (-\infty, x]$ we write $P(X \le x)$ instead of $P(\{X \in E\})$. Similar notation is given below:

$$P(\{X \in (a, b)\}) = P(a < X < b),$$
$$P(\{X \in (a, b]\}) = P(a < X \le b),$$
$$P(\{X \in (x, \infty)\}) = P(X > x),$$
$$P(\{X \in \{x\}\}) = P(X = x).$$

Example 2.5 (Two-Coin Tossing). Consider Example 2.2. Suppose all four outcomes are equally likely. Then

$$P(X = 1) = P(\{(T, H), (H, T)\}) = .5,$$
$$P(X \ge 1) = P(\{(T, H), (H, T), (H, H)\}) = .75. \qquad \square$$

Example 2.6 (Two-Dice Tossing). Consider Example 2.3. Suppose all 36 outcomes are equally likely. Then

$$P(X = 2) = P(\{(0, 2), (1, 1), (2, 0)\}) = \frac{3}{36} = \frac{1}{12},$$

$$P(X > 9) = P(\{(4, 6), (5, 5), (6, 4), (5, 6), (6, 5), (6, 6)\}) = \frac{6}{36} = \frac{1}{6}. \qquad \square$$

But how shall we give $P(X \in E)$ for all subsets E of the real line? The countable additivity of the probability function allows us to do this in a compact way: it suffices to give $P(X \in E)$ only for sets E of the type $(-\infty, x]$ for all $x \in (-\infty, \infty)$. Hence the probabilities of these sets are given a special name.

Definition 2.3 (Cumulative Distribution Function (cdf)). The function

$$F(x) = P(X \le x), \qquad x \in (-\infty, \infty)$$

is called the cumulative distribution function of the random variable X.

We can compute $P(X \in E)$ for other subsets E of the real line using $F(\cdot)$. For example,

$$P(a < X \leq b) = F(b) - F(a),$$
$$P(X > x) = 1 - F(x),$$
$$P(X < x) = \lim_{\epsilon \downarrow 0} F(x - \epsilon) = F(x^-),$$

$$P(X = x) = F(x) - F(x^-).$$

Thus the cumulative distribution function provides all the information about the random variable. The next theorem states the four main properties of the cdf:

Theorem 2.1 (Properties of the cdf).

(1) $F(\cdot)$ is a nondecreasing function, i.e.,

$$x \leq y \quad \Rightarrow \quad F(x) \leq F(y).$$

(2) $F(\cdot)$ is right continuous, i.e.,

$$\lim_{\epsilon \downarrow 0} F(x + \epsilon) = F(x).$$

(3) $\lim_{x \to -\infty} F(x) = F(-\infty) = 0.$

(4) $\lim_{x \to \infty} F(x) = F(\infty) = 1.$

Proof. Suppose $x \leq y$. Then $\{X \leq x\} \subseteq \{X \leq y\}$. Hence $P(\{X \leq x\}) \leq P(\{X \leq y\})$. This establishes property (1). Assertion (2) follows due to the weak inequality $\leq$ in the definition of F. Assertions (3) and (4) are the result of the fact that the random variable X takes values in $(-\infty, \infty)$. $\qquad \square$

Any function satisfying the four properties of Theorem 2.1 is a cdf of some random variable.

Example 2.7 (Two-Coin Tossing). Consider Examples 2.1 and 2.2. The cdf of the random variable X (= number of heads in two coin tosses) is as given below:

$$F(x) = \begin{cases} 0 & \text{if } x < 0, \\ 0.25 & \text{if } 0 \leq x < 1, \\ 0.75 & \text{if } 1 \leq x < 2, \\ 1 & \text{if } x \geq 2. \end{cases}$$

Note that $F(2^-) = .75$, while $F(2^+) = F(2) = 1$. Hence

$$P(X = 2) = F(2) - F(2^-) = .25. \qquad \square$$

Example 2.8 (Lifetimes). Consider Example 2.4. Suppose the cdf of X (= lifespan of the light bulb) is as given below:

$$F(x) = \begin{cases} 0 & \text{if } x < 0, \\ \dfrac{x}{100} & \text{if } 0 \le x \le 100, \\ 1 & \text{if } x > 100. \end{cases}$$

Now

$$P(X < 0) = F(0^-) = 0; \quad P(X > 100) = 1 - F(100) = 0.$$

Hence the lifespan of the light bulb is between 0 to 100 time units (hours, say). What is the probability that the light bulb lasts more than 20 hours but no more than 70 hours? It can be computed as

$$P(20 < X \le 70) = F(70) - F(20) = \frac{70}{100} - \frac{20}{100} = .5.$$

What is the probability that the lifespan of the light bulb is exactly 75 hours? It is given by

$$P(X = 75) = F(75) - F(75^-) = 0.$$

Indeed, the probability that the lifespan is any particular constant is zero!   □

The contrast between the random variables in Examples 2.7 and 2.8 is worth noting. The cdf in Example 2.7 is a step function with jumps at 0, 1, 2, and the corresponding random variable takes only discrete values 0, 1, and 2. Such a random variable is called a *discrete random variable*. The cdf in Example 2.8, on the other hand, has no jumps, and describes a random variable that takes values continuously in an interval (0, 100). Such a random variable is called a *continuous random variable*. We shall study these two types of random variable in detail in the rest of this chapter. From now on we do not explicitly specify the sample space on which the random variable is defined, although the reader should be able to provide one if needed.

2.3. Discrete Random Variables

Discrete random variables, as the name suggests, take values in a countable subset

$$S = \{x_0, x_1, x_2, \ldots\}$$

of the real line. A formal definition follows:

Definition 2.4 (Discrete Random Variable). A random variable is said to be discrete with state space S if its cdf is a step function with jumps at points in S.

Let

$$p_k = P(X = x_k) = F(x_k) - F(x_k^-), \quad x_k \in S.$$

TABLE 2.1. The pmf for Example 2.9.

k	2	3	4	5	6	7	8	9	10	11	12
p_k	$\dfrac{1}{36}$	$\dfrac{2}{36}$	$\dfrac{3}{36}$	$\dfrac{4}{36}$	$\dfrac{5}{36}$	$\dfrac{6}{36}$	$\dfrac{5}{36}$	$\dfrac{4}{36}$	$\dfrac{3}{36}$	$\dfrac{2}{36}$	$\dfrac{1}{36}$

Since

$$F(x) = \sum_{k:x_k \le x} p_k,$$

it is clear that the function $F(\cdot)$ is completely described by the set S and $\{p_0, p_1, p_2, \ldots\}$. Hence we define the following:

Definition 2.5 (Probability Mass Function (pmf)). The function

$$p_k = \mathsf{P}(X = x_k), \qquad k \ge 0,$$

is called the probability mass function of a discrete random variable X with state space $S = \{x_0, x_1, x_2, \ldots\}$.

The next theorem states the main properties of the probability mass function. We omit the proof, since it is trivial.

Theorem 2.2 (Properties of the pmf).

$$p_k \ge 0, \qquad k \ge 0, \tag{2.1}$$

$$\sum_{k=0}^{\infty} p_k = 1. \tag{2.2}$$

Many discrete random variables take values from the set of integers. In such cases, we define $x_k = k$ and

$$p_k = \mathsf{P}(X = k).$$

Example 2.9 (Sum of Two Dice). Consider Example 2.3. The random variable X ($=$ sum of the two dice) is a discrete random variable with state space $\{2, 3, 4, 5, 6, 7, 8, 9, 10, 11, 12\}$ and the pmf shown in Table 2.1. (The pmf is computed as shown in Example 2.6.)

Note that the pmf is symmetric and takes a maximum value of $\frac{1}{6}$ at $k = 7$. This explains why it is important to block the seventh place in the game of Backgammon!   □

In the next section we study several commonly occurring discrete random variables.

2.4. Common Discrete Random Variables

Many of the commonly used discrete random variables arise out of two situations: (i) repeated independent trials, and (ii) urn models. We shall start with repeated trials.

Independent Trials. Consider a random experiment (tossing a coin, testing an item for defects, etc.). Suppose we are interested in a specific event E (getting a head, finding a defect, etc.) that may occur as a result of the experiment. Assume that we can repeat this experiment in an independent fashion as many times as is necessary. Let p be the probability that the event E occurs on a given trial.

A. Bernoulli Random Variable $B(p)$

Suppose we perform one trial of the random experiment and observe whether or not E occurs. Let

$$X = \begin{cases} 1 & \text{if } E \text{ occurs,} \\ 0 & \text{if } E \text{ does not occur.} \end{cases}$$

The random variable X is called the *indicator random variable* of the event E, since it indicates whether or not E has occurred. The pmf of X is given by

$$p_0 = \mathsf{P}(X = 0) = 1 - p, \qquad p_1 = \mathsf{P}(X = 1) = p. \tag{2.3}$$

Note that the above pmf satisfies (2.1) and (2.2).

Definition 2.6 (Bernoulli Distribution). A random variable X with the pmf given in (2.3) is called a Bernoulli random variable with parameter p, and is denoted by $B(p)$. The distribution in (2.3) is called a Bernoulli distribution with parameter p, and is also denoted by $B(p)$.

Example 2.10 (Random Pairing). Consider the random pairing problem in Example 1.16. Let E_i be the event that the ith woman gets paired with her husband. Let X_i be the indicator random variable of the event E_i. Then X_i is a $B(1/n)$ random variable, since

$$p_1 = \mathsf{P}(X_i = 1) = \mathsf{P}(E_i) = \frac{1}{n}. \qquad\qquad \square$$

B. Binomial Random Variable $\text{Bin}(n, p)$

Suppose we conduct n independent trials of an experiment. Define X to be the number of trials that result in event E. The state space of X is $S = \{0, 1, 2, \ldots, n\}$. By using the results of Example 1.34 we see that the pmf of X is given by

$$p_k = \mathsf{P}(X = k) = \binom{n}{k} p^k (1 - p)^{n-k}, \qquad k \in S. \tag{2.4}$$

Does this pmf satisfy (2.1) and (2.2)? Since $0 \le p \le 1$, we see that $p_k \ge 0$ for all $k \in S$. To verify the second property, we use the binomial expansion of $(1 + (1 - p))^n$ to obtain

$$\begin{aligned} \sum_{k=0}^{n} p_k &= \sum_{k=0}^{n} \binom{n}{k} p^k (1 - p)^{n-k} \\ &= (p + (1 - p))^n \\ &= 1. \end{aligned}$$

Definition 2.7 (Binomial Distribution). The random variable X with the pmf given in (2.4) is called a Binomial random variable with parameters n and p, and is denoted by Bin(n, p). The distribution in (2.4) is called a Binomial distribution with parameters n and p, and is also denoted by Bin(n, p).

Example 2.11 (Quality Control). A quality control department is in charge of ensuring the quality of the manufactured items in a factory. Suppose the department follows the following procedure: It inspects 20 items selected randomly from each batch of production. If two or more are found defective, the entire batch is declared defective and is destroyed. Suppose each item is defective with probability $p = .01$. What is the probability that the batch will be rejected?

Let X be the number of defective items in the inspected batch of 20 items. Assuming the items to be independent, we see that X is a Bin$(20, .01)$ random variable. Hence

$$\begin{aligned}
\text{P(batch is rejected)} &= \text{P}(X \geq 2) \\
&= 1 - \text{P}(X = 0) - \text{P}(X = 1) \\
&= 1 - (.99)^{20} - 20(.01)(.99)^{19} \\
&= .0169.
\end{aligned}$$

Now suppose that the manufacturing department can control the value of p, the probability of an item being defective. If the aim is to keep the probability of rejection from the quality control department below .01, what is the largest value of p that can be tolerated?

We must have

$$1 - (1 - p)^{20} - 20p(1 - p)^{19} \leq .01.$$

We can check numerically that this can be achieved as long as

$$p \leq .0076. \qquad \square$$

Example 2.12 (Parity Check in Digital Communication). The modern digital communication systems encode messages into strings of zeros and ones (called bits), and transmit them over communication channels from the sender to the receiver. Due to engineering problems, a small fraction of the zeros and ones gets flipped during the transmission (these are called bit errors), and thus produce a garbled message at the receiving end. One simple method of detecting the occurrence of bit errors is as follows: for every group of seven bits the network introduces an eighth bit (called the parity bit) so that the number of ones in the group of seven bits plus the parity bit (called a byte) is even. This is called the even parity protocol. Thus if a group of seven bits is 1100100, the parity bit is set to 1 and the byte 11001001 is transmitted. Note that the parity bit itself may get flipped during transmission. At the receiving end, the network checks if the number of ones in a byte is even. If it is, the byte is accepted as error-free. What is the probability that bit errors in a byte will go undetected by this protocol?

First, the byte will be accepted as error-free if the number of errors is even (why?). To compute the required probability we need a model of how errors are generated. A simple model is to assume that each bit has a probability p of getting flipped, and that all bit errors are independent of each other. Let X be the number of errors in a byte. Then under these assumptions, X is a Bin(8, p) random variable. Hence, the probability that a byte with errors will be accepted as error-free is given by

$$
\begin{aligned}
\mathsf{P}(X \text{ is even}|X > 0) &= \frac{\mathsf{P}(X \text{ is even}; X > 0)}{\mathsf{P}(X > 0)} \\
&= \frac{\mathsf{P}(X \in \{2, 4, 6, 8\})}{1 - \mathsf{P}(X = 0)} \\
&= \frac{\sum_{k=1}^{4} \binom{8}{2k} p^{2k}(1-p)^{8-2k}}{1 - (1-p)^8}.
\end{aligned}
$$

Note that we need the conditional probability since we are interested in computing the probability that a byte will be accepted as error-free given that it has errors.   □

C. Geometric Random Variable $G(p)$

Suppose we repeat an experiment independently until we observe the event E. Let X be the number of trials needed to observe E for the first time. The state space of X is $S = \{1, 2, 3, \ldots\}$. The pmf of X is given by

$$
\begin{aligned}
p_k &= \mathsf{P}(X = k) \\
&= \mathsf{P}(E^c \text{ occurs on each of the first } k - 1 \text{ trials and } E \text{ occurs on the } k\text{th trial}) \\
&= \mathsf{P}(E^c \text{ on trial } 1) \cdot \mathsf{P}(E^c \text{ on trial } 2) \cdots \mathsf{P}(E^c \text{ on trial } k - 1) \cdot \mathsf{P}(E \text{ on trial } k) \\
&\quad \text{(due to independence of trials)} \\
&= (1-p)^{k-1}p, \qquad k \in S. \tag{2.5}
\end{aligned}
$$

Again, we can see that $p_k \geq 0$ since $0 \leq p \leq 1$. Furthermore, using the formula for the sum of a geometric series, we get

$$
\begin{aligned}
\sum_{k=1}^{\infty} p_k &= \sum_{k=1}^{\infty} (1-p)^{k-1}p \\
&= p \sum_{k=1}^{\infty} (1-p)^{k-1} \\
&= p \sum_{k=0}^{\infty} (1-p)^k \\
&= p \frac{1}{1 - (1-p)} \\
&= \frac{p}{p} = 1!
\end{aligned}
$$

Thus the pmf of X satisfies (2.1) and (2.2).

Definition 2.8 (Geometric Distribution). The random variable X with the pmf given in (2.5) is called a Geometric random variable with parameter p, and is denoted by $G(p)$. The distribution in (2.5) is called a Geometric distribution with parameter p, and is also denoted by $G(p)$.

Example 2.13 (Machine Failure). Suppose a machine is subject to a series of shocks. If the machine has not already failed, the next shock causes its failure with probability .15. What is the probability that the third shock causes the machine to fail?

Suppose the machine fails on the Xth shock. Then X is a $G(.15)$ random variable. The required probability is

$$P(X = 3) = (.85)^2(.15) = .1084. \qquad \square$$

Example 2.14 (Research Project). A research project in genetic science involves conducting a series of experiments in a given sequence. If the kth experiment succeeds, the next experiment can be performed. Otherwise the whole project has to be abandoned. Suppose the experiments are independent and each experiment has a 92% chance of success. What is the probability that the project is abandoned after the sixth experiment?

Suppose the project has an infinite number of experiments and is abandoned after the Xth experiment. Then X is a $G(.08)$ random variable. Hence the required probability is

$$P(X = 6) = (.92)^5(.08) = .00622.$$

What is the probability that a project with eight experiments is concluded successfully? This is given by

$$P(X > 8) = (.92)^8 = .5132. \qquad \square$$

D. Negative Binomial Random Variable $NB(r, p)$

Let X be the number of trials needed if we repeat an experiment independently until the event E occurs r times. The state space of X is $S = \{r, r + 1, r + 2, \ldots\}$. The pmf is given by

$$p_k = P(X = k) = \binom{k - 1}{r - 1} p^r (1 - p)^{k-r}, \qquad k \in S. \tag{2.6}$$

The above pmf can be derived as follows. To get $X = k$, the rth occurrence of the event E must be on the kth trial. This means the other $r - 1$ occurrences must have taken place in the first $k - 1$ trials. Using the independence of trials and Example 1.34, we see that the probability of observing the event E $r - 1$ times in the first $k - 1$ trials is given by

$$\binom{k - 1}{r - 1} p^{r-1} (1 - p)^{k-r}.$$

This, together with the probability p of observing the event E on the kth trial, produces (2.6). We leave it to the reader to verify that the pmf in (2.6) satisfies (2.1) and (2.2).

Definition 2.9 (Negative Binomial Distribution). The random variable X with the pmf given in (2.6) is called a Negative Binomial random variable with parameters r and p, and is denoted by NB(r, p). The distribution in (2.6) is called a Negative Binomial distribution with parameters r and p, and is also denoted by NB(r, p).

Another name for a Negative Binomial distribution is a *Pascal Distribution*.

Example 2.15 (Slot Machine). It costs a quarter to play the slot machine once. After each play the machine either returns nothing, or it returns a nickel, or a dime, or a quarter, or a dollar, with the corresponding probabilities .5, .45, .04, .009, and .001. A player decides to stop playing as soon as the machine produces zero return a total of five times. What is the probability that the player will play exactly 12 times?

Assume that returns on successive plays are independent. Suppose X is the total number of times the person plays. Then X is an NB(5, .5) random variable. Hence the required probability is given by

$$P(X = 12) = \binom{11}{4}(.5)^5(.5)^{12-5} = .0806.$$

□

E. Poisson Random Variable $P(\lambda)$
Consider the Binomial distribution of (2.4). Computing this distribution for large values of n is difficult since the term $\binom{n}{k}$ becomes very large, while the term $p^k(1-p)^{n-k}$ becomes very small. To avoid such difficulties, it is instructive to look at the limit of the Binomial distribution as $n \to \infty$, and $p \to 0$ in such a way that np approaches a fixed number, say $\lambda \in (0, \infty)$. We shall denote this limiting region as D to avoid clutter in the following derivation. We have

$$\lim_D \binom{n}{k} p^k(1-p)^{n-k} = \lim_D \frac{n!}{k!\,(n-k)!} p^k(1-p)^{n-k}$$

$$= \lim_D \frac{(np)^k}{k!} \left(\prod_{r=0}^{k-1}\left(\frac{n-r}{n}\right)\right)\left(1-\frac{np}{n}\right)^{n-k}$$

$$= \lim_D \frac{(np)^k}{k!} \left(\lim_D \prod_{r=0}^{k-1}\left(\frac{n-r}{n}\right)\right)\lim_D \left(1-\frac{np}{n}\right)^{n}\lim_D \left(1-\frac{np}{n}\right)^{-k}$$

$$= \left(\frac{\lambda^k}{k!}\right)(1)\left(e^{-\lambda}\right)(1)$$

$$= e^{-\lambda}\frac{\lambda^k}{k!}.$$

Here we have used

$$\lim_D \left(1-\frac{np}{n}\right)^{n} = \lim_{n\to\infty}\left(1-\frac{\lambda}{n}\right)^{n} = e^{-\lambda}.$$

Thus, in the limit, the Bin(n, p) random variable approaches a random variable with state space $S = \{0, 1, 2, \ldots\}$ and pmf

$$p_k = e^{-\lambda} \frac{\lambda^k}{k!}, \qquad k \in S. \tag{2.7}$$

The above pmf plays an important part in probability models and hence it has been given a special name:

Definition 2.10 (Poisson Distribution). A random variable with the pmf given in (2.7) is called a Poisson random variable with parameter λ, and is denoted by $\mathrm{P}(\lambda)$. The pmf given in (2.7) is called a Poisson distribution with parameter λ, and is also denoted by $\mathrm{P}(\lambda)$.

In practical terms, for large values of n, a Bin(n, p) distribution may be approximated by a $\mathrm{P}(np)$. The approximation works especially well if p is small, say $p < .1$.

Example 2.16 (Counting Accidents). Suppose accidents occur one at a time at a dangerous traffic intersection during a 24-hour day. We divide the day into 1440 minute-long intervals and assume that there can be only 0 or 1 accidents during a 1 minute interval. Let E_k be the event that there is an accident during the kth minute-long interval, $1 \leq k \leq 1440$. Suppose that the events E_k's are mutually independent, and that $\mathrm{P}(E_k) = .001$. Compute the probability that there are exactly k accidents during one day.

Let X be the number of accidents during a 24-hour period. Using the assumptions above, we see that X is a Bin(1440, .001) random variable. Hence

$$P(X = k) = \binom{1440}{k} (.001)^k (.999)^{1440-k}, \qquad 0 \leq k \leq 1440.$$

This is numerically difficult to compute for even moderate values of k. Hence we approximate X as a $\mathrm{P}(1440 * .0001) = \mathrm{P}(1.440)$ random variable. This yields

$$P(X = k) = e^{-1.440} \frac{1.440^k}{k!}, \qquad k \geq 0.$$

For example, for $k = 0$, the binomial formula produces a value .2368, while the Poisson formula produces .2369.   □

Example 2.17 (Service Facility). Customers arrive at a service facility one at a time. Suppose that the total number of arrivals during a 1-hour period is a Poisson random variable with parameter 8. Compute the probability that at least three customers arrive during 1 hour.

Let X be the number of arrivals during 1 hour. Then we are given that X is a $\mathrm{P}(8)$ random variable. The required probability is given by

$$\begin{aligned}
P(X \geq 3) &= 1 - P(X = 0) - P(X = 1) - P(X = 2) \\
&= 1 - e^{-8}(1 + 8 + 64/2) \\
&= 1 - 41e^{-8} = .9862.
\end{aligned}$$

□

Next we turn our attention to discrete random variables based upon urn models.

Urn Models. Consider an urn containing K types of ball. Let N_k be the number of type k balls in the urn, $1 \leq k \leq K$. Let $N = N_1 + N_2 + \cdots + N_K$ be the total number of balls in the urn. A random experiment consists of randomly removing a ball from the urn and noting its type. If this ball is replaced before the next trial we get independent trials, and the Random Variables **A** through **E** described previously will be needed. However, if the ball is not replaced, then the trials are not independent and new random variables will arise.

F. Finite Discrete Random Variable

Suppose we draw one ball from the urn. Let X be its type. Then X is a discrete random variable with finite state space $S = \{1, 2, \ldots, K\}$. Its pmf is given by

$$p_k = \mathsf{P}(X = k) = \frac{N_k}{N}, \qquad k \in S. \tag{2.8}$$

Consider a special case when all N_k are equal. Then the pmf in (2.8) reduces to

$$p_k = \mathsf{P}(X = k) = \frac{1}{K}, \qquad k \in S. \tag{2.9}$$

Definition 2.11 (Discrete Uniform Distribution). The pmf in (2.9) is called a Discrete Uniform distribution over $\{1, 2, \ldots, K\}$.

Discrete uniform random variables arise frequently in statistical sampling experiments that involve picking one item from a group of K items. They also arise in simulation experiments, where we generate uniform discrete random variables using a computer. We illustrate this in the next example.

Example 2.18 (Random Number Generators). A common technique of generating "random" numbers on a digital computer is to use a congruential random number generator. It starts with an initial integer $X_0 \in [1, 2^{31} - 2]$, called the seed. It then recursively generates numbers $X_1, X_2, X_3, \ldots$ using the following formula:

$$X_{n+1} = 95070637 * X_n \bmod(2^{31} - 1)$$

where $x \bmod(y)$ is the remainder when x is divided by y. The integers X_n also lie in $[0, 2^{31} - 1]$ and behave as if they are uniformly distributed over it, although they are generated by a very deterministic scheme! The uniform random numbers thus generated play a very important role in computer simulations of random systems. □

G. Hypergeometric Random Variable

Suppose we pick n balls from the urn without replacement. Let X be the number of type 1 balls drawn in n drawings. For the sake of simplicity we use $M = N_1$. The state space of X is $S = \{\max(0, M + n - N), \ldots, \min(n, M)\}$. The pmf of X is given by

$$p_k = \mathsf{P}(X = k) = \frac{\dbinom{M}{k}\dbinom{N - M}{n - k}}{\dbinom{N}{n}}, \qquad k \in S. \tag{2.10}$$

The above equation can be derived as follows. First of all, there are a total of $\binom{N}{n}$ ways of picking n balls out of the N balls without replacement. To get exactly k balls of type 1, we first pick k balls out of M type 1 balls, of which there are $\binom{M}{k}$ ways, and pick $n - k$ balls of other types from the remaining $N - M$ balls, of which there are $\binom{N-M}{n-k}$ ways. This produces (2.10). Again, we leave it to the reader to verify that (2.1) and (2.2) are satisfied.

Definition 2.12 (Hypergeometric Distribution). The pmf in (2.10) is called a Hypergeometric distribution with parameters N, M, and n.

Example 2.19 (Committees). A school board consists of twelve elected officials: seven democrats, four republicans, and one independent. A subcommittee of four is to be formed to look into the issue of violence in schools. What is the distribution of the number of democrats on the committee, if the four committee members are randomly chosen from the twelve (although in politics things are almost never done randomly!)?

Let X be the number of democrats on the committee. Then X is a hypergeometric random variable with parameters $N = 12$, $M = 7$, $n = 4$. The distribution is given by

$$P(X = 0) = \frac{\binom{7}{0}\binom{5}{4}}{\binom{12}{4}} = \frac{5}{495} = .0101,$$

$$P(X = 1) = \frac{\binom{7}{1}\binom{5}{3}}{\binom{12}{4}} = \frac{70}{495} = .1414,$$

$$P(X = 2) = \frac{\binom{7}{2}\binom{5}{2}}{\binom{12}{4}} = \frac{210}{495} = .4242,$$

$$P(X = 3) = \frac{\binom{7}{3}\binom{5}{1}}{\binom{12}{4}} = \frac{175}{495} = .3535,$$

$$P(X = 4) = \frac{\binom{7}{4}\binom{5}{0}}{\binom{12}{4}} = \frac{35}{495} = .0707.$$

□

2.5. Continuous Random Variables

Continuous random variables, as the name suggests, take values continuously in an interval in the real line. The lifetime of the light bulb in Example 2.8 is a continuous random variable. A formal definition follows:

Definition 2.13 (Continuous Random Variable). A random variable with cdf $F(\cdot)$ is said to be continuous if there exists a function $f(\cdot)$ such that

$$F(x) = \int_{-\infty}^{x} f(u)\,du \tag{2.11}$$

for all $x \in (-\infty, \infty)$.

In particular, if $F(\cdot)$ is a differentiable function, then it is a cdf of a continuous random variable and the function $f(\cdot)$ is given by

$$f(x) = \frac{d}{dx}F(x) = F'(x). \tag{2.12}$$

Furthermore, if $F(\cdot)$ is continuous (i.e., no jumps) and piecewise differentiable, then the above result holds true and f is given by (2.12) in each of the intervals over which F is differentiable. The function $f(\cdot)$ completely determines $F(\cdot)$ by (2.11). Hence it provides an alternate way of describing a continuous random variable, and is given a special name:

Definition 2.14 (Probability Density Function (pdf)). The function $f(\cdot)$ of (2.11) is called the probability density function of X.

Example 2.20 (Lifetimes). Consider the cdf of the lifespan X of the light bulb given in Example 2.8. The cdf is continuous, and piecewise differentiable over $(-\infty, 0)$, $[0, 100]$, and $(100, \infty)$. Hence the random variable X is continuous, and its pdf can be computed by using (2.12) in the three intervals as follows:

$$f(x) = \begin{cases} 0 & \text{if } x < 0, \\ \dfrac{1}{100} & \text{if } 0 \le x \le 100, \\ 0 & \text{if } x > 100. \end{cases}$$

$\square$

The pdf of a continuous random variable is equivalent to the pmf of a discrete random variable, and hence has similar properties. They are given in the following theorem. We omit the proof.

Theorem 2.3. *A function f is a pdf of a continuous random variable if and only if it satisfies the following:*

$$f(x) \ge 0 \qquad \text{for all} \quad x \in (-\infty, \infty), \tag{2.13}$$

$$\int_{-\infty}^{\infty} f(u)\,du = 1. \tag{2.14}$$

Notice that $f(x)$ is not a probability itself, and hence can be more than 1. One intuitive way of thinking about f is the following:

$$P(x < X \leq x + h) = F(x + h) - F(x) \approx f(x)h$$

for sufficiently small values of h. This is just a result of f being the derivative of F.

2.6. Common Continuous Random Variables

In this section we shall study several commonly used continuous random variables.

A. Uniform Random Variable $U(a, b)$
Consider a random variable X with parameters a and b ($-\infty < a < b < \infty$) taking values in the state space $[a, b]$ with the following pdf:

$$f(x) = \begin{cases} 0 & \text{if } x < a, \\ \dfrac{1}{b-a} & \text{if } a \leq x \leq b, \\ 0 & \text{if } x > b. \end{cases} \tag{2.15}$$

The corresponding cdf can be computed using (2.11) as

$$F(x) = \begin{cases} 0 & \text{if } x < a, \\ \dfrac{x-a}{b-a} & \text{if } a \leq x \leq b, \\ 1 & \text{if } x > b. \end{cases} \tag{2.16}$$

We see that for $x \in (a, b)$,

$$P(x < X \leq x + h) \approx f(x)h = \frac{h}{b-a}$$

is independent of x. Thus the random variable is equally likely to take any value in (a, b). Hence it is called a Uniform random variable and is denoted by $U(a, b)$.

Definition 2.15 (Uniform Distribution). The pdf given by (2.15) is called the Uniform Density. The cdf given by (2.16) is called the Uniform Distribution. They are both denoted by $U(a, b)$.

Example 2.21 (Lifetimes). The lifespan of a light bulb with a cdf as given in Example 2.8 and a pdf as in Example 2.20 is a $U(0, 100)$ random variable. □

B. Exponential Random Variable $Exp(\lambda)$
Consider a random variable with parameter $\lambda > 0$ taking values in $[0, \infty)$ with the following pdf:

$$f(x) = \begin{cases} 0 & \text{if } x < 0, \\ \lambda e^{-\lambda x} & \text{if } x \geq 0. \end{cases} \tag{2.17}$$

The corresponding cdf can be computed using (2.11) as

$$F(x) = \begin{cases} 0 & \text{if } x < 0, \\ 1 - e^{-\lambda x} & \text{if } x \geq 0. \end{cases} \tag{2.18}$$

Such a random variable is called an Exponential random variable and is denoted by Exp(λ). Note that if the random variable X has units of time, the parameter λ has units of $(\text{time})^{-1}$ From (2.18) we see that

$$P(X > x) = e^{-\lambda x}, \qquad x \geq 0.$$

Definition 2.16 (Exponential Distribution). The pdf given by (2.17) is called the Exponential Density. The cdf given by (2.18) is called the Exponential Distribution. They are both denoted by Exp(λ).

Example 2.22 (Time to Failure). Suppose a new machine is put into operation at time zero. Its lifetime is known to be an Exp(λ) random variable, with $\lambda = .1/\text{hr}$. What is the probability that the machine will give trouble-free service continuously for 1 day?

To use consistent units we use 24 hours instead of 1 day. Let X be the lifetime of the machine in hours. We compute the required probability as

$$P(X > 24) = 1 - P(X \leq 24) = e^{-(.1)(24)} = e^{-2.4} = .0907.$$

Suppose the machine has not failed by the end of the first day. What is the probability that it will give trouble-free service for the whole of the next day?

The required probability is given by

$$P(X > 48 | X > 24) = \frac{P(X > 48; X > 24)}{P(X > 24)}$$
$$= \frac{P(X > 48)}{P(X > 24)}$$
$$= \frac{e^{-(.1)(48)}}{e^{-(.1)(24)}}$$
$$= e^{-(.1)(24)}$$
$$= .0907.$$

But this is the same as the previous answer. Thus given that the machine has not failed by day 1, it is as good as new! This is called the *memoryless property* and is one of the most important properties of the exponential random variable. We shall see it in greater detail in Chapter 6.   □

C. Erlang Random Variable Erl(k, λ)

Consider a random variable with parameters $k = 1, 2, 3, \ldots$ and $\lambda > 0$ taking values in

$[0, \infty)$ with the following pdf:

$$f(x) = \begin{cases} 0 & \text{if } x < 0, \\ \lambda e^{-\lambda x} \dfrac{(\lambda x)^{k-1}}{(k-1)!} & \text{if } x \geq 0. \end{cases} \tag{2.19}$$

Computing F from f of the above equation by using (2.11) is a tedious exercise in integration by parts, and we omit the details. The final expression is given below:

$$F(x) = \begin{cases} 0 & \text{if } x < 0, \\ 1 - \sum_{r=0}^{k-1} e^{-\lambda x} \dfrac{(\lambda x)^r}{r!} & \text{if } x \geq 0. \end{cases} \tag{2.20}$$

Such a random variable is called an Erlang random variable and is denoted by $\text{Erl}(k, \lambda)$. As in the exponential case, if the random variable X has units of time, the parameter λ has units of $(\text{time})^{-1}$. From (2.20) we see that

$$P(X > x) = e^{-\lambda x} \sum_{r=0}^{k-1} \frac{(\lambda x)^r}{r!}, \qquad x \geq 0.$$

Definition 2.17 (Erlang Distribution). The pdf given by (2.19) is called the Erlang Density. The cdf given by (2.20) is called the Erlang Distribution. They are both denoted by $\text{Erl}(k, \lambda)$.

Example 2.23 (Time to Failure). We shall redo Example 2.22 under the assumption that the lifetime of the machine is an $\text{Erl}(k, \lambda)$ random variable with parameters $k = 2$ and $\lambda = .2/\text{hr}$. From (2.20) we have

$$P(\text{no failure in the first 24 hr}) = P(X > 24)$$
$$= e^{-(.2)(24)}(1 + (.2)(24))$$
$$= .0477$$

and

$$P(\text{no failure in the second 24 hr} \mid \text{no failure in the first 24 hr})$$
$$= P(X > 48 | X > 24)$$
$$= \frac{P(X > 48)}{P(X > 24)}$$
$$= \frac{e^{-(.2)(48)}(1 + (.2)(48))}{e^{-(.2)(24)}(1 + (.2)(24))}$$
$$= .0151.$$

The second probability is lower than the first, indicating that the machine deteriorates with age. This is what we would expect.   □

D. Hyperexponential Random Variable $\text{Hex}(k, \lambda, p)$

Consider a random variable taking values in $[0, \infty)$ with the following parameters: $k \in \{1, 2, 3, \ldots\}$, $\lambda = [\lambda_1, \lambda_2, \ldots, \lambda_k]$, and $p = [p_1, p_2, \ldots, p_k]$. We assume that $\{\lambda_1 > 0,$

$\lambda_2 > 0, \ldots, \lambda_k > 0\}$, $\{p_1 > 0, p_2 > 0, \ldots, p_k > 0\}$, and that $\sum_{r=1}^{k} p_r = 1$. Suppose the pdf is given by

$$f(x) = \begin{cases} 0 & \text{if } x < 0, \\ \sum_{r=1}^{k} p_r \lambda_r e^{-\lambda_r x} & \text{if } x \geq 0. \end{cases} \tag{2.21}$$

Computing F from f of the above equation by using (2.11) we get

$$F(x) = \begin{cases} 0 & \text{if } x < 0, \\ 1 - \sum_{r=1}^{k} p_r e^{-\lambda_r x} & \text{if } x \geq 0. \end{cases} \tag{2.22}$$

Such a random variable is called a Hyperexponential random variable and is denoted by Hex(k, λ, p). As in the exponential case, if the random variable X has units of time, the parameters λ_r have units of (time)$^{-1}$. From (2.20) we see that

$$\mathsf{P}(X > x) = \sum_{r=1}^{k} p_r e^{-\lambda_r x}, \qquad x \geq 0.$$

Definition 2.18 (Hyperexponential Distribution). The pdf given by (2.21) is called the Hyperexponential Density. The cdf given by (2.22) is called the Hyperexponential Distribution. They are both denoted by Hex(k, λ, p).

The density given in (2.21) is seen to be a weighted sum of exponential densities. Since the weights p_r are nonnegative and add up to 1, such a density is called a mixture of exponential densities.

Example 2.24 (Mixture Density). Suppose a factory receives interchangeable components from k vendors, with the components from vendor r having Exp(λ_r) lifetimes. Let p_r be the fraction of the components that come from vendor r, $r = 1, 2, \ldots, k$. Suppose one component is picked at random from all the incoming components. We shall show later that the lifetime of this randomly picked component is Hex(k, λ, p). We need to develop the concepts of multivariate random variables in the next chapter before we can show this rigorously. Intuitively, the density is Exp(λ_r) with probability p_r, hence the "total density" is given by the weighted sum as in (2.21). $\qquad \square$

E. Normal Random Variable $N(\mu, \sigma^2)$
Consider a random variable taking values in $(-\infty, \infty)$ with the following parameters: $\mu \in (-\infty, \infty)$ and $\sigma^2 \in (0, \infty)$. Suppose the pdf is given by

$$f(x) = \frac{1}{\sqrt{2\pi\sigma^2}} \exp\left\{ -\frac{1}{2} \left(\frac{x - \mu}{\sigma} \right)^2 \right\}, \qquad x \in (-\infty, \infty). \tag{2.23}$$

Such a random variable is called a Normal random variable and is denoted by $N(\mu, \sigma^2)$. There is no closed-form expression for the cdf of this random variable. The $N(0, 1)$ random variable is called the standard Normal random variable, and its cdf is denoted by $\Phi(\cdot)$:

$$\Phi(x) = \int_{-\infty}^{x} \frac{1}{\sqrt{2\pi}} \exp\left\{ -\frac{1}{2} u^2 \right\} du. \tag{2.24}$$

Tables of Φ are generally available. The following property of the Φ function is very useful and can be proved by using the symmetry of the density function around $x = \mu$:

$$\Phi(-x) = 1 - \Phi(x).$$

Definition 2.19 (Normal Distribution). The pdf given by (2.23) is called the Normal Density, and is denoted by $N(\mu, \sigma^2)$. The cdf given by (2.24) is called the Standard Normal Distribution. It is denoted by $N(0, 1)$.

The Normal distribution is also called the *Gaussian distribution*. It occurs frequently in statistics as the limiting case of other distributions. In the next section we shall show how the cdf of an $N(\mu, \sigma^2)$ random variable can be computed from Φ of (2.24).

2.7. Functions of Random Variables

Let X be a random variable, and let g be a function. Then $Y = g(X)$ is another random variable, and has its own cdf, and pmf (in case Y is discrete) or pdf (in case it is continuous).

Example 2.25 (Common Functions). Some common examples of functions of random variables are:

(1) $Y = aX + b$, where a and b are constants.

(2) $Y = X^2$.

(3) $Y = X^m$.

(4) $Y = e^{-sX}$, where s is a constant.   □

The cdf of Y is in general difficult to compute. We illustrate with a few examples.

Example 2.26 (Linear Transformation). Let X be a random variable with cdf F_X, and let $Y = aX + b$, where a and b are constants. Thus Y is a linear transformation of X. The cdf F_Y of Y is given by

$$\begin{aligned}
F_Y(y) &= P(Y \le y) \\
&= P(aX + b \le y) \\
&= P(X \le (y - b)/a) \\
&= F_X\left(\frac{y - b}{a}\right).
\end{aligned} \tag{2.25}$$

If X is continuous with pdf f_X, then so is Y, and its pdf f_Y can be computed by differentiating (2.25) as follows:

$$f_Y(y) = \frac{1}{a} f_X\left(\frac{y - b}{a}\right). \tag{2.26}$$

□

We illustrate the use of the above formula in the next two examples.

Example 2.27 (Normalized Normal Random Variable). Let X be an $N(\mu, \sigma^2)$ random variable, and define

$$Y = \frac{X - \mu}{\sigma}.$$

Y is a linear transformation of X with $a = 1/\sigma$ and $b = -\mu/\sigma$. Hence we can use the results of the above example. Using (2.26) and (2.23) we get

$$f_Y(y) = \frac{1}{\sqrt{2\pi}} \exp\left\{-\tfrac{1}{2}y^2\right\}.$$

But this is a pdf of an $N(0, 1)$ random variable. Hence Y must be an $N(0, 1)$ random variable with cdf Φ given in (2.24). For example, suppose X is an $N(2, 16)$ random variable, and we want to compute $P(-4 \leq X \leq 8)$. We know that $Y = (X - 2)/4$ is an $N(0, 1)$ random variable. Hence

$$P(-4 \leq X \leq 8) = P((-4 - 2)/4 \leq (X - 2)/4 \leq (8 - 2)/4)$$
$$= P(-1.5 \leq Y \leq 1.5) = \Phi(1.5) - \Phi(-1.5) = 2\Phi(1.5) - 1. \qquad \square$$

Example 2.28 (Scaled Exponential Random Variable). Let X be an $Exp(\lambda)$ random variable, and let $Y = aX$ be a scaled version of it. Using Example 2.26 we can show that

$$f_Y(y) = \frac{1}{a} f_X\left(\frac{y}{a}\right) = \frac{\lambda}{a} \exp\left\{-\left(\frac{\lambda}{a}\right)y\right\}.$$

Thus f_Y is an $Exp(\lambda/a)$ density and Y is an $Exp(\lambda/a)$ random variable. $\qquad \square$

Example 2.29 (Square Transformation). Let X be a continuous random variable with cdf F_X and pdf f_X, and let $Y = X^2$. Thus Y is a square transformation of X. The cdf F_Y of Y is given by

$$\begin{aligned}
F_Y(y) &= P(Y \leq y) \\
&= P(X^2 \leq y) \\
&= P(-\sqrt{y} \leq X \leq \sqrt{y}) \\
&= F_X(\sqrt{y}) - F_X(-\sqrt{y}).
\end{aligned} \qquad (2.27)$$

The pdf f_Y of Y can be computed by differentiating (2.27) as follows:

$$f_Y(y) = \frac{1}{2\sqrt{y}}\left(f_X(\sqrt{y}) + f_X(-\sqrt{y})\right). \qquad (2.28)$$

$\qquad \square$

Example 2.30 (Exponential Transformation). Let X be a continuous nonnegative random variable with cdf F_X and pdf f_X, and let $Y = \exp\{-sX\}$, where $s > 0$ is a constant. Thus Y is an exponential transformation of X. Then $Y \in (0, 1)$. The cdf F_Y of Y is given by

$$\begin{aligned}
F_Y(y) &= P(Y \leq y) \\
&= P(\exp\{-sX\} \leq y)
\end{aligned}$$

$$= P\left(X \geq -\frac{1}{s}\ln(y)\right)$$

$$= 1 - F_X\left(-\frac{1}{s}\ln(y)\right), \qquad y \in (0, 1). \tag{2.29}$$

The pdf f_Y of Y can be computed by differentiating (2.29) as follows:

$$f_Y(y) = \frac{1}{sy}f_X\left(-\frac{1}{s}\ln(y)\right), \qquad y \in (0, 1). \tag{2.30}$$

$\square$

Example 2.31 (Exponential Transformation). Let X be an Exp(1) random variable, and let $Y = \exp\{-X\}$. Thus Y is an exponential transformation as described in Example 2.30 with $s = 1$. Hence, using (2.30) and (2.17) we get

$$f_Y(y) = \frac{1}{y}\exp\{\ln(y)\} = \frac{y}{y} = 1, \qquad y \in (0, 1).$$

Thus Y is a U(0, 1) random variable. $\square$

2.8.   Expectation of a Discrete Random Variable

"Expectation" is one of the most important concepts in probability. Intuitively, expectation of a random variable is the value that we expect the random variable to take. Although this is vague as stated, the following example shows that we have a pretty good idea of what it means:

Example 2.32 (Intuitive Expected Value). Suppose we toss a fair coin. If we get a tail we win \$1, if we get a head we win \$2. However, there is a fee to play this tossing game. How much fee should we be willing to pay?

Since our winnings are going to be either \$1 or \$2, we should be willing to pay at least \$1 and at most \$2. But exactly how much? Since the chances are 50/50, our intuition says that the amount should be midway, namely, \$1.5. We take this to be the expected value of our winnings on the toss. $\square$

The above example suggests the following definition of the "expected value" of a random variable:

Definition 2.20 (Expected Value). Let X be a discrete random variable with state space $S = \{x_0, x_1, x_2, \ldots\}$. Then the expected value of X is defined as

$$E(X) = \sum_{k=0}^{\infty} x_k P(X = x_k). \tag{2.31}$$

Note that $E(X)$ may not be an element of the state space of the random variable. Thus the expected value of an integer-valued random variable may be noninteger. We illustrate with several examples.

Example 2.33 (Example 2.32 Continued). Let X be the amount we win on the coin tossing experiment of Example 2.32. Then the fairness of the coin implies that $P(X = 1) = .5$ and $P(X = 2) = .5$. The expected value of X is then given by (2.31) as

$$E(X) = 1 * P(X = 1) + 2 * P(X = 2) = 1 * .5 + 2 * .5 = 1.5.$$

This matches the expected value we had intuitively associated with X in Example 2.32. □

Example 2.34 (Bernoulli Random Variable). Let X be a $B(p)$ random variable with pmf as in (2.3). Then, by (2.31), we have

$$E(X) = 0 \cdot P(X = 0) + 1 \cdot P(X = 1) = 0(1 - p) + 1p = p.$$

Thus the expected value of the $B(p)$ random variable is p. □

Example 2.35 (Sum of Two Dice). Let X be the sum of the outcomes from tossing two fair dice as in Example 2.9. Using the pmf given there we get

$$E(X) = \sum_{k=2}^{12} kP(X = k)$$

$$= 2 * \frac{1}{36} + 3 * \frac{2}{36} + 4 * \frac{3}{36} + 5 * \frac{4}{36} + 6 * \frac{5}{36}$$

$$+ 7 * \frac{6}{36} + 8 * \frac{5}{36} + 9 * \frac{4}{36} + 10 * \frac{3}{36} + 11 * \frac{2}{36} + 12 * \frac{1}{36}$$

$$= 7. \tag{2.32}$$

□

Example 2.36 (Binomial Random Variable). Let X be a $Bin(n, p)$ random variable, with pmf given in (2.4). Then

$$E(X) = \sum_{k=0}^{n} kP(X = k)$$

$$= \sum_{k=0}^{n} k \binom{n}{k} p^k (1 - p)^{n-k}$$

$$= \sum_{k=0}^{n} k \frac{n!}{k!(n - k)!} p^k (1 - p)^{n-k}$$

$$= \sum_{k=1}^{n} \frac{n!}{(k - 1)!(n - k)!} p^k (1 - p)^{n-k}$$

$$= np \sum_{k=1}^{n} \frac{(n - 1)!}{(k - 1)!(n - k)!} p^{k-1} (1 - p)^{n-k}$$

$$= np \sum_{k=0}^{n-1} \frac{(n - 1)!}{k!(n - 1 - k)!} p^k (1 - p)^{n-1-k}$$

$$= np(p + 1 - p)^{n-1}$$

$$= np. \tag{2.33}$$

□

Example 2.37 (Poisson Random Variable). Let X be a $P(\lambda)$ random variable with pmf given in (2.7). Then

$$
\begin{aligned}
E(X) &= \sum_{k=0}^{\infty} k P(X = k) \\
&= \sum_{k=0}^{\infty} k e^{-\lambda} \frac{\lambda^k}{k!} \\
&= \lambda \sum_{k=1}^{\infty} e^{-\lambda} \frac{\lambda^{k-1}}{(k-1)!} \\
&= \lambda e^{-\lambda} \sum_{k=0}^{\infty} \frac{\lambda^k}{k!} \\
&= \lambda e^{-\lambda} e^{\lambda} \\
&= \lambda.
\end{aligned}
\tag{2.34}
$$

$\square$

When X is a discrete random variable taking values in $S = \{0, 1, 2, \ldots\}$ we can show that (see Conceptual Problem 2.14)

$$
E(X) = \sum_{k=0}^{\infty} P(X > k).
\tag{2.35}
$$

This formula sometimes simplifies the computation of $E(X)$. We illustrate with the following example:

Example 2.38 (Geometric Random Variable). Let X be a $G(p)$ random variable with pmf given in (2.5). Then,

$$
P(X > k) = (1 - p)^k, \qquad k \geq 0.
$$

Hence, from (2.35)

$$
\begin{aligned}
E(X) &= \sum_{k=0}^{\infty} P(X > k) \\
&= \sum_{k=0}^{\infty} (1 - p)^k \\
&= \frac{1}{1 - (1 - p)} \\
&= \frac{1}{p}.
\end{aligned}
\tag{2.36}
$$

$\square$

For the expected values of other discrete random variables, see Table 2.2 in Section 2.11.

2.9. Expectation of a Continuous Random Variable

In this section we extend the concept of expected value to continuous random variables. This essentially involves replacing the sum in (2.31) by the corresponding integral. The formal definition follows:

Definition 2.21 (Expected Value). Let X be a continuous random variable with pdf $f(\cdot)$. Then the expected value of X is defined as

$$E(X) = \int_{-\infty}^{\infty} x f(x)\, dx. \tag{2.37}$$

Example 2.39 (Uniform Random Variable). Let X be a $U(a, b)$ random variable with pdf given in (2.15). Then

$$
\begin{aligned}
E(X) &= \int_{-\infty}^{\infty} x f(x)\, dx \\
&= \int_{a}^{b} x \frac{1}{b-a}\, dx \\
&= \frac{1}{b-a} \frac{b^2 - a^2}{2} \\
&= \frac{a+b}{2}.
\end{aligned} \tag{2.38}
$$

Since the random variable is uniformly distributed over a to b, it is intuitively clear that its expected value should be the midpoint of the interval (a, b). □

Example 2.40 (Triangular Density). Consider a random variable X with the following density over $(0, 2)$:

$$
f(x) = \begin{cases}
0 & \text{if } x < 0, \\
x & \text{if } 0 \leq x \leq 1, \\
2 - x & \text{if } 1 \leq x \leq 2, \\
0 & \text{if } x > 2.
\end{cases}
$$

Then the expected value of X is given by

$$
\begin{aligned}
E(X) &= \int_{-\infty}^{\infty} x f(x)\, dx \\
&= \int_{0}^{1} x^2\, dx + \int_{1}^{2} x(2-x)\, dx \\
&= \left. \frac{x^3}{3} \right|_{0}^{1} + \left. \left(x^2 - \frac{x^3}{3} \right) \right|_{1}^{2} \\
&= 1.
\end{aligned} \tag{2.39}
$$

This is what we would expect, since the density is symmetric around 1. □

Example 2.41 (Exponential Random Variable). Let X be an Exp(λ) random variable with pdf given in (2.17). Then

$$
\begin{aligned}
\mathsf{E}(X) &= \int_{-\infty}^{\infty} x f(x) \, dx \\
&= \int_{0}^{\infty} x \lambda e^{-\lambda x} \, dx \\
&= \frac{1}{\lambda} \int_{0}^{\infty} \lambda e^{-\lambda x} (\lambda x) \, dx \\
&= \frac{1}{\lambda}.
\end{aligned}
$$

Here the last integral is one since it is an integral of an Erl(2, λ) density.   □

Example 2.42 (Erlang Random Variable). Let X be an Erl(k, λ) random variable with pdf given in (2.19). Then

$$
\begin{aligned}
\mathsf{E}(X) &= \int_{-\infty}^{\infty} x f(x) \, dx \\
&= \int_{0}^{\infty} x \lambda e^{-\lambda x} \frac{(\lambda x)^{k-1}}{(k-1)!} \, dx \\
&= \frac{k}{\lambda} \int_{0}^{\infty} \lambda e^{-\lambda x} \frac{(\lambda x)^{k}}{k!} \, dx \\
&= \frac{k}{\lambda}.
\end{aligned}
\tag{2.40}
$$

Here the last integral is one since it is an integral of an Erl($k + 1, \lambda$) density.   □

Example 2.43 (Normal Random Variable). Let X be a Normal random variable with pdf given in (2.23) with $\mu = 0$ and $\sigma^2 = 1$. Then

$$
\begin{aligned}
\mathsf{E}(X) &= \int_{-\infty}^{\infty} x f(x) \, dx \\
&= \int_{-\infty}^{\infty} x \frac{1}{\sqrt{2\pi}} \exp\left\{ -\tfrac{1}{2} x^2 \right\} \, dx. \\
&= 0.
\end{aligned}
\tag{2.41}
$$

Here the last equality follows because the integrand is an odd function of x. We shall see in the next section how the expected value of an N(μ, σ^2) random variable can be easily computed from this.   □

We can establish the following for a nonnegative continuous random variable (see Conceptual Problem 2.15):

$$
\mathsf{E}(X) = \int_{0}^{\infty} (1 - F(x)) \, dx.
\tag{2.42}
$$

We illustrate the use of (2.42) with the help of an Exp(λ) random variable.

Example 2.44 (Exponential Random Variable). Let X be an $\text{Exp}(\lambda)$ random variable. Then

$$1 - F(x) = \mathsf{P}(X > x) = e^{-\lambda x}.$$

Hence, using (2.42) we get

$$\mathsf{E}(X) = \int_0^\infty e^{-\lambda x}\,dx$$
$$= \frac{1}{\lambda}. \tag{2.43}$$

This agrees with (2.40) (with $k = 1$), as it must.   □

See Table 2.3 for the expected values of the common continuous random variables.

2.10. Expectation of a Function of a Random Variable

Let X be a random variable with cdf F_X, and $Y = g(X)$. In Section 2.7 we saw that Y is also a random variable and showed how its cdf F_Y (or pmf p_Y, or pdf f_Y) can be computed. Since Y is a random variable, it has an expectation. From (2.31) and (2.37), we see that

$$\mathsf{E}(Y) = \begin{cases} \sum_i y_i\, p_Y(y_i) & \text{if } Y \text{ is discrete,} \\[2mm] \displaystyle\int_y y f_Y(y)\,dy & \text{if } Y \text{ is continuous.} \end{cases}$$

Unfortunately, this method requires the pdf or pmf of Y, which is difficult to compute, as we saw in Section 2.7. Is there an easier way? There is indeed a way to avoid computing the distribution of Y. It is called

Theorem 2.4 (Law of Subconscious Statistician).

$$\mathsf{E}(Y) = \mathsf{E}(g(X)) = \begin{cases} \sum_i g(x_i) p_X(x_i) & \text{if } X \text{ is discrete,} \\[2mm] \displaystyle\int_x g(x) f_X(x)\,dx & \text{if } X \text{ is continuous.} \end{cases} \tag{2.44}$$

Proof. Assume that X is discrete with state space $\{x_0, x_1, \ldots\}$. Then Y is necessarily discrete. Let its state space be $\{y_0, y_1, \ldots\}$. We have

$$\mathsf{P}(Y = y_j) = \sum_{i:g(x_i)=y_j} \mathsf{P}(X = x_i). \tag{2.45}$$

Now,

$$\sum_i g(x_i) p_X(x_i) = \sum_j \sum_{i:g(x_i)=y_j} g(x_i)\mathsf{P}(X = x_i)$$
$$= \sum_j y_j \sum_{i:g(x_i)=y_j} \mathsf{P}(X = x_i)$$

$$= \sum_j y_j P(Y = y_j) \quad \text{from (2.45)}$$

$$= E(Y).$$

This proves the theorem in the discrete case. The proof in the continuous case is similar but requires technicalities, hence we omit it. □

We now show that the expectation is linear, i.e.,

$$E(aX + b) = aE(X) + b, \tag{2.46}$$

where a and b are constants. To begin with, assume that X is a continuous random variable with pdf f. Let $Y = aX + b$. Then

$$E(Y) = E(aX + b)$$

$$= \int_{-\infty}^{\infty} (ax + b) f(x) \, dx$$

$$= a \int_{-\infty}^{\infty} x f(x) \, dx + b \int_{-\infty}^{\infty} f(x) \, dx$$

$$= aE(X) + b.$$

Here the second equality follows due to (2.44), the third equality follows due to the linearity property of integrals, and the fourth equality follows due to (2.37) and (2.14). In particular, the above derivation shows that

$$E(c) = c$$

for any constant c. In general we have

$$E(g_1(X) + g_2(X)) = E(g_1(X)) + E(g_2(X)) \tag{2.47}$$

for all random variables X, continuous or discrete.

Example 2.45 (Expectation of $N(\mu, \sigma^2)$). Let X be a standard Normal random variable. Then using the results of Example 2.27 we see that $Y = \sigma X + \mu$ is an $N(\mu, \sigma^2)$ random variable. Hence, using Example 2.43 and (2.46) we get

$$E(Y) = \sigma E(X) + \mu = \sigma \cdot 0 + \mu = \mu. \qquad \qquad □$$

Definition 2.22 (Special Expectations).

$$E(X^n) = n\text{th moment of } X,$$

$$E((X - E(X))^n) = n\text{th central moment of } X,$$

$$E((X - E(X))^2) = \text{variance of } X. \tag{2.48}$$

□

Variance of a random variable, as defined by (2.48) is an important concept. Here we derive an alternate equation for computing it. Using $m = E(X)$ in the definition of variance in

(2.48) we get

$$
\begin{aligned}
\mathrm{Var}(X) &= \mathsf{E}((X-m)^2) \\
&= \mathsf{E}(X^2 - 2Xm + m^2) \\
&= \mathsf{E}(X^2) - \mathsf{E}(2Xm) + m^2 \quad \text{(from (2.47))} \\
&= \mathsf{E}(X^2) - 2m\mathsf{E}(X) + m^2 \\
&= \mathsf{E}(X^2) - 2m^2 + m^2 \\
&= \mathsf{E}(X^2) - m^2.
\end{aligned}
$$

Hence we have

$$
\mathrm{Var}(X) = \mathsf{E}(X^2) - (\mathsf{E}(X))^2. \tag{2.49}
$$

Example 2.46 (Variance of Bin(n, p)). Let X be a Bin(n, p) random variable. We first compute

$$
\begin{aligned}
\mathsf{E}(X^2) &= \sum_{k=0}^{n} k^2 \binom{n}{k} p^k (1-p)^{n-k} \\
&= \sum_{k=0}^{n} (k(k-1) + k) \frac{n!}{k!\,(n-k)!} p^k (1-p)^{n-k} \\
&= \sum_{k=2}^{n} \frac{n!}{(k-2)!\,(n-k)!} p^k (1-p)^{n-k} + \sum_{k=1}^{n} \frac{n!}{(k-1)!\,(n-k)!} p^k (1-p)^{n-k} \\
&= n(n-1)p^2 \sum_{k=2}^{n} \frac{(n-2)!}{(k-2)!\,(n-k)!} p^{k-2} (1-p)^{n-k} \\
&\quad + np \sum_{k=1}^{n} \frac{(n-1)!}{(k-1)!\,(n-k)!} p^{k-1} (1-p)^{n-k} \\
&= n(n-1)p^2 \sum_{k=0}^{n-2} \frac{(n-2)!}{k!\,(n-2-k)!} p^k (1-p)^{n-2-k} \\
&\quad + np \sum_{k=0}^{n-1} \frac{(n-1)!}{k!\,(n-1-k)!} p^k (1-p)^{n-1-k} \\
&= n(n-1)p^2 (p+1-p)^{n-2} + np(p+1-p)^{n-1} \\
&= n(n-1)p^2 + np. \tag{2.50}
\end{aligned}
$$

Substituting in (2.49), and using the results of Example 2.36 we get

$$
\begin{aligned}
\mathrm{Var}(X) &= \mathsf{E}(X^2) - (\mathsf{E}(X))^2 \\
&= n(n-1)p^2 + np - (np)^2 \\
&= np(1-p). \qquad \square
\end{aligned}
$$

See Tables 2.2 and 2.3 in Section 2.11 for the variances of common random variables.

Example 2.47 (Moments of a Uniform Random Variable). Let X be a $U(0, 1)$ random variable. Then

$$E(X^n) = \int_0^1 x^n \, dx$$
$$= \frac{1}{n+1}.$$

□

2.11. Reference Tables

Tables 2.2 and 2.3 give the means and variances of the common discrete and continuous random variables.

TABLE 2.2. Means and variances of discrete random variables.

X	Parameters	Notation	$E(X)$	$Var(X)$
Bernoulli	p	$B(p)$	p	$p(1-p)$
Binomial	n, p	$Bin(n, p)$	np	$np(1-p)$
Geometric	p	$G(p)$	$1/p$	$(1-p)/p^2$
Negative Binomial	r, p	$NB(r, p)$	r/p	$[r(1-p)]/p^2$
Poisson	λ	$P(\lambda)$	λ	λ

TABLE 2.3. Means and variances of continuous random variables.

X	Parameters	Notation	$E(X)$	$Var(X)$
Uniform	a, b	$U(a, b)$	$(a+b)/2$	$(b-a)^2/12$
Exponential	λ	$Exp(\lambda)$	$1/\lambda$	$1/\lambda^2$
Hyperexponential	$\lambda = [\lambda_1, \ldots, \lambda_n]$ $p = [p_1, \ldots, p_n]$	$Hex(n, \lambda, p)$	$\sum_1^n p_i/\lambda_i$	$\sum_1^n 2p_i/\lambda_i^2 - \left(\sum_1^n p_i/\lambda_i\right)^2$
Erlang	k, λ	$Erl(k, \lambda)$	k/λ	k/λ^2
Normal	μ, σ^2	$N(\mu, \sigma^2)$	μ	σ^2

2.12. Problems

CONCEPTUAL PROBLEMS

2.1. Show that:

(1) $\text{Bin}(1, p) = B(p)$.

(2) $\text{NB}(1, p) = G(p)$.

2.2. Show that $\text{Erl}(1, \lambda) = \text{Exp}(\lambda)$.

2.3. Derive the properties of the pmf given in Theorem 2.2.

2.4. Derive the properties of the pdf given in Theorem 2.3.

Definition 2.23 (Mode of a Discrete Distribution). Let X be a discrete random variable on $S = \{0, 1, 2, \ldots\}$, with pmf $\{p_k, k \in S\}$. An integer m is said to be the mode of the pmf (or of X) if

$$p_m \geq p_k \quad \text{for all} \quad k \in S.$$

2.5. Find the mode of a $\text{Bin}(n, p)$ distribution.

2.6. Find the mode of a $\text{P}(\lambda)$ distribution.

Definition 2.24 (Mode of a Continuous Distribution). Let X be a continuous random variable with pdf f. A number m is said to be the mode of the pdf (or of X) if

$$f(m) \geq f(x) \quad \text{for all} \quad x \in S.$$

2.7. Find the mode of an $\text{Erl}(k, \lambda)$ density.

2.8. Find the mode of an $\text{N}(\mu, \sigma^2)$ density.

Definition 2.25 (Median of a Distribution). Let X be a random variable with cdf F. A number m is said to be the median of the cdf (or of X) if

$$F(m^-) \leq .5 \quad \text{and} \quad F(m) \geq .5.$$

2.9. Find the median of an $\text{Exp}(\lambda)$ distribution.

2.10. Find the median of a $\text{G}(p)$ distribution.

2.11. Find the median of an $\text{N}(\mu, \sigma^2)$ distribution.

Definition 2.26 (Hazard Rate of a Distribution). Let X be a nonnegative continuous random variable with pdf f and cdf F. The hazard rate, or failure rate, of X is defined as

$$h(x) = \frac{f(x)}{1 - F(x)} \quad \text{for all } x \text{ such that} \quad F(x) < 1.$$

2.12. Compute the hazard rate of an $\text{Exp}(\lambda)$ random variable.

2.13. Let h be the hazard rate of a random variable with cdf F. Show that

$$1 - F(x) = \exp\left(-\int_0^x h(u)\,du\right).$$

This shows that the hazard rate uniquely determines the cdf.

2.14. Prove (2.35).

2.15. Prove (2.42).

Definition 2.27 (Factorial Moments of a Random Variables). The second factorial moment of a random variable is defined as $E(X(X - 1))$. The rth factorial moment of a random variable is defined as $E(X(X - 1)(X - 2)\ldots(X - r + 1))$.

2.16. Compute the second factorial moment of a $P(\lambda)$ random variable.

2.17. Compute the second factorial moment of a $\text{Bin}(n, p)$ random variable.

2.18. Show that $E(X^2) \geq (E(X))^2$. When does the equality hold?

2.19. Show that $\text{Var}(aX + b) = a^2\,\text{Var}(X)$, where a and b are constants.

2.20. Show that the variance of a standard Normal random variable is 1. (*Hint:* Use the identity

$$\int_{-\infty}^{\infty} x^2 \exp\left\{-\tfrac{1}{2}x^2\right\} dx = \sqrt{2\pi}.)$$

2.21. Using Example 2.27, Conceptual Problems 2.19 and 2.20 show that the variance of an $N(\mu, \sigma^2)$ random variable is σ^2.

2.22. Show that the nth moment of an $\text{Exp}(\lambda)$ random variable is given by $n!/\lambda^n$.

2.23. Let X be a nonnegative random variable with cdf $G(\cdot)$. Let T be a fixed nonnegative constant. Show that

$$E(\min(X, T)) = \int_0^T (1 - G(t))\,dt.$$

2.24. A system consists of n independent components. Each component functions with probability p. The system as a whole functions if at least k components are functioning ($1 \leq k \leq n$). What is the probability that the system functions? (Such a system is called k-out-of-n system.)

COMPUTATIONAL PROBLEMS

2.1. State the sample space and the random function in the following experiments:

(1) Toss three coins and observe the outcome on each coin. $X =$ the number of heads − the number of tails.

(2) Toss two dice and observe the outcome on each die. $X =$ the absolute value of the difference between the two outcomes.

2.2. State the sample space and the random function in the following experiments:

(1) Toss three coins and observe the outcome on each coin. Suppose we win \$3 for every head, and lose \$2 for every tail. Let $X =$ net winnings.

(2) The movement of a coin on a game board is based on the outcome of tossing two dice. If the outcome is not a double, move the sum of the two outcomes, or else move twice the sum of the two outcomes. $X =$ the number of spaces the coin is moved.

2.3. Compute the pmf of the total number of heads obtained on three independent tosses of a fair coin.

2.4. Compute the pmfs of the random variables defined in Computational Problem 2.1. Assume all outcomes in the sample space are equally likely.

2.5. Compute the pmfs of the random variables defined in Computational Problem 2.2. Assume all outcomes in the sample space are equally likely.

2.6. Compute the pmf of the random variable Y defined in Example 2.3. Assume all outcomes in the sample space are equally likely.

2.7. A multiple choice question has five choices, only one of which is correct. A student is given one point if he circles the right answer, and a $-1/4$ point if he circles a wrong answer. If a student does not know the answer and picks a random answer to circle, compute the distribution of the points the student will get for that question.

2.8. Compute the cdf of the random variable in Example 2.19.

2.9. Suppose the weather condition on a given day is classified as dry or wet. Assume that the weather conditions on successive days are independent, and the probability of a wet day is .3. We say that there is a rainy spell on day n if the days $n - 2, n - 1$, and n are wet. Let X_n be the indicator random variable of a rainy spell on day $n, n > 2$. Compute the distribution of X_n.

2.10. A patient is tested for blood sugar every morning. If his sugar level is outside the normal range for 2 days in a row, he is started on a sugar control regime. Suppose the successive sugar levels are independent, and the probability of the sugar level being outside the normal range is p. Let X be the day the patient is put on the sugar control regime. What is the state space of X? Is X a geometric random variable? Compute $P(X = k)$ for $k = 2, 3, 4$.

2.11. Suppose the patient in Computational Problem 2.10 is declared as having no sugar problem if he is not put on a sugar control regime for the first 4 days. Let X be 1 if the patient is declared as problem free, and 0 otherwise. Compute the distribution of X.

2.12. A coin is tossed 10 times in an independent fashion. It is declared to be a fair coin if the number of heads and tails on the 10 trials differ at most by 2. Let X be the indicator of the event that the coin is declared to be fair. If the probability of heads is p, compute the distribution of X.

2.13. Suppose that each incoming customer to a store is female with probability .6 and male with probability .4. Assuming that the store had nine customers in the first hour, what is the probability that the majority of them were females?

2.14. Redo Example 2.12 for the odd parity protocol, i.e., the parity digit is set to one or zero so as to make the total number of ones in the byte an odd number.

2.15. Suppose a family has six children. Compute the distribution of the number of girls in the six children if each child is equally likely to be a boy or a girl.

2.16. Consider the k-out-of-n system of Conceptual Problem 2.24. The probability that a component is functioning at time t is given to be e^{-t}. Compute the probability that the system is functioning at time t for $n = 3, k = 2$.

2.17. Each lottery ticket has a 15-digit number. Each digit of the winning number is decided by a random draw of a ball (with replacement) from a basket containing three hundred balls. Thirty of the balls are labeled 0, thirty are labeled 1, and so on. What is the probability that at least three of the digits of the winning number are seven or more?

2.18. A couple decides to have children until a daughter is born. What is the distribution of the number of children born to the couple if each child is equally likely to be a girl or a boy?

2.19. A couple decides to have children until two sons are born. What is the distribution of the number of children born to the couple if each child is equally likely to be a girl or a boy?

2.20. Consider the machine of Example 2.13. Suppose a shock causes a unit damage to the machine with probability .1 and no damage with probability .9. Successive shocks are independent. Suppose the damages are cumulative and the machine can withstand at most four units of damage. (That is, the machine fails when the fifth unit of damage is inflicted on it.) What is the probability that the kth shock kills the machine?

2.21. A student takes a test repeatedly until he passes it three times in total. If the probability of passing the test on any given attempt is .7 and is independent of the other tests, what is the probability that the student will have to repeat the test exactly seven times?

2.22. Suppose the number of movies produced during a summer season is a $P(8)$ random variable. Each movie ticket is $5. If a person has a budget of $25 for summer movies, what is the probability that he/she will be able to see all the movies during the season?

2.23. In a city of 2.5 million residents, each person has a probability of 10^{-6} of carrying the TB infection, independent of each other. What is the probability that the city has at least five TB patients? Approximate the answer using a Poisson distribution.

2.24. A broker gives a client a choice of 10 mutual funds, six bonds and eight stocks for building a portfolio. Suppose the client chooses exactly five of these financial instruments in a random fashion. What is the pmf of the number of mutual funds picked by the client?

2.25. Suppose the pdf of a continuous random variable is:

$$f(x) = \begin{cases} 0 & \text{if } x < 0, \\ (x-a)^2 & \text{if } 0 \leq x \leq 2a, \\ 0 & \text{if } x > 2a. \end{cases} \tag{2.51}$$

(1) Compute the value of a.

(2) Compute the corresponding cdf.

2.26. For the random variable in Computational Problem 2.25 compute:

(1) $P(X > 2)$;

(2) $P(1 < X \leq 2)$.

2.27. Suppose the pdf of a continuous random variable is:

$$f(x) = \begin{cases} 0 & \text{if } x < 0, \\ x^2 & \text{if } 0 \leq x \leq c, \\ 0 & \text{if } x > c. \end{cases} \tag{2.52}$$

(1) Compute the value of c.

(2) Compute the corresponding cdf.

2.28. For the random variable in Computational Problem 2.27 compute:

(1) $P(X < .5 \text{ or } X > 1)$;

(2) $P(1 \leq X \leq 1.2)$.

2.29. Suppose the pdf of a continuous random variable is:

$$f(x) = \begin{cases} 0 & \text{if } x < 0, \\ cx^2 & \text{if } 0 \leq x \leq 1, \\ 0 & \text{if } x > 1. \end{cases} \tag{2.53}$$

(1) Compute the value of c.

(2) Compute the corresponding cdf.

2.30. For the random variable in Computational Problem 2.29 compute:

(1) $P(X < .5 \text{ or } X > .8)$;

(2) $P(.3 \leq X \leq 2)$.

2.31. The lifetime X of a radioactive atom is exponentially distributed with mean 2138 years. What is its half-life, i.e., what is the value of x such that $P(X > x) = .5$?

2.32. The weight of a ketchup bottle is Erlang distributed with $k = 10$ and $\lambda = .25 \, (\text{oz})^{-1}$. The bottle is declared to be under weight if the weight is under 39 oz. What is the probability that the bottle is declared under weight?

2.33. Consider Example 2.24. Suppose there are two vendors and each provides 50% of the items. The lifetime (in days) of an item from the first vendor is Exp(.1), and that from the second vendor is Exp(.08). Compute the probability that a randomly picked item will last more than 12 days.

2.34. The temperature indicator of an industrial oven indicates that the temperature is 1200° Fahrenheit. However, the actual temperature is distributed as a Normal random variable with parameters $\mu = 1200$ and $\sigma^2 = 400$. What is the probability that the actual temperature is greater than 1230°?

2.35. Suppose X has the density given in Computational Problem 2.25. Compute the density of $2X + 3$.

2.36. Suppose X has the density given in Computational Problem 2.27. Compute the density of X^2.

2.37. Suppose X has the density given in Computational Problem 2.29. Compute the cdf of $(X - 2)^2$.

2.38. Let X be a U(0, 1) random variable. Compute the density of $-\ln(X)$.

2.39. Let X be a U(0, 1) random variable. Compute the density of $a + (b - a)X$, where a and b are constants with $a < b$.

2.40. Let X be a P(λ), and define Y to be 0 if X is even and 1 if X is odd. Compute the pmf of Y.

2.41. The lifetime of a component (in integer number of days) is a G(.2) random variable. The component is replaced as soon as it fails, or upon reaching age 10 days. Compute the pmf of the lifetime of the component when it is replaced.

2.42. The lifetime of a car (in miles) is an Exp(λ) random variable with $1/\lambda = 130,000$ miles. If the car is driven 13,000 miles a year, what is the distribution of the lifetime of the car in years?

2.43. Compute the expected values of the random variables defined in Computational Problem 2.1 assuming all outcomes in the sample space are equally likely.

2.44. Compute the expected values of the random variables defined in Computational Problem 2.2 assuming all outcomes in the sample space are equally likely.

2.45. Compute the expected number of points the student of Computational Problem 2.7 gets on a problem.

2.46. Two vendors offer functionally identical products. The expected lifetime of both the products is 10 months. However, the distribution of the first is Exp(λ), while that of the other is Erl(2, μ). If the aim is to maximize the probability that the lifetime of the product is greater than 8 months, which of the two products should be chosen?

2.47. Compute the expected value of the random variable with pdf as given in Computational Problem 2.25.

2.48. Compute the expected value of the random variable with pdf as given in Computational Problem 2.27.

2.49. Compute the expected value of the random variable with pdf as given in Computational Problem 2.29.

2.50. Compute the variance of the random variable with pdf as given in Computational Problem 2.25.

2.51. Compute the variance of the random variable with pdf as given in Computational Problem 2.27.

2.52. Compute the variance of the random variable with pdf as given in Computational Problem 2.29.

2.53. The lifetime of an item is an exponential random variable with mean 5 days. We replace this item upon failure or at the end of the fourth day whichever occurs first. Compute the expected time of replacement.

2.54. Consider the k-out-of-n system of Computational Problem 2.16. Suppose we visit this system at time $t = 3$ and replace all the failed components, at a cost of $75 each. If the system has failed, it costs us an additional $1000. Compute the expected total cost incurred at time 3.

3

Multivariate Random Variables

3.1. Multivariate Random Variables

In Chapter 2 we defined a random variable as a numerical function from the sample space into the real line. We introduced the concepts of cdf, pdf, and pmf as alternate means of describing a random variable. It is sometimes necessary to define more than one random variable on a probability space to study different aspects of a random phenomenon. We saw an example of this in Example 2.3 in the experiment of tossing two dice, where we defined X as the sum of the outcomes on the two dice and Y as the maximum of the two outcomes. Although we can compute the cdf and pmf for these two random variables separately, it is clear that the two random variables are related to each other. For example, $X = 12$ implies that $Y = 6$. This interdependence is not captured by the two separate cdfs. How do we capture it? That is the subject of this chapter.

Definition 3.1 (Multivariate Random Variable). Let $X_i, i = 1, 2, \ldots, n$, be a random variable with state space S_i. The vector $X = (X_1, X_2, \ldots, X_n)$ is called a multivariate random variable, or $\{X_1, X_2, \ldots, X_n\}$ are called jointly distributed random variables. The set of all values that X can take is called the state space of X and is given by $S = S_1 \times S_2 \times \cdots \times S_n$.

When $n = 2$, $X = (X_1, X_2)$ is called a *bivariate random variable*.

Example 3.1 (Tossing Two Dice). Suppose a random experiment involves tossing two dice and noting the individual outcomes of the two dice. Define the following random variables:

$X_1 =$ maximum of the outcomes of the two dice;
$X_2 =$ minimum of the outcomes of the two dice;

X_3 = sum of the outcomes of the two dice;

X_4 = absolute value of the difference of the two outcomes.

Then (X_1, X_2, X_3, X_4) is a multivariate random variable. (X_3, X_4) is a bivariate random variable.   □

Next we investigate how the random nature of the multivariate random variables can be described mathematically. Let $X = (X_1, X_2, \ldots, X_n)$ be a multivariate random variable with state space S. Following the discussion in Section 2.2 we see that we can describe such a random variable by giving its multivariate cumulative distribution function (cdf) as defined below.

Definition 3.2 (Multivariate cdf). The function

$$F(x_1, x_2, \ldots, x_n) = P(X_1 \leq x_1, X_2 \leq x_2, \ldots, X_n \leq x_n),$$

$$x_i \in (-\infty, \infty) \qquad \text{for} \quad 1 \leq i \leq n,$$

is called the multivariate cdf or joint cdf of $(X_1, X_2, \ldots, X_n)$.

The probabilities of other events associated with the multivariate random variable can be computed in terms of its multivariate cdf. For example, we can show that in the bivariate case

$$P(a_1 < X_1 \leq b_1, a_2 < X_2 \leq b_2) = F(b_1, b_2) - F(a_1, b_2) - F(b_1, a_2) + F(a_1, a_2). \quad (3.1)$$

As in Chapter 2, we study two classes of multivariate random variables, namely discrete and continuous, in the next two sections.

3.2.   Multivariate Discrete Random Variables

We start with the formal definition:

Definition 3.3 (Multivariate Discrete Random Variable). A multivariate random variable $X = (X_1, X_2, \ldots, X_n)$ is said to be discrete if each X_i is a discrete random variable, $1 \leq i \leq n$.

As in the univariate case, a multivariate discrete random variable can be described by its multivariate probability mass function (pmf) defined below.

Definition 3.4 (Multivariate pmf). Let $X = (X_1, X_2, \ldots, X_n)$ be a multivariate random variable with state space $S = S_1 \times S_2 \times \cdots \times S_n$. The function

$$p(x_1, x_2, \ldots, x_n) = P(X_1 = x_1, X_2 = x_2, \ldots, X_n = x_n), \qquad x_i \in S_i \qquad \text{for} \quad 1 \leq i \leq n,$$

is called the multivariate pmf or joint pmf of X.

The multivariate cdf of a multivariate discrete random variable X can be computed from its multivariate pmf, and vice versa. Hence multivariate pmf is an equivalent way of describing a multivariate discrete random variable.

TABLE 3.1. The pmf $P(X_1 = k_1, X_2 = k_2)$ for the two dice example.

$k_2 \downarrow k_1 \rightarrow$	1	2	3	4	5	6
1	$\dfrac{1}{36}$	$\dfrac{2}{36}$	$\dfrac{2}{36}$	$\dfrac{2}{36}$	$\dfrac{2}{36}$	$\dfrac{2}{36}$
2	0	$\dfrac{1}{36}$	$\dfrac{2}{36}$	$\dfrac{2}{36}$	$\dfrac{2}{36}$	$\dfrac{2}{36}$
3	0	0	$\dfrac{1}{36}$	$\dfrac{2}{36}$	$\dfrac{2}{36}$	$\dfrac{2}{36}$
4	0	0	0	$\dfrac{1}{36}$	$\dfrac{2}{36}$	$\dfrac{2}{36}$
5	0	0	0	0	$\dfrac{1}{36}$	$\dfrac{2}{36}$
6	0	0	0	0	0	$\dfrac{1}{36}$

Example 3.2 (Tossing Two Dice). Let (X_1, X_2) be as defined in Example 3.1. The bivariate pmf (i.e., multivariate pmf for a bivariate random variable) of (X_1, X_2) is given in Table 3.1. □

Example 3.3 (Multinomial Random Variable). Consider an experiment that can result in one and only one of r distinct outcomes. Let p_i be the probability of observing outcome i, $1 \leq i \leq r$. We assume that $p_i > 0$ and $p_1 + p_2 + \cdots + p_r = 1$. Suppose this experiment is repeated n times in an independent fashion. Let X_i be the number of times outcome i is observed in these n trials. Then $X = (X_1, X_2, \ldots, X_r)$ is called a multinomial random variable. Its state space is

$$S = \{k = (k_1, k_2, \ldots, k_r) : k_i \text{ are nonnegative integers and } k_1 + k_2 + \cdots + k_r = n\}.$$

Its multivariate pmf is given below:

$$p(k) = \frac{n!}{k_1! k_2! \cdots k_r!} p_1^{k_1} p_2^{k_2} \cdots p_r^{k_r}, \qquad k \in S. \tag{3.2}$$

The above equation follows because $p_1^{k_1} p_2^{k_2} \cdots p_r^{k_r}$ is the probability of observing a specific sequence of n outcomes containing exactly k_i outcomes of type i, $1 \leq i \leq r$, and $(n!/k_1! k_2! \cdots k_r!)$ is the number of such sequences. Equation (3.2) can be seen as a direct extension of the Binomial pmf of (2.4). □

3.3. Multivariate Continuous Random Variables

We start with the formal definition:

Definition 3.5 (Multivariate Continuous Random Variable). A multivariate random variable $X = (X_1, X_2, \ldots, X_n)$ with multivariate cdf $F(\cdot)$ is said to be continuous if there is a function

$f(\cdot)$ such that

$$F(x_1, x_2, \ldots, x_n) = \int_{-\infty}^{x_n} \int_{-\infty}^{x_{n-1}} \cdots \int_{-\infty}^{x_1} f(u_1, u_2, \ldots, u_n) \, du_1 \, du_2 \cdots du_n. \tag{3.3}$$

X is also called jointly continuous and f is called the multivariate pdf or joint pdf of X.

Suppose the multivariate cdf $F(x_1, x_2, \ldots, x_n)$ of $X = (X_1, X_2, \ldots, X_n)$ is a continuous and piecewise differentiable function of $(x_1, x_2, \ldots, x_n)$. Let

$$f(x_1, x_2, \ldots, x_n) = \frac{\partial^n}{\partial x_1 \partial x_2 \cdots \partial x_n} F(x_1, x_2, \ldots, x_n). \tag{3.4}$$

Then X is a multivariate continuous random variable with multivariate pdf $f(\cdot)$ given above if (3.3) is satisfied. We can intuitively interpret the multivariate pdf as follows:

$$P(x_i < X_i \le x_i + h_i, 1 \le i \le n) \approx f(x_1, x_2, \ldots, x_n) h_1 h_2 \cdots h_n.$$

Any function $f(x_1, x_2, \ldots, x_n)$ is a valid multivariate density if it satisfies the following:

$$f(x_1, x_2, \ldots, x_n) \ge 0, \tag{3.5}$$

and

$$\int_{-\infty}^{\infty} \int_{-\infty}^{\infty} \cdots \int_{-\infty}^{\infty} f(u_1, u_2, \ldots, u_n) \, du_1 \, du_2 \cdots du_n = 1. \tag{3.6}$$

Example 3.4. Consider the following function:

$$f(x_1, x_2) = \begin{cases} c(x_1 + x_2) & \text{if } 0 \le x_1, x_2 \le 2, \\ 0 & \text{otherwise.} \end{cases} \tag{3.7}$$

If this is a bivariate density of a random variable (X_1, X_2), what is the value of c? Also compute the joint cdf of (X_1, X_2).

A joint density must satisfy (3.6). Hence we must have

$$\int_0^2 \int_0^2 c(u_1 + u_2) \, du_1 \, du_2 = 8c = 1.$$

Hence

$$c = \tfrac{1}{8}.$$

For $0 \le x_1, x_2 \le 2$, the joint cdf is given by

$$F(x_1, x_2) = \int_0^{x_1} \int_0^{x_2} \tfrac{1}{8}(u_1 + u_2) \, du_1 \, du_2 = \tfrac{1}{16}(x_1 x_2^2 + x_1^2 x_2).$$

We leave it to the reader to specify the joint cdf in the remaining region.   □

3.4. Marginal Distributions

Let $X = (X_1, X_2, \ldots, X_n)$ be a multivariate random variable with joint cdf $F(\cdot)$. Here we define the concept of marginal distributions.

Definition 3.6 (Marginal cdf). The function

$$F_{X_i}(x_i) = P(X_i \leq x_i) \tag{3.8}$$

is called the marginal cdf of X_i.

Thus the marginal cdf of X_i is the same as the cdf of the univariate random variable X_i. The marginal cdf of X_i can be computed from the joint cdf of X as follows:

$$
\begin{aligned}
F_{X_i}(x_i) &= P(X_i \leq x_i) \\
&= P(X_1 < \infty, \ldots, X_{i-1} < \infty, X_i \leq x_i, X_{i+1} < \infty, \ldots, X_n < \infty) \\
&= F(\infty, \ldots, \infty, x_i, \infty, \ldots, \infty). \tag{3.9}
\end{aligned}
$$

We classify the random variable X_i as discrete or continuous based on its marginal cdf, using the definitions of Chapter 2. Then we define the concepts of the marginal pmf of X_i if it is discrete, and marginal pdf of X_i if it is continuous.

Definition 3.7 (Marginal pmf). Suppose $X = (X_1, X_2, \ldots, X_n)$ is a discrete multivariate random variable with state space $S = S_1 \times S_2 \times \cdots \times S_n$. The function

$$p_{X_i}(x) = P(X_i = x), \qquad x \in S_i \tag{3.10}$$

is called the marginal pmf of X_i.

The marginal pmf of X_i can be computed from the joint pmf of X as follows:

$$
\begin{aligned}
p_{X_i}(x) &= P(X_i = x) \\
&= \sum_{x_1} \cdots \sum_{x_{i-1}} \sum_{x_{i+1}} \cdots \sum_{x_n} p(x_1, \ldots, x_{i-1}, x, x_{i+1}, \ldots, x_n). \tag{3.11}
\end{aligned}
$$

Here the last equation follows from the law of total probability. Thus the marginal pmf of X_i is obtained by summing the multivariate pmf over all other coordinates.

Example 3.5 (Tossing Two Dice). Let X_1 and X_2 be as defined in Example 3.1. The joint pmf of (X_1, X_2) is shown in Table 3.1. The marginal pmfs of X_1 and X_2 are given in Table 3.2. The entries in the table are computed by using the following equations:

$$P(X_1 = k_1) = \sum_{k_2=1}^{6} P(X_1 = k_1, X_2 = k_2), \qquad 1 \leq k_1 \leq 6,$$

$$P(X_2 = k_2) = \sum_{k_1=1}^{6} P(X_1 = k_1, X_2 = k_2), \qquad 1 \leq k_2 \leq 6. \qquad \square$$

TABLE 3.2. The marginal pmfs $p_{X_1}(k)$, $p_{X_2}(k)$ for the two dice example.

k	1	2	3	4	5	6
$P(X_1 = k)$	$\dfrac{1}{36}$	$\dfrac{3}{36}$	$\dfrac{5}{36}$	$\dfrac{7}{36}$	$\dfrac{9}{36}$	$\dfrac{11}{36}$
$P(X_2 = k)$	$\dfrac{11}{36}$	$\dfrac{9}{36}$	$\dfrac{7}{36}$	$\dfrac{5}{36}$	$\dfrac{3}{36}$	$\dfrac{1}{36}$

Example 3.6 (Multinomial Random Variable). Consider the Multinomial random variable $X = (X_1, X_2, \ldots, X_r)$ with joint pmf as given in (3.2). Using (3.11) we compute the marginal pmf of X_1 as follows:

$$
\begin{aligned}
p_{X_1}(k_1) &= P(X_1 = k_1) \\
&= \sum_{k_2} \cdots \sum_{k_r} p(k_1, k_2, \ldots, k_r) \\
&= \sum_{k_2} \cdots \sum_{k_r} \frac{n!}{k_1! k_2! \cdots k_r!} p_1^{k_1} p_2^{k_2} \cdots p_r^{k_r} \\
&= \frac{n!}{k_1!(n-k_1)!} p_1^{k_1} \sum_{k_2} \cdots \sum_{k_r} \frac{(n-k_1)!}{k_2! \cdots k_r!} p_2^{k_2} \cdots p_r^{k_r} \\
&= \frac{n!}{k_1!(n-k_1)!} p_1^{k_1} (p_2 + p_3 + \cdots + p_r)^{n-k_1} \\
&= \frac{n!}{k_1!(n-k_1)!} p_1^{k_1} (1 - p_1)^{n-k_1}, \qquad 0 \le k_1 \le n.
\end{aligned}
$$

Here we have used

$$p_1 + p_2 + p_3 + \cdots + p_r = 1$$

and the multinomial expansion formula

$$\sum_{k_1} \cdots \sum_{k_r} \frac{n!}{k_1! \cdots k_r!} x_1^{k_1} x_2^{k_2} \cdots x_r^{k_r} = (x_1 + x_2 + \cdots + x_r)^n,$$

where the sum is taken over all nonnegative integers $(k_1, k_2, \ldots, k_r)$ such that $k_1 + k_2 + \cdots + k_r = n$. We see that X_1 is a $\text{Bin}(n, p_1)$ random variable. In a similar way we can show that the marginal distribution of X_i is $\text{Bin}(n, p_i)$.   □

Definition 3.8 (Marginal pdf). Suppose X_i is a continuous random variable. The function $f_{X_i}(\cdot)$ is called the marginal pdf of X_i if

$$F_{X_i}(x) = \int_{-\infty}^{x} f_{X_i}(u)\, du, \qquad -\infty < x < \infty. \tag{3.12}$$

Suppose X is a multivariate continuous random variable. Then the marginal pdf of X_i can be computed from the joint pdf of X as follows:

$$
\begin{aligned}
f_{X_i}(x) = \int_{x_1} \cdots \int_{x_{i-1}} \int_{x_{i+1}} \cdots \int_{x_n} & f(x_1, \cdots, x_{i-1}, x, x_{i+1}, \cdots, x_n) \\
& \times dx_1 \ldots dx_{i-1}\, dx_{i+1} \cdots dx_n.
\end{aligned}
\tag{3.13}
$$

Here the last equation follows from the law of total probability. Thus the marginal pdf is obtained by integrating the multivariate pdf over all other coordinates.

Example 3.7 Compute the marginal pdf for the bivariate density function given in Example 3.4.

From (3.13) we have, $0 \le x_1 \le 2$,

$$
\begin{aligned}
f_{X_1}(x_1) &= \int_{-\infty}^{\infty} f(x_1, x_2)\, dx_2 \\
&= \int_0^2 \tfrac{1}{8}(x_1 + x_2)\, dx_2 \\
&= \tfrac{1}{4}(1 + x_1).
\end{aligned}
$$

From symmetry we get

$$
f_{X_2}(x_2) = \tfrac{1}{4}(1 + x_2), \qquad 0 \le x_2 \le 2. \qquad \Box
$$

So far we have discussed computing the marginal cdf, pmf, or pdf of X_i from the joint cdf, pmf, or pdf of $(X_1, X_2, \ldots, X_n)$. These results can be extended to computing the joint distributions of (X_i, X_j), $i \ne j$. We give the final result below, without proof. We use $i = 1$, $j = 2$ for concreteness.

$$
F_{X_1 X_2}(x_1, x_2) = F(x_1, x_2, \infty, \ldots, \infty), \tag{3.14}
$$

$$
p_{X_1 X_2}(x_1, x_2) = \sum_{x_3} \sum_{x_4} \cdots \sum_{x_n} p(x_1, x_2, x_3, x_4, \ldots, x_n), \tag{3.15}
$$

$$
f_{X_1 X_2}(x_1, x_2) = \int_{x_3} \int_{x_4} \cdots \int_{x_n} f(x_1, x_2, x_3, x_4, \ldots, x_n)\, dx_3\, dx_4 \cdots dx_n. \tag{3.16}
$$

Example 3.8 (Multinomial Random Variable). Let $(X_1, X_2, \ldots, X_r)$ be a Multinomial random variable with joint pmf as given in Example 3.3. Compute the joint pmf of (X_1, X_2).

Using (3.15) and Example 3.6 we get

$$
\begin{aligned}
p_{X_1 X_2}(k_1, k_2) &= \mathrm{P}(X_1 = k_1;\, X_2 = k_2) \\
&= \sum_{k_3} \cdots \sum_{k_r} \frac{n!}{k_1! k_2! \cdots k_r!} p_1^{k_1} p_2^{k_2} p_3^{k_3} \cdots p_r^{k_r} \\
&= \frac{n!}{k_1! k_2! (n - k_1 - k_2)!} p_1^{k_1} p_2^{k_2} \sum_{k_3} \cdots \sum_{k_r} \frac{(n - k_1 - k_2)!}{k_3! \cdots k_r!} p_3^{k_3} \cdots p_r^{k_r} \\
&= \frac{n!}{k_1! k_2! (n - k_1 - k_2)!} p_1^{k_1} p_2^{k_2} (p_3 + \cdots + p_r)^{n - k_1 - k_2} \\
&= \frac{n!}{k_1! k_2! (n - k_1 - k_2)!} p_1^{k_1} p_2^{k_2} (1 - p_1 - p_2)^{n - k_1 - k_2}, \\
&\qquad\qquad k_1 \ge 0, \qquad k_2 \ge 0, \qquad k_1 + k_2 \le n. \qquad \Box
\end{aligned}
$$

We end this section with a definition of a commonly used term:

Definition 3.9 (Identically Distributed Random Variables). Jointly distributed random variables $(X_1, X_2, \ldots, X_n)$ are said to be identically distributed if their marginal cdfs are identical, i.e.,

$$F_{X_1}(\cdot) = F_{X_2}(\cdot) = \cdots = F_{X_n}(\cdot),$$

or (in the discrete case) their marginal pmfs are identical, i.e.,

$$p_{X_1}(\cdot) = p_{X_2}(\cdot) = \cdots = p_{X_n}(\cdot),$$

or (in the continuous case) their marginal pdfs are identical, i.e.,

$$f_{X_1}(\cdot) = f_{X_2}(\cdot) = \cdots = f_{X_n}(\cdot).$$

3.5. Independence

The concept of independence arises naturally when we study multivariate random variables. We have encountered this concept in connection with events in Section 1.8. Recall that events A and B are said to be independent if

$$P(AB) = P(A)P(B).$$

This suggests that the "independence" of two random random variables should imply

$$P(X_1 \in A; X_2 \in B) = P(X_1 \in A)P(X_2 \in B)$$

for all subsets A and B of real numbers. This prompts the following formal definition:

Definition 3.10 (Independent Random Variables). The jointly distributed random variables $(X_1, X_2, \ldots, X_n)$ are said to be independent if

$$F(x_1, x_2, \ldots, x_n) = F_{X_1}(x_1)F_{X_2}(x_2) \ldots F_{X_n}(x_n),$$

or (in the discrete case)

$$p(x_1, x_2, \ldots, x_n) = p_{X_1}(x_1)p_{X_2}(x_2) \ldots p_{X_n}(x_n),$$

or (in the continuous case)

$$f(x_1, x_2, \ldots, x_n) = f_{X_1}(x_1)f_{X_2}(x_2) \ldots f_{X_n}(x_n).$$

In many applications we shall encounter random variables $X_1, X_2, \ldots, X_n$ that are independent and have identical marginal distributions. Such random variables are called independent and identically distributed random variables, or *iid random variables* for short.

Example 3.9. Consider a bivariate random variable (X_1, X_2) with joint pdf as given in Example 3.4. Are X_1 and X_2 independent?

The marginal pdfs of X_1 and X_2 are given in Example 3.7. From those results we see that

$$f(x_1, x_2) \neq f_{X_1}(x_1) f_{X_2}(x_2).$$

Hence the random variables X_1 and X_2 are not independent. They are, however, identically distributed. □

Example 3.10 (Series System). A system consisting of n components is called a series system if the system fails as soon as any one of the n components fails. Let X_i be the lifetime of component i, and assume that X_i, $1 \leq i \leq n$, are iid random variables with common marginal cdf $F(x) = \mathsf{P}(X_i \leq x)$. Let T be the time when the system fails. Compute the cdf of T.

The series nature of the system implies that

$$T = \min\{X_1, X_2, \dots, X_n\}.$$

If $X_i > t$ for all $1 \leq i \leq n$, then $T > t$ and vice versa. Hence

$$
\begin{aligned}
\mathsf{P}(T > t) &= \mathsf{P}(\min\{X_1, X_2, \dots, X_n\} > t) \\
&= \mathsf{P}(X_1 > t; X_2 > t; \dots; X_n > t) \\
&= \mathsf{P}(X_1 > t)\mathsf{P}(X_2 > t) \cdots \mathsf{P}(X_n > t) \quad \text{(due to independence)} \\
&= (1 - F(t))^n.
\end{aligned}
$$

Hence the cdf of the system lifetime is given by

$$\mathsf{P}(T \leq t) = 1 - (1 - F(t))^n. \qquad \square$$

Example 3.11 (Parallel System). A system consisting of n components is called a parallel system if the system fails as soon as all the n components fail. Let X_i be the lifetime of component i, and assume that X_i, $1 \leq i \leq n$, are iid random variables with common marginal cdf $F(x) = \mathsf{P}(X_i \leq x)$. Let T be the time when the system fails. Compute the cdf of T.

The parallel nature of the system implies that

$$T = \max\{X_1, X_2, \dots, X_n\}.$$

If $X_i \leq t$ for all $1 \leq i \leq n$, then $T \leq t$ and vice versa. Hence the cdf of the system lifetime is given by

$$
\begin{aligned}
\mathsf{P}(T \leq t) &= \mathsf{P}(\max\{X_1, X_2, \dots, X_n\} \leq t) \\
&= \mathsf{P}(X_1 \leq t; X_2 \leq t; \dots; X_n \leq t) \\
&= \mathsf{P}(X_1 \leq t)\mathsf{P}(X_2 \leq t) \cdots \mathsf{P}(X_n \leq t) \quad \text{(due to independence)} \\
&= F(t)^n. \qquad \square
\end{aligned}
$$

3.6. Sums of Random Variables

Following the development in Chapter 2, we now consider functions of multivariate random variables. Let $(X_1, X_2, \cdots, X_n)$ be a multivariate random variable, and let $Z = g(X_1, X_2, \ldots, X_n)$ where g is a real-valued function. Then Z is also a random variable and has a cdf and pdf (if it is continuous) or pmf (if it is discrete). Computing it is, in general, a very difficult problem. In this section we concentrate on one of the most common functions, namely, the sum:

$$Z = X_1 + X_2 + \cdots + X_n.$$

First consider the discrete case. Let $p(x_1, x_2)$ be the bivariate pmf of (X_1, X_2) and let $Z = X_1 + X_2$. Then

$$
\begin{aligned}
p_Z(z) &= P(Z = z) \\
&= P(X_1 + X_2 = z) \\
&= \sum_{x_1} P(X_1 = x_1; X_2 = z - x_1).
\end{aligned}
\tag{3.17}
$$

Here the last equality follows from the law of total probability.

Example 3.12 (Multinomial Random Variable). Let $(X_1, X_2, \ldots, X_r)$ be a Multinomial random variable with joint pmf as given in Example 3.3. Compute the pmf of $Z = X_1 + X_2$. The joint pmf of (X_1, X_2) is given in Example 3.8. Using (3.17) we get, for $0 \le k \le n$,

$$
\begin{aligned}
p_Z(k) &= \sum_{k_1=0}^{k} P(X_1 = k_1; X_2 = k - k_1) \\
&= \sum_{k_1=0}^{k} \frac{n!}{k_1!\,(k - k_1)!\,(n - k)!} p_1^{k_1} p_2^{k-k_1} (1 - p_1 - p_2)^{n-k} \\
&= \frac{n!}{k!\,(n - k)!}(1 - p_1 - p_2)^{n-k} \sum_{k_1=0}^{k} \frac{k!}{k_1!\,(k - k_1)!} p_1^{k_1} p_2^{k-k_1} \\
&= \frac{n!}{k!\,(n - k)!}(1 - p_1 - p_2)^{n-k}(p_1 + p_2)^k.
\end{aligned}
$$

We have used the binomial expansion of $(p_1 + p_2)^k$ to evaluate the last sum. The last expression implies that $X_1 + X_2$ is a $\text{Bin}(n, p_1 + p_2)$ random variable. $\square$

Next consider the continuous case. Let $f(x_1, x_2)$ be the bivariate pmf of (X_1, X_2) and let $Z = X_1 + X_2$. Then we have

$$
\begin{aligned}
F_Z(z) &= P(Z \le z) \\
&= P(X_1 + X_2 \le z)
\end{aligned}
$$

$$= \int\int_{x_1+x_2 \le z} f(x_1, x_2)\, dx_1\, dx_2$$

$$= \int_{-\infty}^{\infty} \int_{-\infty}^{z-x_2} f(x_1, x_2)\, dx_1\, dx_2.$$

The pdf of Z can be computed by taking derivatives as follows:

$$f_Z(z) = \frac{d}{dz} F_Z(z)$$

$$= \frac{d}{dz} \int_{-\infty}^{\infty} \int_{-\infty}^{z-x_2} f(x_1, x_2)\, dx_1\, dx_2$$

$$= \int_{-\infty}^{\infty} f(z - x_2, x_2)\, dx_2. \tag{3.18}$$

Example 3.13. Consider a bivariate random variable (X_1, X_2) with joint pdf as given in Example 3.4. Compute the density of $Z = X_1 + X_2$.

From Example 3.4 we have the joint pdf of (X_1, X_2) as follows:

$$f(x_1, x_2) = \begin{cases} \frac{1}{8}(x_1 + x_2) & \text{if } 0 \le x_1, x_2 \le 2, \\ 0 & \text{otherwise.} \end{cases} \tag{3.19}$$

Since $0 \le X_1, X_2 \le 2$, we see that $0 \le Z \le 4$. Also, if $Z = z$, then we must have $\max(0, z-2) \le x_2 \le \min(2, z)$ in order to keep $0 \le x_1, x_2 \le 2$. Substituting in (3.18) we get, for $0 \le z \le 4$,

$$f_Z(z) = \int_{-\infty}^{\infty} f(z - x_2, x_2)\, dx_2$$

$$= \int_{\max(0,z-2)}^{\min(2,z)} \frac{1}{8}(z - x_2 + x_2)\, dx_2$$

$$= \frac{1}{8}(z)(\min(2, z) - \max(0, z - 2))$$

$$= \begin{cases} \frac{1}{8}z^2 & \text{if } 0 \le z \le 2, \\ \frac{1}{8}z(4 - z) & \text{if } 2 \le z \le 4. \end{cases} \qquad \square$$

When (X_1, X_2) are discrete *independent* random variables, (3.17) reduces to

$$p_Z(z) = \sum_{x_1} p_{X_1}(x_1) p_{X_2}(z - x_1) \tag{3.20}$$

$$= \sum_{x_2} p_{X_1}(z - x_2) p_{X_2}(x_2). \tag{3.21}$$

The pmf p_Z is called a discrete *convolution* of p_{X_1} and p_{X_2}.

Example 3.14 (Sums of Binomials). Let X_i be a Bin(n_i, p) random variable, $i = 1, 2$. Suppose X_1 and X_2 are independent. Compute the pmf of $Z = X_1 + X_2$.

Using (3.20) we get

$$p_Z(k) = \sum_{k_2=0}^{\min(k,n_2)} p_{X_1}(k - k_2) p_{X_2}(k_2)$$

$$= \sum_{k_2=0}^{\min(k,n_2)} \frac{n_1!}{(k - k_2)!(n_1 - k + k_2)!} p^{k-k_2}(1 - p)^{n_1-k+k_2} \frac{n_2!}{k_2!(n_2 - k_2)!} p^{k_2}(1 - p)^{n_2-k_2}$$

$$= p^k(1 - p)^{n_1+n_2-k} \sum_{k_2=0}^{\min(k,n_2)} \binom{n_1}{k - k_2}\binom{n_2}{k_2}$$

$$= \binom{n_1 + n_2}{k} p^k(1 - p)^{n_1+n_2-k}, \tag{3.22}$$

where the last equality was obtained from the well-known Binomial identity:

$$\sum_{r=0}^{\min(k,n_2)} \binom{n_1}{k - r}\binom{n_2}{r} = \binom{n_1 + n_2}{k}.$$

Equation (3.22) implies that Z is a Bin$(n_1 + n_2, p)$ random variable. □

Recall that a B(p) random variable is the same as a Bin$(1, p)$ random variable. Hence the above result can be used to show the following: Let $X_i, 1 \leq i \leq n$, be iid B(p) random variables. Then $X_1 + X_2 + \cdots + X_n$ is a Bin(n, p) random variable.

Example 3.15 (Sums of Poissons). Let X_i be a P(λ_i) random variable, $i = 1, 2$. Suppose X_1 and X_2 are independent. Compute the pmf of $Z = X_1 + X_2$.

Using (3.20) we get

$$p_Z(k) = \sum_{k_2=0}^{k} e^{-\lambda_1} \frac{\lambda_1^{k-k_2}}{(k - k_2)!} e^{-\lambda_2} \frac{\lambda_2^{k_2}}{(k_2)!}$$

$$= e^{-\lambda_1-\lambda_2} \frac{1}{k!} \sum_{k_2=0}^{k} \frac{k!}{(k - k_2)! k_2!} \lambda_1^{k-k_2} \lambda_2^{k_2}$$

$$= e^{-\lambda_1-\lambda_2} \frac{(\lambda_1 + \lambda_2)^k}{k!}. \tag{3.23}$$

The last equality follows from the binomial expansion

$$\sum_{r=0}^{k} \frac{k!}{(k - r)! r!} \lambda_1^{k-r} \lambda_2^r = (\lambda_1 + \lambda_2)^k.$$

Thus, (3.23) implies that the sum of independent Poisson random variables P(λ_1) and P(λ_2), is a P$(\lambda_1 + \lambda_2)$ random variable. This can be extended to sums of more than two independent Poisson random variables. □

When (X_1, X_2) are continuous *independent* random variables, (3.18) reduces to

$$f_Z(z) = \int_{-\infty}^{\infty} f_{X_1}(z - x_2) f_{X_2}(x_2)\, dx_2 \tag{3.24}$$

$$= \int_{-\infty}^{\infty} f_{X_1}(x_1) f_{X_2}(z - x_1)\, dx_1. \tag{3.25}$$

The pdf f_Z is called a *convolution* of f_{X_1} and f_{X_2}.

Example 3.16 (Sums of Exponentials). Suppose X_1 and X_2 are iid $\text{Exp}(\lambda)$ random variables. Compute the pdf of $Z = X_1 + X_2$.

Using (3.24) we get

$$f_Z(z) = \int_{-\infty}^{\infty} f_{X_1}(z - x_2) f_{X_2}(x_2)\, dx_2$$

$$= \int_0^z \lambda e^{-\lambda(z - x_2)} \lambda e^{-\lambda x_2}\, dx_2$$

$$= \lambda^2 e^{-\lambda z} \int_0^z dx_2$$

$$= e^{-\lambda z}(\lambda^2 z)$$

$$= \lambda e^{-\lambda z} \frac{(\lambda z)^1}{1!}.$$

The last expression can be recognized as the pdf of an $\text{Erl}(2, \lambda)$ random variable. Thus the sum of two iid $\text{Exp}(\lambda)$ random variables is an $\text{Erl}(2, \lambda)$ random variable. Similarly we can show that the sum of n iid $\text{Exp}(\lambda)$ random variables is an $\text{Erl}(n, \lambda)$ random variable.   □

3.7. Expectations

We studied the expectations of univariate random variables and of functions of univariate random variables in Section 2.10. In particular, we saw in Theorem 2.4 that if X is a univariate random variable, then

$$\mathsf{E}(g(X)) = \begin{cases} \sum_i g(x_i) p_X(x_i) & \text{if } X \text{ is discrete,} \\[2ex] \displaystyle\int_x g(x) f_X(x)\, dx & \text{if } X \text{ is continuous.} \end{cases} \tag{3.26}$$

Now suppose $X = (X_1, X_2, \ldots, X_n)$ is a multivariate random variable. Let $g(x_1, x_2, \ldots, x_n)$ be a real-valued function of $(x_1, x_2, \ldots, x_n)$. Then $Z = g(X_1, X_2, \ldots, X_n)$ is a univariate random variable, and hence has an expectation. We can compute this expectation by first computing the pdf or pmf of Z, but this is in general difficult. Fortunately, Theorem 2.4 extends to the multivariate case. We state it below without proof.

Theorem 3.1 (Expectation of a Function of a Multivariate Random Variable). *Let* $X = (X_1, X_2, \ldots, X_n)$ *be a multivariate random variable. Let* $g(x_1, x_2, \ldots, x_n)$ *be a real-valued function of* $(x_1, x_2, \ldots, x_n)$. *Then*

$$E(g(X_1, X_2, \ldots, X_n))$$
$$= \begin{cases} \sum_{x_1} \cdots \sum_{x_n} g(x_1, x_2, \ldots, x_n) p(x_1, x_2, \ldots, x_n) & \text{if } X \text{ is discrete,} \\ \int_{x_1} \cdots \int_{x_n} g(x_1, x_2, \ldots, x_n) f(x_1, x_2, \ldots, x_n) \, dx_1 \, dx_2 \ldots dx_n & \text{if } X \text{ is continuous.} \end{cases}$$
$$(3.27)$$

The above theorem makes it easy to compute the expectations of functions of multivariate random variables. As a special case, consider the function

$$g(x_1, x_2, \ldots, x_n) = x_1.$$

Then, assuming $(X_1, X_2, \ldots, X_n)$ to be continuous we get

$$E(g(X_1, X_2, \ldots, X_n)) = \int_{x_1} \cdots \int_{x_n} x_1 f(x_1, x_2, \ldots, x_n) \, dx_1 \, dx_2 \cdots dx_n$$
$$= \int_{x_1} x_1 \int_{x_2} \cdots \int_{x_n} f(x_1, x_2, \ldots, x_n) \, dx_1 \, dx_2 \cdots dx_n$$
$$= \int_{x_1} x_1 f_{X_1}(x_1),$$

where the last equation follows from (3.13). This is consistent with the definition of the expectation of X_1. In the discrete case we get

$$E(X_1) = \sum_{x_1} x_1 p_{X_1}(x_1).$$

An important consequence of Theorem 3.1 is given in the following theorem:

Theorem 3.2 (Expectation of a Sum). *Let* $X = (X_1, X_2, \ldots, X_n)$ *be a multivariate random variable. Then*

$$E(X_1 + X_2 + \cdots + X_n) = E(X_1) + E(X_2) + \cdots + E(X_n). \qquad (3.28)$$

Proof. We consider the continuous case. The discrete case follows similarly. From Theorem 3.1 we get

$$E(X_1 + X_2 + \cdots + X_n)$$
$$= \int_{x_1} \int_{x_2} \cdots \int_{x_n} (x_1 + x_2 + \cdots + x_n) f(x_1, x_2, \ldots, x_n) \, dx_1 \, dx_2 \cdots dx_n$$
$$= \int_{x_1} \int_{x_2} \cdots \int_{x_n} x_1 f(x_1, x_2, \ldots, x_n) \, dx_1 \, dx_2 \cdots dx_n$$
$$+ \int_{x_1} \int_{x_2} \cdots \int_{x_n} x_2 f(x_1, x_2, \ldots, x_n) \, dx_1 \, dx_2 \cdots dx_n + \cdots$$

$$+ \int_{x_1} \int_{x_2} \cdots \int_{x_n} x_n f(x_1, x_2, \ldots, x_n) \, dx_1 \, dx_2 \cdots dx_n$$

$$= \int_{x_1} x_1 f_{X_1}(x_1) + \int_{x_2} x_2 f_{X_2}(x_2) + \cdots + \int_{x_n} x_n f_{X_n}(x_n)$$

$$= E(X_1) + E(X_2) + \cdots + E(X_n).$$

This completes the proof. □

Thus the expected value of a sum of a fixed number of random variables is the sum of the expected values of those random variables. This holds even if the random variables are not independent! In fact this result provides an important tool for the computation of the expected values of random variables.

Example 3.17 (Random Pairing). Consider the random pairing experiment discussed in Example 1.16. Compute the expected number of women that get paired with their own husbands.

Let X_i be 1 if the ith woman gets paired with her own husband. Now X_i is a B($1/n$) random variable (see Example 2.10). Hence

$$E(X_i) = \frac{1}{n}.$$

The number of women who get paired with their own husbands is given by

$$X = X_1 + X_2 + \cdots + X_n.$$

Note that $X_1, \ldots, X_n$ are dependent random variables. However, (3.28) remains valid. Hence we get

$$E(X) = E(X_1) + E(X_2) + \cdots + E(X_n)$$
$$= \frac{1}{n} + \frac{1}{n} + \cdots + \frac{1}{n}$$
$$= 1.$$

Thus regardless of the number of couples involved in the game, on average, one couple will get paired correctly! □

Example 3.18 (Expectation of Binomial). Let X_i, $1 \leq i \leq n$, be iid B(p) random variables. We have argued earlier that $X = X_1 + X_2 + \cdots + X_n$ is a Bin(n, p) random variable. Hence

$$E(X) = E(X_1) + E(X_2) + \cdots + E(X_n) = np.$$

The above equation follows because $E(X_i) = p$ for all $1 \leq i \leq n$. □

Example 3.19 (Expectation of an Erlang). Let $X_i, 1 \leq i \leq n$, be iid Exp(λ) random variables. Using Example 3.16 we can show that $X = X_1 + X_2 + \cdots + X_n$ is an Erl(n, λ) random variable. Hence

$$E(X) = E(X_1) + E(X_2) + \cdots + E(X_n) = \frac{n}{\lambda}.$$

The above equation follows because $E(X_i) = 1/\lambda$ for all $1 \leq i \leq n$. □

Theorem 3.2 is about the expectation of a sum of random variables. The next theorem is about the expectation of products of random variables.

Theorem 3.3 (Expectation of a Product). *Let* $(X_1, X_2, \ldots, X_n)$ *be independent random variables. Then*

$$E(g_1(X_1)g_2(X_2)\cdots g_n(X_n)) = E(g_1(X_1))E(g_2(X_2))\cdots E(g_n(X_n)). \tag{3.29}$$

Proof. We consider the continuous case. Independence implies that

$$f(x_1, x_2, \ldots, x_n) = f_{X_1}(x_1)f_{X_2}(x_2)\cdots f_{X_n}(x_n).$$

From Theorem 3.1 we get

$$E(g_1(X_1)g_2(X_2)\cdots g_n(X_n))$$
$$= \int_{x_1}\int_{x_2}\cdots\int_{x_n} g_1(x_1)g_2(x_2)\cdots g_n(x_n)f(x_1, x_2, \ldots, x_n)\,dx_1\,dx_2\cdots dx_n$$
$$= \int_{x_1}\int_{x_2}\cdots\int_{x_n} g_1(x_1)g_2(x_2)\cdots g_n(x_n)f_{X_1}(x_1)f_{X_2}(x_2)\cdots f_{X_n}(x_n)\,dx_1\,dx_2\cdots dx_n$$
$$= \int_{x_1} g_1(x_1)f_{X_1}(x_1)\,dx_1\int_{x_2} g_2(x_2)f_{X_2}(x_2)\,dx_2\cdots\int_{x_n} g_n(x_n)f_{X_n}(x_n)\,dx_n$$
$$= E(g_1(X_1))E(g_2(X_2))\cdots E(g_n(X_n))$$

as desired. The discrete case follows similarly.   □

We state an important consequence of the above theorem. Suppose $X_1, X_2, \ldots, X_n$ are independent random variables. Then from Theorem 3.3 we have

$$E(X_1X_2\cdots X_n) = E(X_1)E(X_2)\cdots E(X_n).$$

Note that the expected value of a sum of random variables is the sum of their expected values (from Theorem 3.2), regardless of independence. However, the expectation of a product of random variables is the product of their expected values if they are independent. Another consequence of independence is given in the following theorem:

Theorem 3.4 (Variance of a Sum of Random Variables). *Let* $(X_1, X_2, \ldots, X_n)$ *be independent random variables. Then*

$$\text{Var}\left(\sum_{i=1}^{n} X_i\right) = \sum_{i=1}^{n} \text{Var}(X_i). \tag{3.30}$$

Proof. We shall prove (3.30) for the case of $n = 2$. The general case follows similarly:

$$\text{Var}(X_1 + X_2) = E([X_1 + X_2 - E(X_1 + X_2)]^2)$$
$$= E([X_1 + X_2 - E(X_1) - E(X_2)]^2)$$
$$= E([X_1 - E(X_1) + X_2 - E(X_2)]^2)$$
$$= E([X_1 - E(X_1)]^2 + [X_2 - E(X_2)]^2$$

$$+ 2[X_1 - E(X_1)][X_2 - E(X_2)])$$
$$= E([X_1 - E(X_1)]^2) + E([X_2 - E(X_2)]^2)$$
$$+ 2E([X_1 - E(X_1)][X_2 - E(X_2)])$$
$$= \text{Var}(X_1) + \text{Var}(X_2) + 2E([X_1 - E(X_1)])E([X_2 - E(X_2)]).$$

(due to independence)

$$= \text{Var}(X_1) + \text{Var}(X_2).$$

The last equality follows because

$$E([X_i - E(X_i)]) = E(X_i) - E(X_i) = 0, \qquad i = 1, 2.$$

This proves the theorem. $\qquad\qquad\qquad\qquad\qquad\qquad\qquad\qquad\qquad$ □

Example 3.20 (Variance of a Binomial Random Variable). Compute the variance of a Bin(n, p) random variable.

Let X_i, $1 \leq i \leq n$, be iid B(p) random variables. Then $X = X_1 + X_2 + \cdots + X_n$ is a Bin(n, p) random variable. Using (3.30) we get

$$\text{Var}(X) = \text{Var}\left(\sum_{i=1}^{n} X_i \right)$$

$$= \sum_{i=1}^{n} \text{Var}(X_i) \quad \text{(due to the independence of } X_i\text{'s)}$$

$$= \sum_{i=1}^{n} p(1 - p) \quad \text{(since } X_i \text{ are iid B}(p))$$

$$= np(1 - p). \tag{3.31}$$

This is much easier than computing the variance by direct computation. $\qquad\qquad$ □

3.8. Problems

CONCEPTUAL PROBLEMS

3.1. Derive (3.1).

3.2. A coin is tossed repeatedly in an independent fashion. The probability of observing a head on a given toss is p. Let T_1 be the first toss on which a head is observed, and let T_2 be the first toss on which a tail is observed. Thus if the sequence of outcomes is $HHHTHTT\ldots$, $T_1 = 1$ and $T_2 = 4$. Compute the joint pmf of (T_1, T_2).

3.3. A system consisting of n components is called a k-out-of-n system if it functions as long as at least k of the n components function. Let X_i be the lifetime of component i, and assume that X_i, $1 \leq i \leq n$, are iid random variables with common marginal cdf $F(x) = P(X_i \leq x)$. Let T be the time when the system fails. Compute the cdf of T.

3.4. Let (X_1, X_2) be two iid $\text{Exp}(\lambda)$ random variables. Show that $\min\{X_1, X_2\}$ is an $\text{Exp}(2\lambda)$ random variable. (*Hint*: See Example 3.10).

3.5. Let X_i be a $\text{P}(\lambda_i)$ random variable, $i = 1, 2, \ldots, n$. Suppose $X_1, X_2, \ldots, X_n$ are independent. Using Example 3.15 and induction, show that $X_1 + X_2 + \cdots + X_n$ is a $\text{P}(\lambda_1 + \cdots + \lambda_n)$ random variable.

3.6. Suppose $X_1, X_2, \ldots, X_n$ are iid $\text{Exp}(\lambda)$ random variables. Using Example 3.16 and induction, show that $X_1 + X_2 + \cdots + X_n$ is an $\text{Erl}(n, \lambda)$ random variable.

3.7. Suppose $X_1, X_2, \ldots, X_r$ are iid $\text{G}(p)$ random variables. Show that $X_1 + X_2 + \cdots + X_r$ is an $\text{NB}(r, p)$ random variable.

3.8. Let $(X_1, X_2, \ldots, X_r)$ be a multinomial random variable with joint pmf as given in Example 3.3. Let A be a strict subset of $\{1, 2, \ldots, r\}$. Show that $Z = \sum_{i \in A} X_i$ is a $\text{Bin}(n, \sum_{i \in A} p_i)$ random variable. (*Hint*: See Example 3.12.)

3.9. Let X_1 and X_2 be two independent discrete random variables taking values in $\{0, \pm1 \pm 2, \ldots\}$. Let $X = X_1 - X_2$. Show that

$$p_X(k) = \sum_{k_2 = -\infty}^{\infty} p_{X_1}(k + k_2) p_{X_2}(k_2).$$

3.10. Suppose a coin is tossed independently n times. Let H be the number of heads and T the number of tails obtained in the n tosses. Compute the distribution of $H - T$.

3.11. Let X_1 and X_2 be two independent continuous random variables. Let $X = X_1 - X_2$. Show that

$$f_X(x) = \int_{x_2 = -\infty}^{\infty} f_{X_1}(x + x_2) f_{X_2}(x_2) \, dx_2$$

3.12. Show that

$$E(a_1 X_1 + a_2 X_2 + \cdots + a_n X_n) = a_1 E(X_1) + a_2 E(X_2) + \cdots + a_n E(X_n),$$

where $a_1, a_2, \ldots, a_n$ are constants.

3.13. Suppose $X_1, X_2, \ldots, X_n$ are independent random variables, and $a_1, a_2, \ldots a_n$, and c are constants. Show that

$$\text{Var}(a_1 X_1 + a_2 X_2 + \cdots + a_n X_n + c) = a_1^2 \text{Var}(X_1) + a_2^2 \text{Var}(X_2) + \cdots + a_n^2 \text{Var}(X_n).$$

3.14. Consider the multinomial experiment described in Example 3.3. Compute the expected number of outcomes that do not occur in the n trials. (*Hint*: Let

$$X_i = \begin{cases} 1 & \text{if no type } i \text{ outcome occurs in } n \text{ trials,} \\ 0 & \text{otherwise.} \end{cases}$$

Then $\sum_{i=1}^{r} X_i$ is the number of outcomes that do not occur in the n trials.)

3.15. Using the result of Conceptual Problem 3.6 compute the variance of the $\text{Erl}(n, \lambda)$ random variable.

3.16. Let $X_1, X_2, \ldots$ be iid continuous random variables. We say that a record occurs at time k if

$$X_k = \max\{X_1, X_2, \ldots, X_k\}.$$

Let N be the number of records that occur up to time n. Show that $\mathsf{E}(N) = \sum_{i=1}^{n} 1/i$. (*Hint*: Let N_k be 1 if a record occurs at time k and 0 otherwise. Show that N_k is a B($1/k$) random variable and use $N = N_1 + N_2 + \cdots + N_n$.)

COMPUTATIONAL PROBLEMS

3.1. Compute the joint pmf of the bivariate random variable (X_3, X_4) defined in Example 3.1.

3.2. Compute the joint pmf of the bivariate random variable (X_1, X_3) defined in Example 3.1.

3.3. Suppose a class of 24 students consists of 10 juniors, 10 seniors, and 4 graduate students. A group of six students is selected at random from this class. Compute the joint pmf of the number of juniors, seniors, and graduate students in this group.

3.4. A hand of six cards is dealt from a perfectly shuffled deck of 52 playing cards. Compute the joint pmf of the number of hearts, diamonds, spades and clubs in this hand.

3.5. Compute the joint pmf of hearts and diamonds in the hand described in Computational Problem 3.4.

3.6. Consider a bivariate random variable with the following joint pmf:

$$f(x_1, x_2) = \begin{cases} cx_1^2 x_2^2 & \text{if } 0 \le x_1, x_2 \le 1, \\ 0 & \text{otherwise.} \end{cases}$$

Compute the value of c.

3.7. Consider a bivariate random variable with the following joint pmf:

$$f(x_1, x_2) = \begin{cases} cx_1^2 x_2^2 & \text{if } x_1 \ge 0, x_2 \ge 0, x_1 + x_2 \le 1, \\ 0 & \text{otherwise.} \end{cases}$$

Compute the value of c.

3.8. Consider a bivariate random variable with the following joint pmf:

$$f(x_1, x_2) = \begin{cases} x_1 + x_2 & \text{if } a \le x_1, x_2 \le 2a, \\ 0 & \text{otherwise.} \end{cases}$$

Compute the value of a.

3.9. Consider a bivariate random variable with the following joint pmf:

$$f(x_1, x_2) = \begin{cases} cx_1 x_2 & \text{if } x_1 \ge 0, x_2 \ge 0, x_1 + x_2 \le 1, \\ 0 & \text{otherwise.} \end{cases}$$

Compute the value of c.

3.10. Compute the joint cdf associated with the joint pdf of Computational Problem 3.6.

3.11. Compute the joint cdf associated with the joint pdf of Computational Problem 3.7.

3.12. Compute the joint cdf associated with the joint pdf of Computational Problem 3.8.

3.13. Compute the joint cdf associated with the joint pdf of Computational Problem 3.9.

3.14. A multivariate random variable (X_1, X_2, X_3) takes the values in the following set S with equal probability:

$$S = \{(1, 1, 1), (3, 0, 0), (0, 3, 0), (0, 0, 3), (1, 2, 3)\}.$$

Compute the marginal pmf of X_i, $i = 1, 2, 3$.

3.15. Compute the marginal pmf of hearts in the hand described in Computational Problem 3.4.

3.16. Compute the marginal pdfs of X_1 and X_2 if (X_1, X_2) has the joint pdf given in Computational Problem 3.6.

3.17. Compute the marginal pdfs of X_1 and X_2 if (X_1, X_2) has the joint pdf given in Computational Problem 3.7.

3.18. Compute the marginal pdfs of X_1 and X_2 if (X_1, X_2) has the joint pdf given in Computational Problem 3.8.

3.19. Compute the marginal pdfs of X_1 and X_2 if (X_1, X_2) has the joint pdf given in Computational Problem 3.9.

3.20. Consider a bivariate random variable (X_1, X_2) with the joint pdf given in Computational Problem 3.9. Are X_1 and X_2 independent random variables?

3.21. Suppose a machine has three independent components with Exp(.1) lifetimes. Compute the expected lifetime of the machine if it needs all three components to function properly.

3.22. A rope has three strands, all of which share a load. The lifetime of each strand (in months) is a U(10, 12) random variable, regardless of the load it is handling. The rope breaks as soon as all three strands break. Compute the cdf of the lifetime of the rope assuming that the strand lifetimes are independent.

3.23. Let U_1, U_2 be two iid U(0, 1) random variables. Compute the pdf of $U_1 + U_2$.

3.24. The stock of Random Products, Inc., is currently valued at \$10 per share. Suppose each week the stock rises by \$1 with probability .6, or falls by \$1 with probability .4. Assuming that the stock movement is independent from week to week, compute the distribution of the stock at the end of the third week.

3.25. A fair die is tossed 10 times independently. Compute the expected number of outcomes that do not occur at all. (*Hint*: See Conceptual Problem 3.14).

3.26. A statistical experiment consists of starting 10 machines at time 0, and noting the number of working machines after 10 hours. If the lifetimes (in hours) of these machines are an iid Exp(.125) random variable, compute the mean and variance of the number of machines still working after 10 hours.

3.27. An urn contains four balls: red, green, orange, and blue. Balls are drawn with replacement repeatedly from this urn. Compute the expected number of drawings needed to observe at least one ball of each color.

3.28. Consider the multinomial experiment described in Example 3.3, with the following modification. The trials continue until each outcome is observed at least once. Compute the expected number of trials conducted if all outcomes are equally likely.

3.29. Compute the following for a bivariate random variable (X_1, X_2) with pdf as given in Computational Problem 3.6:

(1) $E(X_1), E(X_2)$;

(2) $Var(X_1), Var(X_2)$.

3.30. Compute the following for a bivariate random variable (X_1, X_2) with pdf as given in Computational Problem 3.7:

(1) $E(X_1), E(X_2)$;

(2) $Var(X_1), Var(X_2)$.

(*Hint*: Use the following integral:

$$\int_0^1 x^m (1-x)^n \, dx = m! \, n! / (m+n+1)! .)$$

3.31. Compute the following for a bivariate random variable (X_1, X_2) with pdf as given in Computational Problem 3.8:

(1) $E(X_1), E(X_2)$;

(2) $Var(X_1), Var(X_2)$.

3.32. Compute the following for a bivariate random variable (X_1, X_2) with pdf as given in Computational Problem 3.9:

(1) $E(X_1), E(X_2)$;

(2) $Var(X_1), Var(X_2)$.

4

Conditional Probability and Expectations

4.1. Introduction

Let (X_1, X_2) be a bivariate random variable. In Chapter 3 we saw that it is described by the bivariate cumulative distribution function (cdf)

$$F(x_1, x_2) = \mathsf{P}(X_1 \le x_1, X_2 \le x_2), \qquad x_i \in (-\infty, \infty), \qquad i = 1, 2.$$

Using this joint distribution we can compute the probabilities of the type $\mathsf{P}(X_1 \in A; X_2 \in B)$ where A and B are sets of real numbers. In many applications we know that $X_2 \in B$, and we need the probability that $X_1 \in A$ given this knowledge. Using the definition of the conditional probability from Chapter 1, this can be computed as

$$\mathsf{P}(X_1 \in A | X_2 \in B) = \frac{\mathsf{P}(X_1 \in A; X_2 \in B)}{\mathsf{P}(X_2 \in B)}. \tag{4.1}$$

For example, we have

$$\mathsf{P}(X_1 \le x_1 | X_2 \le x_2) = \frac{\mathsf{P}(X_1 \le x_1; X_2 \le x_2)}{\mathsf{P}(X_2 \le x_2)}$$
$$= \frac{F(x_1, x_2)}{F_{X_2}(x_2)},$$

where $F_{X_2}(x_2)$ is the marginal cdf of X_2. Note that if X_1 and X_2 are independent random variables

$$F(x_1, x_2) = F_{X_1}(x_1) F_{X_2}(x_2),$$

and hence

$$\mathsf{P}(X_1 \le x_1 | X_2 \le x_2) = F_{X_1}(x_1),$$

and

$$P(X_2 \le x_2 | X_1 \le x_1) = F_{X_2}(x_2).$$

This makes intuitive sense. In order to be able to compute general conditional probabilities in (4.1) we need the concepts of conditional pmf and conditional pdf. They are introduced in the next two sections.

4.2. Conditional Probability Mass Function (pmf)

In this section we consider the discrete case. Let (X_1, X_2) be a jointly distributed discrete random variable with joint pmf $p(x_1, x_2)$. We formally define the concepts of conditional pmf below.

Definition 4.1 (Conditional pmf). The conditional probability $P(X_1 = x_1 | X_2 = x_2)$ is called the conditional pmf of X_1 given $X_2 = x_2$, and is denoted by $p_{X_1 | X_2}(x_1 | x_2)$.

By symmetry, the conditional pmf of X_2 given $X_1 = x_1$ is given by

$$p_{X_2 | X_1}(x_2 | x_1) = P(X_2 = x_2 | X_1 = x_1).$$

Note that the conditional pmf has all the properties of the usual pmf as given in Theorem 2.2. The next theorem shows how to compute the conditional pmf.

Theorem 4.1 (Conditional pmf). *If* $p_{X_2}(x_2) > 0$

$$p_{X_1 | X_2}(x_1 | x_2) = \frac{p(x_1, x_2)}{p_{X_2}(x_2)}, \tag{4.2}$$

where $p_{X_2}(x_2)$ *is the marginal pmf of* X_2.

Proof. Assume $p_{X_2}(x_2) = P(X_2 = x_2) > 0$. We have

$$\begin{aligned}
p_{X_1 | X_2}(x_1 | x_2) &= P(X_1 = x_1 | X_2 = x_2) \\
&= \frac{P(X_1 = x_1; X_2 = x_2)}{P(X_2 = x_2)} \\
&= \frac{p(x_1, x_2)}{p_{X_2}(x_2)}.
\end{aligned}$$

This yields (4.2).   □

We illustrate these concepts with several examples.

Example 4.1 (Tossing Two Dice). Consider the dice-tossing experiment of Examples 3.1, 3.2, and 3.5. Let X_1 be the maximum of the outcomes of the two dice and let X_2 be the

TABLE 4.1. The conditional pmf $p_{X_1|X_2}(x_1|3)$ for the two dice example.

x_1	1	2	3	4	5	6
$p_{X_1|X_2}(x_1|3)$	0	0	$\dfrac{1}{7}$	$\dfrac{2}{7}$	$\dfrac{2}{7}$	$\dfrac{2}{7}$

minimum of the outcomes of the two dice. Compute the conditional pmf of X_1 given $X_2 = 3$.

Using (4.2) we get

$$p_{X_1|X_2}(x_1|3) = \frac{p(x_1, 3)}{p_{X_2}(3)}. \tag{4.3}$$

For example, using the joint pmf from Example 3.2 and the marginal pmf from Example 3.5 we get

$$p_{X_1|X_2}(4|3) = \frac{p(4, 3)}{p_{X_2}(3)} = \frac{2/36}{7/36} = \frac{2}{7}.$$

The conditional pmf $p_{X_1|X_2}(x_1|3)$ is shown in Table 4.1. $\square$

Example 4.2 (Poisson Random Variables). Let X_i be a P(λ_i) random variable for $i = 1, 2$. Assume that X_1 and X_2 are independent. Compute the conditional pmf of X_1 given that $X_1 + X_2 = k$.

From Example 3.15 we see that $Z = X_1 + X_2$ is a P($\lambda_1 + \lambda_2$) random variable. We want to compute $p_{X_1|Z}(i|k)$ for $0 \le i \le k$. We have

$$
\begin{aligned}
p_{X_1|Z}(i|k) &= P(X_1 = i|Z = k) \\
&= \frac{P(X_1 = i; Z = k)}{P(Z = k)} \\
&= \frac{P(X_1 = i; X_2 = k - i)}{P(Z = k)} \\
&= \frac{P(X_1 = i)P(X_2 = k - i)}{P(Z = k)} \\
&\quad \text{(due to independence of } X_1 \text{ and } X_2) \\
&= \frac{e^{-\lambda_1}(\lambda_1^i/i!)\, e^{-\lambda_2}\left[\lambda_2^{k-i}/(k - i)!\right]}{e^{-\lambda_1-\lambda_2}\left[(\lambda_1 + \lambda_2)^k/k!\right]} \\
&= \frac{k!}{i!\,(k - i)!}\left(\frac{\lambda_1}{\lambda_1 + \lambda_2}\right)^i \left(\frac{\lambda_2}{\lambda_1 + \lambda_2}\right)^{k-i}
\end{aligned}
$$

Thus, given $X_1 + X_2 = k$, X_1 is a Bin($k, \lambda_1/(\lambda_1 + \lambda_2)$) random variable. $\square$

Conditional distributions can be defined in connection with more than two variables in a similar way. Instead of giving a formal definition, we illustrate the concept with an example.

Example 4.3 (Multinomial Random Variable). Consider the Multinomial random variable $X = (X_1, X_2, \ldots, X_r)$ with joint pmf as given in Equation 3.2. Compute the conditional joint pmf of $(X_1, X_2, \ldots, X_{r-1})$ given $X_r = k_r$.

From Example 3.6 we know that the marginal distribution of X_r is Bin(n, p_r). Hence we have

$$p_{X_1, X_2, \ldots, X_{r-1}|X_r}(k_1, k_2, \ldots, k_{r-1}|k_r)$$
$$= P(X_1 = k_1, X_2 = k_2, \ldots, X_{r-1} = k_{r-1}|X_r = k_r)$$
$$= \frac{P(X_1 = k_1, X_2 = k_2, \ldots, X_{r-1} = k_{r-1}, X_r = k_r)}{P(X_r = k_r)}$$
$$= \frac{[n!/(k_1! k_2! \cdots k_r!)] p_1^{k_1} p_2^{k_2} \cdots p_r^{k_r}}{[n!/(k_r!(n - k_r)!)] p_r^{k_r}(1 - p_r)^{n-k_r}}$$
$$= \frac{(n - k_r)!}{k_1! k_2! \cdots k_{r-1}!} \left(\frac{p_1}{1 - p_r}\right)^{k_1} \left(\frac{p_2}{1 - p_r}\right)^{k_2} \cdots \left(\frac{p_{r-1}}{1 - p_r}\right)^{k_{r-1}}. \quad (4.4)$$

Here we have used

$$n - k_r = k_1 + k_2 + \cdots + k_{r-1}.$$

From (4.4) we see that conditioned on $X_r = k_r$ $(X_1, X_2, \ldots, X_{r-1})$ is a multinomial random variable with parameters $n - k_r$ and $(p_1/(1 - p_r), p_2/(1 - p_r), \ldots, p_{r-1}/(1 - p_r))$. $\quad\square$

4.3. Conditional Probability Density Function (pdf)

In this section we consider the continuous case. Let (X_1, X_2) be a jointly distributed continuous random variable with joint pdf $f(x_1, x_2)$, and marginal pdfs $f_{X_1}(x_1)$ and $f_{X_2}(x_2)$. (See Section 3.4.) We formally define the concept of conditional pdf.

Definition 4.2 (Conditional pdf). The conditional pdf of X_1 given $X_2 = x_2$ is denoted by $f_{X_1|X_2}(x_1|x_2)$ and is defined to be

$$f_{X_1|X_2}(x_1|x_2) = \frac{f(x_1, x_2)}{f_{X_2}(x_2)} \quad (4.5)$$

assuming $f_{X_2}(x_2) > 0$.

Notice that we cannot define the conditional pdf as $P(X_1 = x_1|X_2 = x_2)$ since the events $X_1 = x_1$ and $X_2 = x_2$ both occur with zero probabilities due to the continuous nature of X_1 and X_2. However, the conditional pdf has the following intuitive "infinitesimal" interpretation:

$$P(x_1 \le X_1 \le x_1 + h|X_2 = x_2) \approx f_{X_1|X_2}(x_1|x_2)h$$

for "small" values of h. Thus for every x_2 for which $f_{X_2}(x_2) > 0$, the conditional pdf $f_{X_1|X_2}(\cdot|x_2)$ is a probability density function.

Example 4.4. Consider the bivariate random variable (X_1, X_2) of Example 3.4 with the following joint pdf:

$$f(x_1, x_2) = \begin{cases} \frac{1}{8}(x_1 + x_2) & \text{if } 0 \le x_1, x_2 \le 2, \\ 0 & \text{otherwise.} \end{cases}$$

Compute the conditional pdf of X_1 given $X_2 = x_2$.

From Example 3.7 we have the following expression for the marginal pdf of X_2:

$$f_{X_2}(x_2) = \tfrac{1}{4}(1 + x_2), \qquad 0 \le x_2 \le 2.$$

Substituting in (4.5) we get

$$f_{X_1|X_2}(x_1|x_2) = \frac{(x_1 + x_2)}{2(1 + x_2)}, \qquad 0 \le x_1, x_2 \le 2. \qquad \Box$$

4.4. Computing Probabilities by Conditioning

In Section 1.6 we studied the Law of Total Probability which we restate below: For any event E

$$P(E) = \sum_{n=1}^{\infty} P(E|E_n)P(E_n), \tag{4.6}$$

where $E_1, E_2, E_3, \dots$ are mutually exclusive and exhaustive events. We can use this law in conjunction with random variables as follows. Let X be a discrete random variable taking values $x_0, x_1, x_2, \dots$. Then the events $E_i = \{X = x_i\}, i = 0, 1, 2, \dots$, are mutually exclusive and exhaustive. Hence, substituting in (4.6) we get

$$P(E) = \sum_{i=0}^{\infty} P(E|X = x_i)P(X = x_i). \tag{4.7}$$

When X is a continuous random variable with pdf $f(\cdot)$, the sum in the above equation becomes an integral to yield

$$P(E) = \int_{-\infty}^{\infty} P(E|X = x)f(x)\,dx. \tag{4.8}$$

These two equations provide a very useful tool of computing probabilities of events when computing the conditional probabilities is easier. We illustrate with several examples below.

Example 4.5 (Splitting a Poisson Random Variable). The number of voters that turn out to vote on an election day is a Poisson random variable with parameter λ. Each voter casts exactly one vote. Suppose each voter votes for the democratic candidate with probability p.

Voters behave independently of each other. Compute the pmf of the number of votes received by the democratic candidate.

Let X be the total numbers of votes cast. Then X is a $P(\lambda)$ random variable. Let Y be the number of votes received by the democratic candidate. If the number of votes cast is n, then the independence of voters implies that Y is a $\text{Bin}(n, p)$ random variable. Using this conditional information we get

$$
\begin{aligned}
P(Y = k) &= \sum_{n=0}^{\infty} P(Y = k \mid X = n) P(X = n) \\
&= \sum_{n=k}^{\infty} \binom{n}{k} p^k (1-p)^{n-k} e^{-\lambda} \frac{\lambda^n}{n!} \\
&= e^{-\lambda} \frac{p^k}{k!} \sum_{n=k}^{\infty} \frac{n!}{(n-k)!} (1-p)^{n-k} \frac{\lambda^n}{n!} \\
&= e^{-\lambda} \frac{(\lambda p)^k}{k!} \sum_{n=k}^{\infty} \frac{(\lambda(1-p))^{n-k}}{(n-k)!} \\
&= e^{-\lambda} \frac{(\lambda p)^k}{k!} \sum_{n=0}^{\infty} \frac{(\lambda(1-p))^n}{n!} \\
&= e^{-\lambda} \frac{(\lambda p)^k}{k!} e^{-\lambda(1-p)} \\
&= e^{-\lambda p} \frac{(\lambda p)^k}{k!}.
\end{aligned}
\tag{4.9}
$$

Thus Y is seen to be a $P(\lambda p)$ random variable. As a numerical example, suppose that the number of votes cast in a given precinct is a Poisson random variable with mean 1000, and about 40% of the voters vote democratic. Thus the number of votes cast is a Poisson random variable with parameter $\lambda = 1000$, and each voter votes democratic with probability $p = .4$. Hence the number of votes received by the democratic candidate is a Poisson random variable with parameter (and mean) $(1000)(.4) = 400$.

Since the voters are split into those who voted democratic and those who did not, we say the Poisson random variable X is *split* into Y and $X - Y$. □

Example 4.6 (Probability of First Failure). Personal computers are available from two vendors. The lifetime of a computer from vendor i is an $\text{Exp}(\lambda_i)$ random variable, $i = 1, 2$. Suppose we purchase two computers, one from each vendor. What is the probability that the computer from vendor 1 fails before that from vendor 2?

Let X_i be the lifetime of the computer from vendor i, $i = 1, 2$. We are given that X_i is an $\text{Exp}(\lambda_i)$ random variable, and we want to compute $P(X_1 < X_2)$. Conditioning on X_1 we get

$$
P(X_1 < X_2) = \int_0^{\infty} P(X_1 < X_2 \mid X_1 = x) \lambda_1 e^{-\lambda_1 x} \, dx
$$

$$= \int_0^\infty P(x < X_2 | X_1 = x) \lambda_1 e^{-\lambda_1 x} \, dx$$

$$= \int_0^\infty P(X_2 > x) \lambda_1 e^{-\lambda_1 x} \, dx$$

$$= \int_0^\infty e^{-\lambda_2 x} \lambda_1 e^{-\lambda_1 x} \, dx$$

$$= \lambda_1 \int_0^\infty e^{-(\lambda_1 + \lambda_2)x} \, dx$$

$$= \frac{\lambda_1}{\lambda_1 + \lambda_2}.$$

As a numerical example, suppose the mean lifetime of the computer from vendor 1 is 20 months, while that from vendor 2 is 25 months. Then $\lambda_1 = \frac{1}{20} = .05$ (months)$^{-1}$, and $\lambda_2 = \frac{1}{25} = .04$ (months)$^{-1}$. Hence the required probability is

$$\frac{\lambda_1}{\lambda_1 + \lambda_2} = \frac{.05}{.05 + .04} = \frac{5}{9}.$$ □

Example 4.7 (Mixtures). Suppose that an item is available from n vendors. The lifetime of an item from vendor i has a cdf $F_i(\cdot)$, $1 \leq i \leq n$. If an item in use comes from vendor i with probability p_i, what is the probability that a randomly picked item will last at most t amount of time?

Let X be the vendor that supplied the item, and let T be the lifetime of the item. Then

$$P(X = i) = p_i, \quad i = 1, 2, \ldots, n,$$

and

$$P(T \leq t | X = i) = F_i(t), \quad i = 1, 2, \ldots, n.$$

Hence the cdf of T is given by

$$F(t) = P(T \leq t) = \sum_{i=1}^n P(T \leq t | X = i) P(X = i) = \sum_{i=1}^n F_i(t) p_i.$$

The cdf F of T is said to be a *mixture* of F_i, $i = 1, 2, \ldots, n$.

As a numerical example, suppose there are two vendors. The lifetimes of the item from them are exponentially distributed with means 20 months and 25 months, respectively. Suppose vendor 1 supplies 40% of the items and vendor 2 supplies the rest. What is the probability that an arbitrary item will fail within 2 years? We have $n = 2$, $p_1 = .4$, $p_2 = .6$, $F_1(t) = 1 - e^{-.05t}$, and $F_2(t) = 1 - e^{-.04t}$, where t is in months. Hence we get

$$P(T \leq 24) = .4(1 - e^{-.05*24}) + .6(1 - e^{-.04*24}) = .6498.$$ □

Example 4.8 (Compound Random Variables). A random variable Y, with parameter X, is said to be a compound random variable if X itself is a random variable. For example, suppose X is uniformly distributed over $(0, 1)$, and Y is a Geometric random variable with parameter X. Compute $P(Y = k)$.

We have

$$P(Y = k | X = p) = (1 - p)^{k-1} p, \qquad k = 1, 2, 3, \ldots,$$

and

$$f_X(p) = 1, \qquad 0 \le p \le 1.$$

Hence we get

$$P(Y = k) = \int_0^1 P(Y = k | X = p) f_X(p) \, dp$$

$$= \int_0^1 (1 - p)^{k-1} p \, dp$$

$$= \frac{1}{k(k + 1)}.$$

Note that the pmf of Y is not at all like a Geometric pmf, although for a given $X = p$, the conditional pmf of Y is $G(p)$! □

4.5. Conditional Expectations

The concept of conditional pmf was defined in (4.2), and that of conditional pdf was defined in (4.5). We saw that the conditional pmf and pdf have all the properties of the usual pmf and pdf. Hence it should be possible to define expected values using the conditional pmf and pdf. They are called conditional expected values and are defined formally below.

Definition 4.3 (Conditional Expectation). The conditional expectation of X, given $Y = y$, is given by

$$E(X | Y = y) = \sum_x x p_{X|Y}(x|y) \tag{4.10}$$

in the discrete case, and by

$$E(X | Y = y) = \int_{-\infty}^{\infty} x f_{X|Y}(x|y) \, dx \tag{4.11}$$

in the continuous case.

We illustrate the above concept with several examples.

Example 4.9 (Tossing Two Dice). Let X_1 and X_2 be as in Example 4.1. Compute $E(X_1 | X_2 = 3)$.

The conditional pmf of X_1 given $X_2 = 3$ is given in Example 4.1. Using that we get

$$\begin{aligned} \mathsf{E}(X_1|X_2 = 3) &= \sum_{i=1}^{6} i p_{X_1|X_2}(i|3) \\ &= 1 \cdot 0 + 2 \cdot 0 + 3 \cdot \tfrac{1}{7} + 4 \cdot \tfrac{2}{7} + 5 \cdot \tfrac{2}{7} + 6 \cdot \tfrac{2}{7} \\ &= \frac{33}{7}. \end{aligned}$$

$\square$

Example 4.10 (Poisson Random Variables). Let X_i be a $P(\lambda_i)$ random variable for $i = 1, 2$. Assume that X_1 and X_2 are independent. Compute $\mathsf{E}(X_1|X_1 + X_2 = k)$.

In Example 4.2 we saw that, given $X_1 + X_2 = k$, X_1 is a $\text{Bin}(k, \lambda_1/(\lambda_1 + \lambda_2))$ random variable. Hence

$$\mathsf{E}(X_1|X_1 + X_2 = k) = k \frac{\lambda_1}{\lambda_1 + \lambda_2}.$$

$\square$

Example 4.11. Consider the bivariate random variable (X_1, X_2) of Example 3.4. Compute $\mathsf{E}(X_1| X_2 = x_2)$ for $0 \leq x_2 \leq 2$.

The conditional pmf of X_1 given $X_2 = x_2$ is given in Example 4.4. Using that we get

$$\begin{aligned} \mathsf{E}(X_1|X_2 = x_2) &= \int_{-\infty}^{\infty} x_1 f_{X_1|X_2}(x_1|x_2)\, dx_1 \\ &= \int_{0}^{2} x_1 \frac{(x_1 + x_2)}{2(1 + x_2)}\, dx_1 \\ &= \frac{4 + 3x_2}{3 + 3x_2}. \end{aligned}$$

$\square$

4.6. Computing Expectations by Conditioning

In Section 4.4 we saw how the probabilities of events can be computed by conditioning. In this section we shall see how conditioning can be used to compute the expected values as well. The main result is given in the following theorem:

Theorem 4.2 (Expectation via Conditioning). *Let (X, Y) be a bivariate random variable. Then*

$$\mathsf{E}(X) = \sum_{y} \mathsf{E}(X|Y = y) p_Y(y) \tag{4.12}$$

if Y is discrete, and

$$\mathsf{E}(X) = \int_{-\infty}^{\infty} \mathsf{E}(X|Y = y) f_Y(y)\, dy \tag{4.13}$$

if Y is continuous.

Proof. We shall consider the case of discrete (X, Y) with joint pmf $p_{X,Y}(x, y)$. The continuous case is similar. Substituting for $\mathsf{E}(X|Y = y)$ from (4.10) we get

$$\sum_y \mathsf{E}(X|Y = y)p_Y(y) = \sum_y \sum_x x p_{X|Y}(x|y)p_Y(y)$$

$$= \sum_x x \sum_y p_{X|Y}(x|y)p_Y(y)$$

$$= \sum_x x \sum_y \frac{p_{X,Y}(x, y)}{p_Y(y)} p_Y(y) \quad \text{(from (4.2))}$$

$$= \sum_x x \sum_y p_{X,Y}(x, y)$$

$$= \sum_x x p_X(x) \quad \text{(from (3.11))}$$

$$= \mathsf{E}(X).$$

This proves the theorem. □

Theorem 4.2 can be interpreted in another useful way. Consider a discrete bivariate random variable (X, Y). Define

$$g(y) = \mathsf{E}(X|Y = y).$$

Since g is a well-defined function, we can talk about $g(Y)$. Note that $g(Y)$ is a random variable that takes value $g(y)$ with probability $p_Y(y)$. (This statement is precise when g is a one-to-one function.) The random variable $g(Y)$ is more transparently written as $\mathsf{E}(X|Y)$. Indeed $\mathsf{E}(X|Y)$ is a well-defined random variable for the continuous case as well. Theorem 4.2 then says

$$\mathsf{E}(g(Y)) = \sum_y g(y)p_Y(y) = \sum_y \mathsf{E}(X|Y = y)p_Y(y) = \mathsf{E}(X).$$

This is written in a succinct manner as

$$\mathsf{E}(\mathsf{E}(X|Y)) = \mathsf{E}(X). \tag{4.14}$$

This equation is valid in general.

Theorem 4.2 and (4.14) provide important computational tools for computing expectations of random variables. We illustrate its usefulness with several examples.

Example 4.12 (Tossing Two Dice). Consider the random variables X_1 and X_2 of Example 4.1. Following the computations of Example 4.9 we create Table 4.2. Thus $\mathsf{E}(X_1|X_2)$ is a random variable that takes value $\frac{41}{11}$ with probability $\frac{11}{36}$, etc. Using (4.14) we get

$$\mathsf{E}(\mathsf{E}(X_1|X_2)) = \frac{41}{11} \cdot \frac{11}{36} + \cdots + \frac{6}{1} \cdot \frac{1}{36} = \frac{150}{36} = \mathsf{E}(X_1). \qquad □$$

Example 4.13. Consider the bivariate random variable (X_1, X_2) of Example 3.4. Compute $\mathsf{E}(X_1)$ using Theorem 4.2.

TABLE 4.2. $E(X_1|X_2 = k)$ and $p_{X_2}(k)$ for the two dice example.

k	1	2	3	4	5	6
$E(X_1|X_2 = k)$	$\dfrac{41}{11}$	$\dfrac{38}{9}$	$\dfrac{33}{7}$	$\dfrac{26}{5}$	$\dfrac{17}{3}$	$\dfrac{6}{1}$
$p_{X_2}(k)$	$\dfrac{11}{36}$	$\dfrac{9}{36}$	$\dfrac{7}{36}$	$\dfrac{5}{36}$	$\dfrac{3}{36}$	$\dfrac{1}{36}$

The conditional expected value of X_1 given $X_2 = x_2$ is given in Example 4.4. Using that in (4.13) we get

$$E(X_1) = \int_{-\infty}^{\infty} E(X_1|X_2 = x_2) f_{X_2}(x_2) \, dx_2$$
$$= \int_0^2 \frac{4 + 3x_2}{3 + 3x_2} \tfrac{1}{4}(1 + x_2) \, dx_2$$
$$= \frac{7}{6}.$$

$\square$

Example 4.14 (Splitting a Poisson Random Variable). Consider the voting problem described in Example 4.5. X, the total number of votes cast, is a $P(\lambda)$ random variable. Each vote goes to a democratic candidate with probability p. Let Y be the number of votes received by the democratic candidate. Compute $E(Y)$.

In Example 4.5 we showed that Y is a $P(\lambda p)$ random variable, and hence $E(Y) = \lambda p$. Here we derive the same result by explicitly using (4.14). We know that, given $X = k$, Y is a Bin(k, p) random variable. Hence

$$E(Y|X = k) = pk.$$

Thus we can write

$$E(Y|X) = pX.$$

This explicitly brings out the fact that $E(Y|X)$ is a random variable that takes value pk whenever X takes value k. Hence, using the fact that $E(X) = \lambda$, we get

$$E(Y) = E(E(Y|X)) = E(pX) = pE(X) = p\lambda.$$

This matches with our earlier result, as it must. $\square$

Example 4.15 (Expectation of Random Sums). Let $\{X_n, n = 1, 2, 3, \ldots\}$ be a sequence of iid random variables, and let N be a nonnegative integer valued random variable that is independent of $\{X_n, n = 1, 2, 3, \ldots\}$. We are interested in

$$Z = \sum_{n=1}^{N} X_n.$$

Thus the number of the terms in the sum is random, and Z is called a random sum. If $N = 0$, then we define $Z = 0$. If $N = 3$, $Z = X_1 + X_2 + X_3$, etc. Such random sums occur often in applications. For example, let X_n be the toll paid by the nth car exiting a turnpike at a given exit, and let N be the total number of cars that exit it during 1 day. Then the total toll collected during the day is Z. Here we compute $E(Z)$.

We shall use

$$E(Z) = E(E(Z|N)). \tag{4.15}$$

First we need to compute $E(Z|N)$. To do this we compute

$$
\begin{aligned}
E(Z|N=k) &= E\left(\sum_{n=1}^{N} X_n | N = k\right) \\
&= E\left(\sum_{n=1}^{k} X_n\right) \quad \text{(due to independence of } N \text{ and } X_n\text{'s)} \\
&= \sum_{n=1}^{k} E(X_n) \\
&= kE(X),
\end{aligned}
$$

where $E(X) = E(X_n)$ for all n, since the X_n's are iid. Hence

$$E(Z|N) = NE(X).$$

Substituting in (4.15), and noting that $E(X)$ is a constant, we get

$$
\begin{aligned}
E(Z) &= E(E(Z|N)) \\
&= E(NE(X)) \\
&= E(N)E(X). \tag{4.16}
\end{aligned}
$$

This makes intuitive sense: If the expected toll paid by each car is $E(X) = \$2.50$ and the expected number of cars is $E(N) = 2000$, the expected total toll collection is $E(Z) = \$2.50 * 2000 = \$5000$. □

Example 4.16 (Mean of Geometric Random Variable). Suppose a coin that lands heads with probability p on any given toss is tossed repeatedly in an independent fashion. Let N be the number of tosses needed to obtain the first head. Compute $E(N)$.

Let X be the outcome on the first toss. If $X = H$, i.e., if the first toss yields a head, then clearly, $N = 1$. Hence

$$E(N|X = H) = 1.$$

On the other hand, if the first toss is a tail, i.e., $X = T$, then we have used up one toss already and the expected number of tosses from now on, to obtain the first head, is still

$E(N)$, due to the independence of the coin tosses. Hence

$$E(N|X = T) = 1 + E(N).$$

Combining these two equations we get

$$E(N) = E(N|X = H)P(X = H) + E(N|X = T)P(X = T)$$
$$= (1)(p) + (1 + E(N))(1 - p).$$

Solving for $E(N)$ we get

$$E(N) = \frac{1}{p}.$$

This agrees with the result in Example 2.38, as it must.   □

Example 4.17 (Two Heads in a Row). In the coin-tossing experiment of Example 4.16 above, let N be the number of tosses needed to observe two consecutive heads for the first time. Compute $E(N)$.

In this case we do not have an obvious candidate random variable to condition on. We need to choose one so that the conditional expected values will be easy to compute. We choose X to be the number of tosses needed to observe the first tail. Clearly, X is a geometric random variable with

$$P(X = k) = p^{k-1}(1 - p), \qquad k = 1, 2, 3, \ldots.$$

Using the same argument as in the previous example we get

$$E(N|X = 1) = 1 + E(N),$$
$$E(N|X = 2) = 2 + E(N),$$
$$E(N|X = k) = 2, \qquad k \geq 3.$$

Hence, using (4.12), we get

$$E(N) = \sum_{k=1}^{\infty} E(N|X = k)P(X = k)$$

$$= (1 + E(N))(1 - p) + (2 + E(N))p(1 - p) + \sum_{k=3}^{\infty}(2)p^{k-1}(1 - p)$$

$$= 1 + p + (1 - p^2)E(N).$$

Solving for $E(N)$, we get

$$E(N) = \frac{1 + p}{p^2}.$$

Note that computing this expected value by computing the pmf of N would be very difficult.   □

4.7.   Problems

CONCEPTUAL PROBLEMS

4.1. An experiment consists of first rolling a six-sided fair die, and if the outcome is k, $1 \leq k \leq 6$, tossing a fair coin k times. Let X be the number of heads and Y the number of tails obtained. Compute the joint distribution of (X, Y).

4.2. A six-sided die is tossed twice in an independent fashion. Let X_i, $i = 1, 2$, be the outcome of the ith toss. Compute the conditional pmf of X_1 given that $X_1 + X_2 = k$, $2 \leq k \leq 12$.

4.3. A coin is tossed repeatedly and independently. Let p be the probability that a head appears on any given toss. Let T_i be the toss number when the ith head is observed ($i = 1, 2, \ldots$). Compute the conditional pmf of T_1 given that $T_2 = k$.

4.4. Consider the experiment of Conceptual Problem 4.3. Let N_k be the number of heads obtained in the first k tosses. Compute the conditional pmf of T_1 given $N_k = 1$ for a fixed k.

4.5. Consider the Multinomial random variable $X = (X_1, X_2, \ldots, X_r)$ with joint pmf as given in Equation 3.2. Let $1 \leq k < r$ be a fixed integer. Compute the conditional joint pmf of $(X_1, X_2, \ldots, X_k)$ given $X_{k+1} + \cdots + X_r = m$.

4.6. Let X and Y be two iid Exp(λ) random variables. Compute the conditional pdf of X given $X + Y = z$.

4.7. Let X and Y be two iid N(μ, σ^2) random variables. Compute the conditional pdf of X given $X + Y = z$.

4.8. Let $\{X_n, n = 1, 2, \ldots\}$ be a set of iid Exp(λ) random variables, and let N be an independent G(p) random variable. Show that

$$Z = X_1 + X_2 + \cdots + X_N$$

is an Exp(λp) random variable.

4.9. Let U be a U(0, 1) random variable. Given $U = u$, X is a Bin(n, u) random variable. Show that

$$P(X = k) = \frac{1}{n + 1}, \qquad 0 \leq k \leq n.$$

4.10. Let U be a U(0, 1) random variable. Given $U = u$, X is an Exp(u) random variable. Compute the pdf of X.

4.11. Let U be an Exp(1) random variable. Given $U = u$, X is a P(u) random variable. Compute the pmf of X.

4.12. Using the technique of Example 4.17 show that the expected number of tosses needed to get k heads in a row is

$$\frac{1}{p^k}\frac{1-p^k}{1-p}.$$

4.13. A six-sided fair die is tossed repeatedly and independently. Compute the expected number of tosses necessary to get the sum of two consecutive tosses to be 7.
(*Hint:* Let u_i be the expected number of tosses needed if the current toss yields i, $1 \le i \le 6$. Derive equations for u_i and solve them. Show that the required answer is given by

$$\frac{1}{6}\sum_1^6 u_i.)$$

4.14. A six-sided fair die is tossed repeatedly and independently. Compute the expected number of tosses needed to obtain a cumulative score of six or more.
(*Hint:* Let m_i be the expected number of tosses needed to get a cumulative score of i or more. Conditioning on the outcome of the first toss, derive an equation for m_i in terms of m_j, $j < i$. Solve these equations to compute m_6, the desired answer.)

4.15. A fair coin is tossed repeatedly and independently until a total of k heads is obtained. Compute the expected number of tosses needed, assuming that the probability of a head on any given toss is p.

COMPUTATIONAL PROBLEMS

4.1. A bivariate random variable (X, Y) has the following pmf:

$$p(0, 0) = .1, \qquad p(1, 0) = .2,$$
$$p(0, 1) = .2, \qquad p(1, 1) = .1,$$
$$p(0, 2) = .2, \qquad p(1, 2) = .2.$$

(1) Compute the conditional pmf of Y given $X = 0$.

(2) Compute the conditional pmf of Y given $X = 1$.

4.2. In Computational Problem 4.1 compute the following:

(1) The conditional pmf of X given $Y = 0$.

(2) The conditional pmf of X given $Y = 1$.

(3) The conditional pmf of X given $Y = 2$.

4.3. A bivariate random variable (X, Y) has the following pmf:

$$p(1, 0) = .1, \qquad p(1, 2) = .25, \qquad p(1, 4) = .05,$$
$$p(3, 0) = .25, \qquad p(3, 2) = .1, \qquad p(3, 4) = .05,$$
$$p(5, 0) = .05, \qquad p(5, 2) = .05, \qquad p(5, 4) = .1.$$

Compute the conditional pmf of Y given $X = k$ for $k = 1, 3, 5$.

4.4. In Computational Problem 4.3 compute the conditional pmf of X given $Y = k$ for $k = 0, 2, 4$.

4.5. A trivariate random variable (X, Y, Z) takes the values in the following set with equal probabilities: $\{(1, 1, 1), (2, 2, 2), (3, 3, 3), (1, 2, 3), (2, 3, 1), (3, 2, 1)\}$. Compute:

(1) The conditional joint pmf of (X, Y) given $Z = 3$.

(2) The conditional pmf of X given $Y = 1$.

(3) The conditional pmf of Z given $X = 2$.

4.6. A class consists of 10 juniors and 10 seniors and four graduate students. A group of six students is selected at random from this class. Compute the conditional pmf of the number of juniors in this group given that it contains exactly one graduate student.

4.7. A hand of six cards is dealt from a perfectly shuffled deck of 52 cards. Compute the conditional pmf of the number of hearts in this hand, given that it includes exactly two diamonds.

4.8. An urn contains eight red balls, six white balls and four blue balls. Balls are drawn from this urn without replacement until a blue ball is drawn. Compute the conditional pmf of the number of white balls drawn given that the first blue ball appears on the seventh draw.

4.9. Do Computational Problem 4.8 if the balls are drawn with replacement.

4.10. Compute the conditional pdf of X_1 given $X_2 = x_2$ if (X_1, X_2) has the joint pdf given in Computational Problem 3.6.

4.11. Compute the conditional pdf of X_2 given $X_1 = x_1$ if (X_1, X_2) has the joint pdf given in Computational Problem 3.7.

4.12. Compute the conditional pdf of X_1 given $X_2 = x_2$ if (X_1, X_2) has the joint pdf given in Computational Problem 3.8.

4.13. Compute the conditional pdf of X_2 given $X_1 = x_1$ if (X_1, X_2) has the joint pdf given in Computational Problem 3.9.

4.14. Compute the conditional pdf of X_2 given $X_1 = x_1$ if (X_1, X_2) has the joint pdf given in Computational Problem 3.6.

4.15. Compute the conditional pdf of X_1 given $X_2 = x_2$ if (X_1, X_2) has the joint pdf given in Computational Problem 3.7.

4.16. Compute the conditional pdf of X_2 given $X_1 = x_1$ if (X_1, X_2) has the joint pdf given in Computational Problem 3.8.

4.17. Compute the conditional pdf of X_1 given $X_2 = x_2$ if (X_1, X_2) has the joint pdf given in Computational Problem 3.9.

4.18. A one-foot long stick is cut at two points that are placed U_1 and U_2 feet from one end of the stick. Suppose U_1 is U(0, .5), and U_2 is U(.5, 1), and U_1 and U_2 are independent. Compute the probability that a triangle can be made from the three pieces so obtained.

4.19. The lifetimes of two car batteries (Brand A and B) are independent exponential random variables with means 12 hours and 10 hours, respectively. What is the probability that Brand B battery outlasts Brand A battery?

4.20. The run times of two computer programs, A and B, are independent random variables. The run time of program A is an exponential random variable with mean 3 minutes, while that of program B is an Erl(2, λ) random variable with mean 3 minutes. If both programs are being executed on two identical computers, what is the probability that Program B will finish first?

4.21. Two cars, 1 and 2, are competing in a 500 mile race. The speed of car 1 is uniformly distributed over 120 to 130 mph, while that of car 2 is uniformly distributed over 115 to 135 mph. What is the probability that car 1 wins the race?

4.22. In Computational Problem 4.1 compute $E(X|Y = k)$ for $0 \leq k \leq 2$.

4.23. In Computational Problem 4.1 compute $E(Y|X = k)$ for $k = 0, 1$.

4.24. In Computational Problem 4.3 compute $E(X|Y = k)$ for $k = 0, 2, 4$.

4.25. In Computational Problem 4.3 compute $E(Y|X = k)$ for $k = 1, 3, 5$.

4.26. Compute the conditional expected value of X_1 given $X_2 = x_2$ if (X_1, X_2) has the joint pdf given in Computational Problem 3.6.

4.27. Compute the conditional expected value of X_2 given $X_1 = x_1$ if (X_1, X_2) has the joint pdf given in Computational Problem 3.7.

4.28. Compute the conditional expected value of X_1 given $X_2 = x_2$ if (X_1, X_2) has the joint pdf given in Computational Problem 3.8.

4.29. Compute the conditional expected value of X_2 given $X_1 = x_1$ if (X_1, X_2) has the joint pdf given in Computational Problem 3.9.

4.30. In Example 4.15 suppose the X_n's are Exp(2) random variables and N is a P(10) random variable. Compute the mean of Z.

4.31. In Example 4.15 suppose the X_n's are Erl(2, 5) random variables and N is a G(0.10) random variable. Compute the mean of Z.

4.32. A six-sided die is tossed once and its outcome is noted. It is then tossed repeatedly and independently until an outcome is observed that is at least as large as that on the first toss. Compute the expected number of tosses needed, not including the first one. (*Hint:* Condition on the outcome of the first toss.)

5

Discrete-Time Markov Models

5.1. What Is a Stochastic Process?

Consider a system that evolves randomly in time. Suppose this system is observed at times $n = 0, 1, 2, 3, \ldots$. Let X_n be the (random) state of the system at time n. The sequence of random variables $\{X_0, X_1, X_2, \ldots\}$ is called a (discrete-time) *stochastic process* and is written as $\{X_n, n \geq 0\}$. Let S be the set of values that X_n can take for any n. Then S is called the *state space* of the stochastic process $\{X_n, n \geq 0\}$. We illustrate with several examples below:

Example 5.1.
(a) X_n = temperature (in degrees Fahrenheit) recorded at Raleigh-Durham Airport at 12:00 noon on the nth day. The state space of the stochastic process $\{X_n, n \geq 0\}$ can be taken to be $S = (-50, 150)$. Note that this implies that the temperature never goes below $-50°$, or over 150°F. In theory the temperature could go all the way down to absolute zero, and all the way up to infinity!
(b) X_n = outcome of the nth toss of a six-sided die. The state space of $\{X_n, n \geq 0\}$ is $\{1, 2, 3, 4, 5, 6\}$.
(c) X_n = inflation rate in the US at the end of the nth week. Theoretically the inflation rate can take any real value, positive or negative. Hence the state space of $\{X_n, n \geq 0\}$ can be taken to be $(-\infty, \infty)$.
(d) X_n = Dow Jones index at the end of the nth trading day. The state space of the stochastic process $\{X_n, n \geq 0\}$ is $[0, \infty)$.
(e) X_n = number of claims submitted to an insurance company during the nth week. The state space of $\{X_n, n \geq 0\}$ is $\{0, 1, 2, 3, \ldots\}$.
(f) X_n = inventory of cars at a dealership at the end of the nth working day. The state space of the stochastic process $\{X_n, n \geq 0\}$ is $\{0, 1, 2, 3, \ldots\}$.

(g) X_n = number of PCs sold at an outlet store during the nth week. The state space of the stochastic process $\{X_n, n \geq 0\}$ is the same as that in (f).

(h) X_n = number of reported accidents in Chapel Hill during the nth day. The state space of the stochastic process $\{X_n, n \geq 0\}$ is the same as that in (f).   □

In this book we shall develop tools to study systems evolving randomly in time, and described by a stochastic process $\{X_n, n \geq 0\}$. What do we mean by "study" a system? The answer to this question depends on the system that we are interested in. Hence we illustrate the idea with an example.

Example 5.2.
(a) Consider the stochastic process in Example 5.1(a). The "study" of this system may involve predicting the temperature on day 10, that is, predicting X_{10}. However, predicting X_{10} is itself ambiguous: it may mean predicting the expected value of X_{10}, or the cdf of X_{10}. Another "study" may involve predicting how long the temperature will remain above 90° Fahrenheit, assuming the current temperature is above 90°. This involves computing the mean or distribution of the random time T when the temperature first dips below 90°.

(b) Consider the system described by the stochastic process of Example 5.1(e). The study of this system may involve computing the mean of the total number of claims submitted to the company during the first 10 weeks, i.e., $E(X_1 + X_2 + \cdots + X_{10})$. We may be interested in knowing if there is a long-term average weekly rate of claim submissions. This involves checking if $E(X_n)$ goes to any limit as n becomes large.

(c) Consider the system described by the stochastic process of Example 5.1(f). Suppose it costs \$20 per day to keep a car on the lot. The study of the system may involve computing the total expected cost over the first 3 days, i.e., $E(20X_1 + 20X_2 + 20X_3)$. We may also be interested in long-run average daily inventory cost, quantified by the limit of $E(20(X_1 + X_2 + \cdots + X_n)/n)$ as n becomes large.   □

The above example shows a variety of questions that can arise in the study of a system being modeled by a stochastic process $\{X_n, n \geq 0\}$. We urge the reader to think up plausible questions that may arise in studying the other systems mentioned in Example 5.1.

In order to answer the kinds of questions mentioned above, it is clear that we must be able to compute the joint distribution of $(X_1, X_2, \ldots, X_n)$ for any n. Then we can use the theory developed in the earlier chapters to answer these questions. How do we compute this joint distribution? We need more information about the stochastic process before it can be done. If the stochastic process has a simple (yet rich enough to capture the reality) structure, then the computation is easy. Over the years scientists have developed special classes of stochastic processes which can be used to describe a wide variety of useful systems and for which it is easy to do probabilistic computations. The first such class is called Discrete-Time Markov Chains. We shall study it in this chapter.

5.2.   Discrete-Time Markov Chains

Consider a system that is observed at times $0, 1, 2, \ldots$. Let X_n be the state of the system at time n for $n = 0, 1, 2, \ldots$. Suppose we are currently at time $n = 10$, say. That is, we have observed $X_0, X_1, \ldots, X_{10}$. The question is: can we predict, in a probabilistic way, the state of the system at time 11? In general, X_{11} depends (in a possibly random fashion) on $X_0, X_1, \ldots, X_{10}$. Considerable simplification occurs if, given the complete history $X_0, X_1, \ldots, X_{10}$, the next state X_{11} depends only upon X_{10}. That is, as far as predicting X_{11} is concerned, the knowledge of $X_0, X_1, \ldots, X_9$ is redundant if X_{10} is known. If the system has this property at all times n (and not just at $n = 10$), it is said to have a *Markov property*. (This is in honor of Markov, who, in the 1900s, first studied the stochastic processes arising out of such systems.) We start with a formal definition below.

Definition 5.1 (Markov Chain). A stochastic process $\{X_n, n \geq 0\}$ on state space S is said to be a Discrete-Time Markov Chain (DTMC) if, for all i and j in S,

$$P(X_{n+1} = j | X_n = i, X_{n-1}, \ldots, X_0) = P(X_{n+1} = j | X_n = i). \tag{5.1}$$

A DTMC $\{X_n, n \geq 0\}$ is said to be time homogeneous if, for all $n = 0, 1, \ldots,$

$$P(X_{n+1} = j | X_n = i) = P(X_1 = j | X_0 = i). \tag{5.2}$$

Note that (5.1) implies that the conditional probability on the left-hand side is the same no matter what values $X_0, X_1, \ldots, X_{n-1}$ take. Sometimes this property is described in words as follows: given the present state of the system (namely X_n), the future state of the DTMC (namely X_{n+1}) is independent of its past (namely $X_0, X_1, \ldots, X_{n-1}$). The quantity $P(X_{n+1} = j | X_n = i)$ is called a *one-step transition probability* of the DTMC at time n. Equation (5.2) implies that, for time homogeneous DTMCs, the one-step transition probability depends on i and j but is the same at all times n. Hence the terminology, *time homogeneous*.

In this chapter we shall consider only time homogeneous DTMCs with *finite* state space $S = \{1, 2, \ldots, N\}$. We shall always mean time homogeneous DTMC when we say DTMC. For such DTMCs we introduce a shorthand notation for the one-step transition probability:

$$p_{i,j} = P(X_{n+1} = j | X_n = i), \qquad i, j = 1, 2, \ldots, N. \tag{5.3}$$

Note the absence of n in the notation. This is because the right-hand side is independent of n for time homogeneous DTMCs. Note that there are N^2 one-step transition probabilities $p_{i,j}$. It is convenient to arrange them in an $N \times N$ matrix form as shown below.

$$P = \begin{bmatrix} p_{1.1} & p_{1.2} & p_{1.3} & \cdots & p_{1.N} \\ p_{2.1} & p_{2.2} & p_{2.3} & \cdots & p_{2.N} \\ p_{3.1} & p_{3.2} & p_{3.3} & \cdots & p_{3.N} \\ \vdots & \vdots & \vdots & \ddots & \vdots \\ p_{N.1} & p_{N.2} & p_{N.3} & \cdots & p_{N.N} \end{bmatrix}. \tag{5.4}$$

The matrix P in the above equation is called the *one-step transition probability matrix*, or transition matrix for short, of the DTMC. Note that the rows correspond to the starting state and the columns correspond to the ending state of a transition. Thus the probability of going from state 2 to state 3 in one step is stored in row number 2 and column number 3.

The information about the transition probabilities can also be represented in a graphical fashion by constructing a *transition diagram* of the DTMC. A transition diagram is a directed graph with N nodes, one node for each state of the DTMC. There is a directed arc going from node i to node j in the graph if $p_{i,j}$ is positive; in this case the value of $p_{i,j}$ is written next to the arc for easy reference. We can use the transition diagram as a tool to visualize the dynamics of the DTMC as follows: imagine a particle on a given node, say i, at time n. At time $n + 1$, the particle moves to node 2 with probability $p_{i,2}$, or node 3 with probability $p_{i,3}$, etc. X_n can then be thought of as the position (node index) of the particle at time n.

Example 5.3 (Transition Matrix and Transition Diagram). Suppose $\{X_n, n \geq 0\}$ is a DTMC with state space $\{1, 2, 3\}$ and transition matrix

$$P = \begin{bmatrix} .20 & .30 & .50 \\ .10 & .00 & .90 \\ .55 & .00 & .45 \end{bmatrix}. \tag{5.5}$$

If the DTMC is in state 3 at time 17, what is the probability that it will be in state 1 at time 18? The required probability is $p_{3,1}$, and is given by the element in the third row and the first column of the matrix P. Hence the answer is .55.

If the DTMC is in state 2 at time 9, what is the probability that it will be in state 3 at time 10? The required probability can be read from the element in the second row and third column of P. It is $p_{2,3} = .90$.

The transition diagram for this DTMC is shown in Figure 5.1. Note that it has no arc from node 3 to 2, representing the fact that $p_{3,2} = 0$.   □

Next we present two main characteristics of a transition probability matrix in the following theorem:

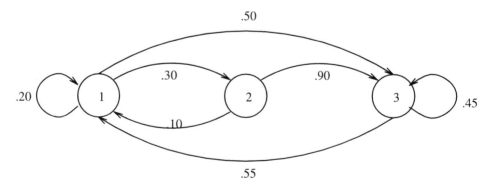

FIGURE 5.1. Transition diagram of the DTMC in Example 5.3.

Theorem 5.1 (Properties of a Transition Probability Matrix). *Let $P = [p_{i,j}]$ be an $N \times N$ transition probability matrix of a DTMC $\{X_n, n \geq 0\}$ with state space $S = \{1, 2, \ldots, N\}$. Then:*

(1) $p_{i,j} \geq 0, \quad 1 \leq i, j \leq N;$

(2) $\sum_{j=1}^{N} p_{i,j} = 1, \quad 1 \leq i \leq N.$

Proof. The nonnegativity of $p_{i,j}$ follows since it is a (conditional) probability. To prove the second assertion, we have

$$\sum_{j=1}^{N} p_{i,j} = \sum_{j=1}^{N} \mathsf{P}(X_{n+1} = j | X_n = i)$$
$$= \mathsf{P}(X_{n+1} \in S | X_n = i). \tag{5.6}$$

Since X_{n+1} must take some value in the state space S, regardless of the value of X_n, it follows that the last quantity is 1. Hence the theorem follows. $\quad\square$

Any square matrix possessing the two properties of the above theorem is called a *stochastic matrix* and can be thought of as a transition probability matrix of a DTMC.

5.3. Examples of Markov Models

Discrete-Time Markov Chains appear as appropriate models in many fields: biological systems, inventory systems, queueing systems, computer systems, telecommunication systems, manufacturing systems, manpower systems, economic systems, and so on. The following examples will give some evidence of this diversity. In each of these examples we derive the transition probability matrix for the appropriate DTMC.

Example 5.4 (Machine Reliability). The Depend-On-Us company manufactures a machine that is either up or down. If it is up at the beginning of a day, then it is up at the beginning of the next day with probability .98 (regardless of the history of the machine); or it fails with probability .02. Once the machine goes down, the company sends a repair person to repair it. If the machine is down at the beginning of a day, it is down at the beginning of the next day with probability .03 (regardless of the history of the machine); or the repair completes and the machine is up with probability .97. A repaired machine is as good as new. Model the evolution of the machine as a DTMC.

Let X_n be the state of the machine at the beginning of day n, defined as follows:

$$X_n = \begin{cases} 0 & \text{if the machine is down at the beginning of day } n, \\ 1 & \text{if the machine is up at the beginning of day } n. \end{cases}$$

The description of the system shows that $\{X_n, n \geq 0\}$ is a DTMC with state space $\{0, 1\}$ and the following transition probability matrix:

$$P = \begin{bmatrix} .03 & .97 \\ .02 & .98 \end{bmatrix}. \tag{5.7}$$

Now suppose the company maintains two such machines that are identical, behave independently of each other, and each has its own repair person. Let Y_n be the number of machines in the "up" state at the beginning of day n. Is $\{Y_n, n \geq 0\}$ a DTMC?

First we identify the state space of $\{Y_n, n \geq 0\}$ to be $\{0, 1, 2\}$. Next we see if the Markov property holds, that is, we check if $P(Y_{n+1} = j | Y_n = i, Y_{n-1}, \ldots, Y_0)$ depends only on i and j for $i, j = 0, 1, 2$. For example, consider the case $Y_n = i = 1$ and $Y_{n+1} = j = 0$. Thus, one machine is up (and one down) at time n. Since both machines are identical, it does not matter which is up and which is down. In order to move to state 0 at time $n + 1$, the down machine must stay down, and the up machine must go down at the beginning of the next day. Since the machines are independent, the probability of this happening is $.03 * .02$, independent of the history of the two machines. Hence we get

$$P(Y_{n+1} = 0 | Y_n = 1, Y_{n-1}, \ldots, Y_0) = .03 * .02 = .0006 = p_{1,0}.$$

Proceeding in this fashion we construct the following transition probability matrix:

$$P = \begin{bmatrix} .0009 & .0582 & .9409 \\ .0006 & .0488 & .9506 \\ .0004 & .0392 & .9604 \end{bmatrix}. \tag{5.8}$$

$\square$

Example 5.5 (Weather Model). The weather in the city of Heavenly is classified as sunny, cloudy, or rainy. Suppose that tomorrow's weather depends only on today's weather as follows: If it is sunny today, it is cloudy tomorrow with probability .3 and rainy with probability .2; if it is cloudy today, it is sunny tomorrow with probability .5 and rainy with probability .3; and finally, if it is rainy today, it is sunny tomorrow with probability .4 and cloudy with probability .5. Model the weather process as a DTMC.

Let X_n be the weather condition in Heavenly on day n, defined as follows:

$$X_n = \begin{cases} 1 & \text{if it is sunny on day } n, \\ 2 & \text{if it is cloudy on day } n, \\ 3 & \text{if it is rainy on day } n. \end{cases}$$

Then we are told that $\{X_n, n \geq 0\}$ is a DTMC with state space $\{1, 2, 3\}$. We next compute its transition matrix. We are given that $p_{1,2} = .3$ and $p_{1,3} = .2$. We are not explicitly given $p_{1,1}$. We use

$$p_{1,1} + p_{1,2} + p_{1,3} = 1$$

to obtain $p_{1,1} = .5$. Similarly we can obtain $p_{2,2}$ and $p_{3,3}$. This yields the following transition probability matrix:

$$P = \begin{bmatrix} .50 & .30 & .20 \\ .50 & .20 & .30 \\ .40 & .50 & .10 \end{bmatrix}. \tag{5.9}$$

$\square$

Example 5.6 (Inventory System). Computers-R-Us stocks a wide variety of PCs for retail sales. It is open for business Monday through Friday 8:00 a.m. to 5:00 p.m. It uses the following operating policy to control the inventory of Pentium PCs: At 5:00 p.m. Friday, the store clerk checks to see how many Pentium PCs are still in stock. If the number is less than 2, then he orders enough PCs to bring the total in stock up to 5 at the beginning of the business day Monday. If the number in stock is 2 or more no action is taken. The demand for the Pentium PCs during the week is a Poisson random variable with mean 3. Any demand that cannot be immediately satisfied is lost. Develop a stochastic model of the inventory of the Pentium PCs at Computers-R-Us.

Let X_n be the number of Pentium PCs in stock in Computers-R-Us at 8:00 a.m. Monday of the nth week. Let D_n be the number of Pentium PCs demanded during the nth week. Then the number of PCs left in the store at the end of the week is $\max(X_n - D_n, 0)$. If $X_n - D_n \geq 2$, then there are 2 or more Pentium PCs left in the store at 5:00 p.m. Friday of the nth week. Hence no more PCs will be ordered that weekend, and we will have $X_{n+1} = X_n - D_n$. On the other hand, if $X_n - D_n \leq 1$, there is 1 or 0 Pentiums left in the store at the end of the week. Hence enough will be ordered over the weekend so that $X_{n+1} = 5$. Putting these observations together we get

$$X_{n+1} = \begin{cases} X_n - D_n & \text{if } X_n - D_n \geq 2, \\ 5 & \text{if } X_n - D_n \leq 1. \end{cases}$$

It then follows that the state space of $\{X_n, n \geq 0\}$ is $\{2, 3, 4, 5\}$. Now assume that the demands from week to week are independent of each other and of the inventory in the store. Then we can show that $\{X_n, n \geq 0\}$ is a DTMC. We have, for $j = 2, 3, 4$ and $i = j, j + 1, \ldots, 5$:

$$\begin{aligned} \mathsf{P}(X_{n+1} = j | X_n = i, X_{n-1}, \ldots, X_0) \\ = \mathsf{P}(X_n - D_n = j | X_n = i, X_{n-1}, \ldots, X_0) \\ = \mathsf{P}(i - D_n = j | X_n = i, X_{n-1}, \ldots, X_0) \\ = \mathsf{P}(D_n = i - j). \end{aligned}$$

Similarly,

$$\mathsf{P}(X_{n+1} = 5 | X_n = i) = \mathsf{P}(X_n - D_n \leq 1 | X_n = i) = \mathsf{P}(D_n \geq i - 1)$$

for $i = 2, 3, 4$. Using the fact that D_n is a $P(3)$ random variable, we get Table 5.1. Using the data in this table we can compute the transition probability matrix of the DTMC $\{X_n, n \geq 0\}$

TABLE 5.1. Demand distribution for Example 5.6.

$k \to$	0	1	2	3	4
$P(D_n = k)$	.0498	.1494	.2240	.2240	.1680
$P(D_n \geq k)$	1.000	.9502	.8008	.5768	.3528

as follows:

$$
P = \begin{bmatrix}
.0498 & 0 & 0 & .9502 \\
.1494 & .0498 & 0 & .8008 \\
.2240 & .1494 & .0498 & .5768 \\
.2240 & .2240 & .1494 & .4026
\end{bmatrix}.
\tag{5.10}
$$

Note the entry for $p_{5.5}$. The inventory can move from 5 to 5 if the demand is either 0 or at least 4. Hence we have

$$
p_{5.5} = P(D_n = 0) + P(D_n \geq 4).
$$
☐

Example 5.7 (Manufacturing). The Gadgets-R-Us company has a manufacturing setup consisting of two distinct machines, each producing one component per hour. Each component can be tested instantly to be identified as defective or nondefective. Let a_i be the probability that a component produced by machine i is nondefective, $i = 1, 2$. (Obviously $1 - a_i$ is the probability that a component produced by machine i is defective.) The defective components are discarded and the nondefective components produced by each machine are stored in two separate bins. When a component is present in each bin, the two are instantly assembled together and shipped out. Each bin can hold at most two components. When a bin is full the corresponding machine is turned off. It is turned on again when the bin has space for at least one component. Model this system by a DTMC.

Let A_n be the number of components in the bin of machine 1, and let B_n be the number of components in the bin of machine 2, at the end of hour n, after accounting for the production and assembly during the hour. The bin size restrictions imply that $0 \leq A_n, B_n \leq 2$. Note that both bins cannot be nonempty simultaneously, since the assembly is instantaneous. Thus, $A_n > 0$ implies that $B_n = 0$, and $B_n > 0$ implies that $A_n = 0$. Let $X_n = A_n - B_n$. Then X_n can take values in $S = \{-2, -1, 0, 1, 2\}$. Note that we can recover A_n and B_n from X_n as follows:

$$
\begin{aligned}
X_n \leq 0 &\Rightarrow A_n = 0; & B_n = -X_n, \\
X_n \geq 0 &\Rightarrow A_n = X_n; & B_n = 0.
\end{aligned}
$$

If the successive components are independent of each other, then $\{X_n, n \geq 0\}$ is a DTMC. For example, suppose the history $\{X_0, \ldots, X_{n-1}\}$ is known, and $X_n = -1$. That is, the bin for machine 1 is empty, while the bin for machine 2 has one component in it. During the next hour each machine will produce one new component. If both these components are nondefective, then they will be assembled instantaneously. If both are defective, they will be discarded. In either case, X_{n+1} will remain -1. If the newly produced component 1 is nondefective, and the newly produced component 2 is defective, then the new component 2 will be discarded, the old component from bin 2 will be assembled with the newly produced nondefective component 1, and shipped out. Hence X_{n+1} will become 0. If the new component 1 is defective, while new component 2 is nondefective, X_{n+1} will become -2. Bin 2 will now be full, and machine 2 will be turned off. In the next hour only machine 1 will produce an item. From this analysis we get

$$P(X_{n+1} = j | X_n = -1, X_{n-1}, \ldots, X_0) = \begin{cases} a_1(1 - a_2) & \text{if } j = 0, \\ a_1 a_2 + (1 - a_1)(1 - a_2) & \text{if } j = -1, \\ (1 - a_1)a_2 & \text{if } j = -2, \\ 0 & \text{otherwise.} \end{cases}$$

Proceeding in this fashion, we can show that $\{X_n, n \geq 0\}$ is a DTMC on $\{-2, -1, 0, 1, 2\}$ with the following transition probability matrix (we use $a = (1 - a_1)a_2$, $b = a_1 a_2 + (1 - a_1)(1 - a_2)$, and $c = a_1(1 - a_2)$ for compactness):

$$P = \begin{bmatrix} 1 - a_1 & a_1 & 0 & 0 & 0 \\ a & b & c & 0 & 0 \\ 0 & a & b & c & 0 \\ 0 & 0 & a & b & c \\ 0 & 0 & 0 & a_2 & 1 - a_2 \end{bmatrix}. \tag{5.11}$$

□

Example 5.8 (Manpower Planning). Paper-Pushers, Inc. is an insurance company that employees 100 workers organized into four grades, labeled 1, 2, 3, and 4. For the sake of simplicity we shall assume that workers may get promoted from one grade to another, or leave the company, only at the beginning of a week. A worker in grade 1 at the beginning of a week gets promoted to grade 2 with probability .03, or leaves the company with probability .02, or continues in the same grade, at the beginning of the next week. A worker in grade 2 at the beginning of a week gets promoted to grade 3 with probability .01, or leaves the company with probability .008, or continues in the same grade, at the beginning of the next week. A worker in grade 3 at the beginning of a week gets promoted to grade 4 with probability .005, or leaves the company with probability .02, or continues in the same grade, at the beginning of the next week. A worker in grade 4 at the beginning of a week either leaves the company with probability .01, or continue in the same grade, at the beginning of the next week. If a worker leaves the company, he is instantly replaced by a new one in grade 1. Model worker movement in this company using a DTMC.

We shall assume that all worker promotions are decided in an independent manner. This simplifies our model considerably. Instead of keeping track of all 100 workers, we keep track of a single worker, say worker number k, $k = 1, 2, \ldots, 100$. We think of k as a worker id, and if and when this worker leaves the company, it gets assigned to the new replacement. Let X_n^k be the grade that the kth worker is in at the beginning of the nth week. Now if we assume that the worker promotions are determined independently of the history of the worker so far (meaning the time spent in a given grade does not affect one's chances of promotion), we see that, for $k = 1, 2, \ldots, 100$, $\{X_n^k, n \geq 0\}$ is a DTMC with state space $\{1, 2, 3, 4\}$. We illustrate the computation of the transition probabilities with an example: Suppose $X_n^k = 3$, i.e., the employee number k is in grade 3 at the beginning of week n. If he gets promoted (which happens with probability .005), we see that $X_{n+1}^k = 4$. Hence $P(X_{n+1}^k = 4 | X_n^k = 3) = .005$. If he leaves the company (which happens with probability .02), he is replaced by a new employee in grade 1, carrying the id k, making $X_{n+1}^k = 1$. Hence $P(X_{n+1}^k = 1 | X_n^k = 3) = .02$. With the remaining probability, .975, the employee continues in the same grade in the next week, making $X_{n+1}^k = 3$. Hence $P(X_{n+1}^k = 3 | X_n^k = 3) = .975$. Proceeding in a similar way we obtain the following transition probability matrix:

$$P = \begin{bmatrix} .9700 & .0300 & 0 & 0 \\ .0080 & .9820 & .0100 & 0 \\ .0200 & 0 & .9750 & .0050 \\ .0100 & 0 & 0 & .9900 \end{bmatrix}. \tag{5.12}$$

Note that the 100 DTMCs $\{X_n^1, n \geq 0\}$ through $\{X_n^{100}, n \geq 0\}$ are independent of each other and have the same transition probability matrix.   □

Example 5.9 (Stock Market). The common stock of the Gadgets-R-Us company is traded in the stock market. The chief financial officer of Gadgets-R-Us buys and sells the stock in his own company so that the price never drops below \$2, and never goes above \$10. For simplicity we assume that X_n, the stock price at the end of day n, takes only integer values, i.e., the state space of $\{X_n, n \geq 0\}$ is $\{2, 3, \ldots, 9, 10\}$. Let I_{n+1} be the potential movement in the stock price on day $n + 1$, in the absence of any intervention from the chief financial officer. Thus, we have

$$X_{n+1} = \begin{cases} 2 & \text{if } X_n + I_{n+1} \leq 2, \\ X_n + I_{n+1} & \text{if } 2 < X_n + I_{n+1} < 10, \\ 10 & \text{if } X_n + I_{n+1} \geq 10. \end{cases}$$

Statistical analysis of the past data suggests that the potential movements $\{I_n, n \geq 1\}$ form an iid sequence of random variables with common pmf given by

$$P(I_n = k) = .2, \qquad k = -2, -1, 0, 1, 2.$$

This implies that $\{X_n, n \geq 0\}$ is a DTMC on $\{2, 3, \ldots, 9, 10\}$. We illustrate the computation of the transition probabilities with three cases:

$$P(X_{n+1} = 2 | X_n = 3) = P(X_n + I_{n+1} \leq 2 | X_n = 3)$$
$$= P(I_{n+1} \leq -1)$$
$$= .4.$$

$$P(X_{n+1} = 6 | X_n = 5) = P(X_n + I_{n+1} = 6 | X_n = 5)$$
$$= P(I_{n+1} = 1)$$
$$= .2.$$

$$P(X_{n+1} = 10 | X_n = 10) = P(X_n + I_{n+1} \geq 10 | X_n = 10)$$
$$= P(I_{n+1} \geq 0)$$
$$= .6.$$

Proceeding in this fashion we get the following transition probability matrix for the DTMC $\{X_n, n \geq 0\}$:

$$P = \begin{bmatrix} .6 & .2 & .2 & 0 & 0 & 0 & 0 & 0 & 0 \\ .4 & .2 & .2 & .2 & 0 & 0 & 0 & 0 & 0 \\ .2 & .2 & .2 & .2 & .2 & 0 & 0 & 0 & 0 \\ 0 & .2 & .2 & .2 & .2 & .2 & 0 & 0 & 0 \\ 0 & 0 & .2 & .2 & .2 & .2 & .2 & 0 & 0 \\ 0 & 0 & 0 & .2 & .2 & .2 & .2 & .2 & 0 \\ 0 & 0 & 0 & 0 & .2 & .2 & .2 & .2 & .2 \\ 0 & 0 & 0 & 0 & 0 & .2 & .2 & .2 & .4 \\ 0 & 0 & 0 & 0 & 0 & 0 & .2 & .2 & .6 \end{bmatrix}. \tag{5.13}$$

$\square$

Example 5.10 (Telecommunications). The Tel-All Switch Corporation manufactures switching equipment for communication networks. Communication networks move data from switch to switch at lightning-fast speed in the form of packets, i.e., strings of zeros and ones (called bits). The Tel-All switches handle data packets of constant lengths, i.e., the same number of bits in each packet. At a conceptual level we can think of the switch as a storage device where packets arrive from the network users according to a random process. They are stored in a buffer with capacity to store K packets, and are removed from the buffer one-by-one according to a prespecified protocol. Under one such protocol, time is slotted into intervals of fixed length, say a microsecond. If there is a packet in the buffer at the beginning of a slot it is removed instantaneously. If there is no packet at the beginning of a slot no packet is removed during the slot even if more packets may arrive during the slot. If a packet arrives during a slot and there is no space for it, it is discarded. Model this as a DTMC.

Let A_n be the number of packets that arrive at the switch during one slot. (Some of these may be discarded.) Let X_n be the number of packets in the buffer at the end of the nth slot. Now, if $X_n = 0$, then there are no packets available for transmission at the beginning of the $(n + 1)$ st slot. Hence all the packets that arrive during that slot, namely A_{n+1}, are in the buffer at the end of slot, unless $A_{n+1} > K$, in which case the buffer is full at the end of the $(n + 1)$ st slot. Hence $X_{n+1} = \min\{A_{n+1}, K\}$. In case $X_n > 0$, one packet is removed at the beginning of the $(n + 1)$ st slot and A_{n+1} packets are added during that slot, subject to the capacity limitation. Combining these cases we get

$$X_{n+1} = \begin{cases} \min\{A_{n+1}, K\} & \text{if } X_n = 0, \\ \min\{X_n + A_{n+1} - 1, K\} & \text{if } 0 < X_n \leq K. \end{cases}$$

Assume that $\{A_n, n \geq 1\}$ is a sequence of iid random variables with common pmf

$$P(A_n = k) = a_k, \quad k \geq 0.$$

Under this assumption $\{X_n, n \geq 0\}$ is a DTMC on state space $\{0, 1, 2, \ldots, K\}$. The transition probabilities can be computed as follows: for $0 \leq j < K$,

$$
\begin{aligned}
P(X_{n+1} = j | X_n = 0) &= P(\min\{A_{n+1}, K\} = j | X_n = 0) \\
&= P(A_{n+1} = j) \\
&= a_j.
\end{aligned}
$$

$$
\begin{aligned}
P(X_{n+1} = K | X_n = 0) &= P(\min\{A_{n+1}, K\} = K | X_n = 0) \\
&= P(A_{n+1} \geq K) \\
&= \sum_{k=K}^{\infty} a_k.
\end{aligned}
$$

Similarly, for $1 \leq i \leq K$, and $i - 1 \leq j < K$,

$$
\begin{aligned}
P(X_{n+1} = j | X_n = i) &= P(\min\{X_n + A_{n+1} - 1, K\} = j | X_n = i) \\
&= P(A_{n+1} = j - i + 1) \\
&= a_{j-i+1}.
\end{aligned}
$$

Finally, for $1 \leq i \leq K$,

$$
\begin{aligned}
P(X_{n+1} = K | X_n = i) &= P(\min\{X_n + A_{n+1} - 1, K\} = K | X_n = i) \\
&= P(A_{n+1} \geq K - i + 1) \\
&= \sum_{k=K-i+1}^{\infty} a_k.
\end{aligned}
$$

Combining all these cases, and using the notation

$$b_j = \sum_{k=j}^{\infty} a_k$$

we get the following transition probability matrix:

$$
P = \begin{bmatrix}
a_0 & a_1 & \cdots & a_{K-1} & b_K \\
a_0 & a_1 & \cdots & a_{K-1} & b_K \\
0 & a_0 & \cdots & a_{K-2} & b_{K-1} \\
\vdots & \vdots & \ddots & \vdots & \vdots \\
0 & 0 & \cdots & a_0 & b_1
\end{bmatrix}. \tag{5.14}
$$

$\square$

Armed with the collection of the examples above, we set out to "analyze" them in the next section.

5.4.   Transient Distributions

Let $\{X_n, n \geq 0\}$ be a time-homogeneous DTMC on state space $S = \{1, 2, \ldots, N\}$, with transition probability matrix P, and initial distribution $a = [a_1, \ldots, a_N]$, where

$$a_i = \mathsf{P}(X_0 = i), \qquad 1 \leq i \leq N$$

In this section we concentrate on the *transient distribution*, i.e., the distribution of X_n for a fixed $n \geq 0$. In other words, we are interested in $\mathsf{P}(X_n = j)$ for all $j \in S$, and $n \geq 0$. We have

$$\mathsf{P}(X_n = j) = \sum_{i=1}^{N} \mathsf{P}(X_n = j | X_0 = i)\mathsf{P}(X_0 = i)$$

$$= \sum_{i=1}^{N} a_i \mathsf{P}(X_n = j | X_0 = i). \tag{5.15}$$

Thus it suffices to study the conditional probability $\mathsf{P}(X_n = j | X_0 = i)$. This quantity is called the *n-step transition probability* of the DTMC. We use the following notation:

$$a_j^{(n)} = \mathsf{P}(X_n = j), \tag{5.16}$$

$$p_{i,j}^{(n)} = \mathsf{P}(X_n = j | X_0 = i). \tag{5.17}$$

Analogous to the one-step transition probability matrix $P = [p_{i,j}]$ we build an *n-step transition probability matrix* as follows:

$$P^{(n)} = \begin{bmatrix} p_{1,1}^{(n)} & p_{1,2}^{(n)} & p_{1,3}^{(n)} & \cdots & p_{1,N}^{(n)} \\ p_{2,1}^{(n)} & p_{2,2}^{(n)} & p_{2,3}^{(n)} & \cdots & p_{2,N}^{(n)} \\ p_{3,1}^{(n)} & p_{3,2}^{(n)} & p_{3,3}^{(n)} & \cdots & p_{3,N}^{(n)} \\ \vdots & \vdots & \vdots & \ddots & \vdots \\ p_{N,1}^{(n)} & p_{N,2}^{(n)} & p_{N,3}^{(n)} & \cdots & p_{N,N}^{(n)} \end{bmatrix}. \tag{5.18}$$

We discuss two cases, $P^{(0)}$ and $P^{(1)}$, below. We have

$$p_{i,j}^{(0)} = \mathsf{P}(X_0 = j | X_0 = i) = \begin{cases} 1 & \text{if } i = j, \\ 0 & \text{if } i \neq j. \end{cases}$$

This implies that

$$P^{(0)} = I,$$

an $N \times N$ identity matrix. From (5.17) and (5.3), we see that

$$p_{i,j}^{(1)} = \mathsf{P}(X_1 = j | X_0 = i) = p_{i,j}. \tag{5.19}$$

Hence, from (5.19), (5.4), and (5.18) we get

$$P^{(1)} = P.$$

Now, construct the transient pmf vector

$$a^{(n)} = [a_1^{(n)}, a_2^{(n)}, \ldots, a_N^{(n)}],$$ (5.20)

so that (5.15) can be written in matrix form as

$$a^{(n)} = a * P^{(n)}.$$ (5.21)

Note that

$$a^{(0)} = a * P^{(0)} = a * I = a,$$

the initial distribution of the DTMC.

In this section we develop methods of computing the n-step transition probability matrix $P^{(n)}$. The main result is given in the following theorem:

Theorem 5.2 (n-Step Transition Probability Matrix).

$$P^{(n)} = P^n,$$ (5.22)

where P^n is the nth power of the matrix P.

Proof. Since $P^0 = I$ and $P^1 = P$, the theorem is true for $n = 0, 1$. Hence, let $n \geq 2$. We have

$$p_{i.j}^{(n)} = \mathsf{P}(X_n = j | X_0 = i)$$

$$= \sum_{k=1}^{N} \mathsf{P}(X_n = j | X_{n-1} = k, X_0 = i) \mathsf{P}(X_{n-1} = k | X_0 = i)$$

$$= \sum_{k=1}^{N} p_{i.k}^{(n-1)} \mathsf{P}(X_n = j | X_{n-1} = k, X_0 = i) \quad \text{(From (5.17))}$$

$$= \sum_{k=1}^{N} p_{i.k}^{(n-1)} \mathsf{P}(X_n = j | X_{n-1} = k) \quad \text{(due to the Markov property)}$$

$$= \sum_{k=1}^{N} p_{i.k}^{(n-1)} \mathsf{P}(X_1 = j | X_0 = k) \quad \text{(due to time-homogeneity)}$$

$$= \sum_{k=1}^{N} p_{i.k}^{(n-1)} p_{k.j}.$$ (5.23)

The last sum can be recognized as a matrix multiplication operation, and the above equation, which is valid for all $1 \leq i, j \leq N$, can be written in a more succinct fashion in matrix terminology as

$$P^{(n)} = P^{(n-1)} * P.$$ (5.24)

Using the above equation for $n = 2$ we get

$$P^{(2)} = P^{(1)} * P = P * P.$$

Similarly, for $n = 3$ we get

$$P^{(3)} = P^{(2)} * P = P * P * P.$$

In general using P^n as the nth power of the matrix P, we get (5.22).   □

From the above theorem we get the following two corollaries:

Corollary 5.1.

$$a^{(n)} = a * P^n.$$

Proof. Left to the reader as Conceptual Problem 5.5.   □

Corollary 5.2.

$$P(X_{n+m} = j | X_n = i, X_{n-1}, \ldots, X_0)$$
$$= P(X_{n+m} = j | X_n = i) = p_{ij}^{(m)}.$$

Proof. Use induction on m.   □

A more general version of (5.23) is given in the following theorem:

Theorem 5.3 (Chapman–Kolmogorov Equations). *The n-step transition probabilities satisfy the following equation, called the Chapman–Kolmogorov equation:*

$$p_{i,j}^{(n+m)} = \sum_{k=1}^{N} p_{i,k}^{(n)} p_{k,j}^{(m)}. \qquad (5.25)$$

Proof.

$$P(X_{n+m} = j | X_0 = i) = \sum_{k=1}^{N} P(X_{n+m} = j | X_n = k, X_0 = i) P(X_n = k | X_0 = i)$$

$$= \sum_{k=1}^{N} P(X_{n+m} = j | X_n = k) P(X_n = k | X_0 = i)$$

(due to Corollary 5.2)

$$= \sum_{k=1}^{N} P(X_m = j | X_0 = k) P(X_n = k | X_0 = i)$$

(due to time-homogeneity)

$$= \sum_{k=1}^{N} P(X_n = k | X_0 = i) P(X_m = j | X_0 = k)$$

$$= \sum_{k=1}^{N} p_{i,k}^{(n)} p_{k,j}^{(m)}. \qquad (5.26)$$

□

In matrix form, (5.25) can be expressed as

$$P^{(n+m)} = P^{(n)} * P^{(m)}. \tag{5.27}$$

Interchanging the roles of n and m we get

$$P^{(n+m)} = P^{(m)} * P^{(n)}.$$

The above equations imply that the matrices $P^{(n)}$ and $P^{(m)}$ commute for all n and m. This is an unusual property for matrices. Theorem 5.2 makes it especially easy to compute the transient distributions in DTMCs, since it reduces the computations to matrix powers and multiplications. Several matrix oriented computer packages are available that make these computations easy to perform.

Example 5.11. Consider the 3-state DTMC with the transition matrix as given in Example 5.3 and the initial distribution $a = [.2 \quad .3 \quad .5]$. Compute the pmf of X_3.

Using Corollary 5.1 we see that the pmf of X_3 can be computed as

$$a^{(3)} = a * P^3$$

$$= [.2 \quad .3 \quad .5] * \begin{bmatrix} .20 & .30 & .50 \\ .10 & .00 & .90 \\ .55 & .00 & .45 \end{bmatrix}^3$$

$$= [.3626 \quad .1207 \quad .5167]. \qquad \square$$

Example 5.12 (Weather Model). Mr and Mrs Happy have planned to celebrate their 25th wedding anniversary in Honeymooners' Paradise, a popular resort in the city of Heavenly. Counting today as the first day, they are supposed to be there on the seventh and eighth day. They are thinking about buying a vacation insurance which promises to reimburse them for the entire vacation package cost of $2500 if it rains on both days, and nothing is reimbursed otherwise. The insurance costs $200. Suppose the weather in the city of Heavenly changes according to the model in Example 5.5. Assuming that it is sunny today in the city of Heavenly, should Mr and Mrs Happy buy the insurance?

Let R be the reimbursement the couple gets from the insurance company. Letting X_n be the weather on the nth day, we see that $X_1 = 1$, and

$$R = \begin{cases} 2500 & \text{if } X_7 = X_8 = 3, \\ 0 & \text{otherwise.} \end{cases}$$

Hence

$$\begin{aligned} \mathsf{E}(R) &= 2500\mathsf{P}(X_7 = X_8 = 3 | X_1 = 1) \\ &= 2500\mathsf{P}(X_8 = 3 | X_7 = 3, X_1 = 1)\mathsf{P}(X_7 = 3 | X_1 = 1) \end{aligned}$$

$$= 2500 P(X_1 = 3|X_0 = 3)P(X_6 = 3|X_0 = 1)$$

$$= (2500)(.10) \left[\begin{bmatrix} .50 & .30 & .20 \\ .50 & .20 & .30 \\ .40 & .50 & .10 \end{bmatrix}^6 \right]_{1.3}$$

$$= 52.52.$$

Since this is less than \$200, the insurance is not worth it. □

Example 5.13 (Manufacturing). Consider the manufacturing operation described in Example 5.7. Suppose that both bins are empty at time 0 at the beginning of an 8-hour shift. What is the probability that both bins are empty at the end of the 8-hour shift? (Assume $a_1 = .99$ and $a_2 = .995$.)

Let $\{X_n, n \geq 0\}$ be the DTMC described in Example 5.7. We see that we are interested in $p_{0,0}^{(8)}$. Using the data given above, we see that the transition probability matrix is given by

$$P = \begin{bmatrix} .0100 & .9900 & 0 & 0 & 0 \\ .00995 & .9851 & .00495 & 0 & 0 \\ 0 & .00995 & .9851 & .00495 & 0 \\ 0 & 0 & .00995 & .9851 & .00495 \\ 0 & 0 & 0 & .9950 & .0050 \end{bmatrix}. \tag{5.28}$$

Computing P^8 we get

$$p_{0,0}^{(8)} = .88938.$$

What is the probability that machine 2 is shut down at the end of the shift? It is given by

$$p_{0,-2}^{(8)} = .0006533.$$ □

Example 5.14 (Telecommunications). Consider the model of the Tel-All data switch as described in Example 5.10. Let Y_n be the number of packets *lost* during the nth time slot. Show how to compute $E(Y_n)$ assuming that the buffer is initially empty. Tabulate $E(Y_n)$ as a function of n if the buffer size is seven packets and the number of packets that arrive at the switch during one time slot is a Poisson random variable with mean 1, packet arrivals in successive time slots being iid.

Let X_n be the number of packets in the buffer at the end of the nth time slot. Then $\{X_n, n \geq 0\}$ is a DTMC as described in Example 5.10. Let A_n be the number of packet arrivals during the nth slot. Then

$$Y_{n+1} = \begin{cases} \max\{0, A_{n+1} - K\} & \text{if } X_n = 0, \\ \max\{0, X_n - 1 + A_{n+1} - K\} & \text{if } X_n > 0. \end{cases}$$

TABLE 5.2. Data for Example 5.14.

$r \to$	0	1	2	3	4	5	6	7
$P(A_n = r)$	.3679	.3679	.1839	.0613	.0153	.0031	.0005	.0001
$P(A_n \geq r)$	1.000	.6321	.2642	.0803	.0190	.0037	.0006	.0001

Hence

$$E(Y_{n+1}) = E(\max\{0, A_{n+1} - K\})p_{0.0}^{(n)}$$
$$+ \sum_{k=1}^{K} E(\max\{0, k - 1 + A_{n+1} - K\})p_{0.k}^{(n)}.$$

Using $a_r = P(A_n = r)$ we get

$$E(Y_{n+1}) = p_{0.0}^{(n)} \sum_{r=K}^{\infty}(r - K)a_r$$
$$+ \sum_{k=1}^{K} p_{0.k}^{(n)} \sum_{r=K+1-k}^{\infty} (r - K - 1 + k)a_r. \qquad (5.29)$$

We are given that A_n is $P(1)$. The pmf and complementary cdf of A_n is as given in Table 5.2. Using the data given there, and using the analysis of Example 5.10, we see that $\{X_n, n \geq 0\}$ is a DTMC with state space $\{0, 1, \ldots, 7\}$ and transition probability matrix as given below:

$$P = \begin{bmatrix} .3679 & .3679 & .1839 & .0613 & .0153 & .0031 & .0005 & .0001 \\ .3679 & .3679 & .1839 & .0613 & .0153 & .0031 & .0005 & .0001 \\ 0 & .3679 & .3679 & .1839 & .0613 & .0153 & .0031 & .0006 \\ 0 & 0 & .3679 & .3679 & .1839 & .0613 & .0153 & .0037 \\ 0 & 0 & 0 & .3679 & .3679 & .1839 & .0613 & .0190 \\ 0 & 0 & 0 & 0 & .3679 & .3679 & .1839 & .0803 \\ 0 & 0 & 0 & 0 & 0 & .3679 & .3679 & .2642 \\ 0 & 0 & 0 & 0 & 0 & 0 & .3679 & .6321 \end{bmatrix}. \qquad (5.30)$$

The matrix P^n is computed numerically from P, and the n-step transition probabilities $p_{0.k}^{(n)}$ are obtained from it by using Theorem 5.2. These are then used in (5.29) to compute the expected number of packets lost in each slot. The final answer is given in the following table:

TABLE 5.3. Expected packet loss in Example 5.14.

$n \to$	1	2	5	10	20	30	40	80
$E(Y_n)$	.0000	.0003	.0063	.0249	.0505	.0612	.0654	.0681

Note that even if the switch removes one packet per time slot, and one packet arrives at the switch on the average per time slot, the losses are not zero. This is due to the randomness in the arrival process. Another interesting feature to note is that, as n becomes large, $E(Y_n)$ seems to tend to a limiting value. More on this in Section 5.6. $\square$

5.5. Occupancy Times

Let $\{X_n, n \geq 0\}$ be a time-homogeneous DTMC on state space $S = \{1, 2, \ldots, N\}$, with transition probability matrix P, and initial distribution $a = [a_1, \ldots, a_N]$. In this section we study occupancy times, i.e., the expected amount of time the DTMC spends in a given state during a given interval of time. Since the DTMC undergoes one transition per unit time, the occupancy time is the same as the expected number of times it visits a given state in a finite number of transitions. We define this quantity formally below.

Let $N_j(n)$ be the number of times the DTMC visits state j over the time span $\{0, 1, \ldots, n\}$, and let

$$m_{i,j}(n) = E(N_j(n)|X_0 = i).$$

The quantity $m_{i,j}(n)$ is called the *occupancy time* up to n of state j starting from state i. Let

$$M(n) = \begin{bmatrix} m_{1,1}(n) & m_{1,2}(n) & m_{1,3}(n) & \cdots & m_{1,N}(n) \\ m_{2,1}(n) & m_{2,2}(n) & m_{2,3}(n) & \cdots & m_{2,N}(n) \\ m_{3,1}(n) & m_{3,2}(n) & m_{3,3}(n) & \cdots & m_{3,N}(n) \\ \vdots & \vdots & \vdots & \ddots & \vdots \\ m_{N,1}(n) & m_{N,2}(n) & m_{N,3}(n) & \cdots & m_{N,N}(n) \end{bmatrix} \tag{5.31}$$

be the *occupancy times matrix*. The next theorem gives a simple method of computing the occupancy times.

Theorem 5.4 (Occupancy Times). *Let $\{X_n, n \geq 0\}$ be a time-homogeneous DTMC on state space $S = \{1, 2, \ldots, N\}$, with transition probability matrix P. The occupancy times matrix is given by*

$$M(n) = \sum_{r=0}^{n} P^r. \tag{5.32}$$

Proof. Fix i and j. Let $Z_n = 1$ if $X_n = j$, and $Z_n = 0$ if $X_n \neq j$. Then,

$$N_j(n) = Z_0 + Z_1 + \cdots + Z_n.$$

Hence

$$m_{i,j}(n) = E(N_j(n)|X_0 = i)$$
$$= E(Z_0 + Z_1 + \cdots + Z_n|X_0 = i)$$

$$= \sum_{r=0}^{n} E(Z_r | X_0 = i)$$

$$= \sum_{r=0}^{n} P(Z_r = 1 | X_0 = i)$$

$$= \sum_{r=0}^{n} P(X_r = j | X_0 = i)$$

$$= \sum_{r=0}^{n} p_{i.j}^{(r)}. \tag{5.33}$$

We get (5.32) by writing the above equation in matrix form. □

Example 5.15 (Three-State DTMC). Consider the DTMC in Example 5.3. Compute the occupancy time matrix $M(10)$.

We have, from Theorem 5.4,

$$M(10) = \sum_{r=0}^{10} \begin{bmatrix} .20 & .30 & .50 \\ .10 & .00 & .90 \\ .55 & .00 & .45 \end{bmatrix}^r$$

$$= \begin{bmatrix} 4.5317 & 1.2484 & 5.2198 \\ 3.5553 & 1.9555 & 5.4892 \\ 3.8583 & 1.0464 & 6.0953 \end{bmatrix}.$$

Thus the expected number of visits to state 1 starting from state 1 over the first 10 transitions is 4.5317. Note that this includes the visit at time 0. □

Example 5.16 (Weather Model). Consider the three state weather model described in Example 5.5. Suppose it is sunny in Heavenly today. Compute the expected number of rainy days in the week starting today.

Let X_n be the weather in Heavenly on the nth day. Then, from Example 5.5, $\{X_n, n \geq 0\}$ is a DTMC with transition matrix P given in (5.9). The required quantity is given by $m_{1.3}(6)$ (why 6 and not 7?). The occupancy matrix $M(6)$ is given by

$$M(6) = \sum_{r=0}^{6} \begin{bmatrix} .5 & .3 & .2 \\ .5 & .2 & .3 \\ .4 & .5 & .1 \end{bmatrix}^r$$

$$= \begin{bmatrix} 3.8960 & 1.8538 & 1.2503 \\ 2.8876 & 2.7781 & 1.3343 \\ 2.8036 & 2.0218 & 2.1747 \end{bmatrix}.$$

Hence $m_{1.3}(6) = 1.25$. □

5.6. Limiting Behavior

Let $\{X_n, n \geq 0\}$ be a DTMC on state space $S = \{1, 2, \ldots, N\}$ with transition probability matrix P. In this section we study the limiting behavior of X_n as n tends to infinity. We start with the most obvious question:

Does the pmf of X_n approach a limit as n tends to infinity?

If it does, we call it the *limiting or steady-state distribution* and denote it by

$$\pi = [\pi_1, \pi_2, \ldots, \pi_N], \tag{5.34}$$

where

$$\pi_j = \lim_{n \to \infty} P(X_n = j), \qquad j \in S. \tag{5.35}$$

The next question is a natural followup:

If the limiting distribution exists, is it unique?

This question makes sense, since it is conceivable that the limit may depend upon the starting state, or the initial distribution of the DTMC. Finally, the question of practical importance is:

If there is a unique limiting distribution, how do we compute it?

It so happens that the answers to the first two questions are involved, but the answer to the last question is easy. Hence we give that first in the following theorem.

Theorem 5.5 (Limiting Distributions). *If a limiting distribution π exists, it satisfies*

$$\pi_j = \sum_{i=1}^{N} \pi_i p_{i,j}, \qquad j \in S, \tag{5.36}$$

and

$$\sum_{j=1}^{N} \pi_j = 1. \tag{5.37}$$

Proof. Conditioning on X_n and using the law of total probability, we get

$$P(X_{n+1} = j) = \sum_{i=1}^{N} P(X_n = i)p_{i,j}, \qquad j \in S. \tag{5.38}$$

Now, let n tend to infinity on both the right- and left-hand sides. Then, assuming that the limiting distribution exists, we see that

$$\lim_{n \to \infty} P(X_n = j) = \lim_{n \to \infty} P(X_{n+1} = j) = \pi_j.$$

Substituting in (5.38) we get (5.36). Equation (5.37) follows since π is a pmf.   □

Equations (5.36) can be written in matrix form as follows:

$$\pi = \pi P, \tag{5.39}$$

and are called *the balance equations* or *the steady-state equations*. Equation (5.37) is called the *normalizing equation*. We illustrate the above theorem with an example.

Example 5.17 (A DTMC with a Unique Limiting Distribution). Suppose $\{X_n, n \geq 0\}$ is a DTMC with state space $\{1, 2, 3\}$ and the following transition matrix (see Example 5.3):

$$P = \begin{bmatrix} .20 & .30 & .50 \\ .10 & .00 & .90 \\ .55 & .00 & .45 \end{bmatrix}. \tag{5.40}$$

We give below the n step transition probability matrix for various values of n:

$$n = 2: \quad P^2 = \begin{bmatrix} .3450 & .0600 & .5950 \\ .5150 & .0300 & .4550 \\ .3575 & .1650 & .4775 \end{bmatrix},$$

$$n = 4: \quad P^4 = \begin{bmatrix} .3626 & .1207 & .5167 \\ .3558 & .1069 & .5373 \\ .3790 & .1052 & .5158 \end{bmatrix},$$

$$n = 10: \quad P^{10} = \begin{bmatrix} .3704 & .1111 & .5185 \\ .3703 & .1111 & .5186 \\ .3704 & .1111 & .5185 \end{bmatrix},$$

$$n \geq 11: \quad P^n = \begin{bmatrix} .3704 & .1111 & .5185 \\ .3704 & .1111 & .5185 \\ .3704 & .1111 & .5185 \end{bmatrix}.$$

From this we see that the pmf of X_n approaches

$$\pi = [.3704, \quad .1111, \quad .5185].$$

It can be checked that π satisfies (5.36) and (5.37). Furthermore, all the rows of P^n are the same in the limit, implying that the limiting distribution of X_n is the same regardless of the initial distribution. □

Example 5.18 (A DTMC with No Limiting Distribution). Consider a DTMC $\{X_n, n \geq 0\}$ with state space $\{1, 2, 3\}$ and transition matrix as given below:

$$P = \begin{bmatrix} 0 & 1 & 0 \\ .10 & 0 & .90 \\ 0 & 1 & 0 \end{bmatrix}. \tag{5.41}$$

We can check numerically that

$$P^{2n} = \begin{bmatrix} .1000 & 0 & .9000 \\ 0 & 1.0000 & 0 \\ .1000 & 0 & .9000 \end{bmatrix}, \quad n \geq 1,$$

$$P^{2n-1} = \begin{bmatrix} 0 & 1.0000 & 0 \\ .1000 & 0 & .9000 \\ 0 & 1.0000 & 0 \end{bmatrix}, \quad n \geq 1.$$

Let a be the initial distribution of the DTMC. We see that the pmf of $X_n, n \geq 1$, is $[.1(a_1 + a_3), \ a_2, \ .9(a_1 + a_3)]$ if n is even, and $[.1a_2, \ a_1 + a_3, \ .9a_2]$ if n is odd. Thus the pmf of X_n does not approach a limit. It fluctuates betwen two pmfs that depend on the initial distribution. The DTMC has no limiting distribution. □

Although there is no limiting distribution for the DTMC of the above example, we can still solve the balance equations and the normalizing equation. The solution is unique and is given by $[.05 \ .50 \ .45]$. To see what this solution means, suppose the initial distribution of the DTMC is $[.05 \ .50 \ .45]$. Then, using (5.38) for $n = 0$, we can show that the pmf of X_1 is also given by $[.05 \ .50 \ .45]$. Proceeding this way, we see that the pmf of X_n is $[.05 \ .50 \ .45]$ for all $n \geq 0$. Thus the pmf of X_n remains the same for all n, if the initial distribution is chosen to be $[.05 \ .50 \ .45]$. We call this initial distribution a *stationary distribution*. Formally:

Definition 5.2 (Stationary Distribution). A distribution

$$\pi^* = [\pi_1^*, \ \pi_2^*, \ldots, \ \pi_N^*] \tag{5.42}$$

is called a stationary distribution if

$$P(X_0 = i) = \pi_i^* \quad \text{for all} \quad 1 \leq i \leq N \Rightarrow$$
$$P(X_n = i) = \pi_i^* \quad \text{for all} \quad 1 \leq i \leq N, \quad \text{and} \quad n \geq 0.$$

The questions about the limiting distribution (namely, existence, uniqueness, and method of computation) can be asked about the stationary distribution as well. We have a slightly stronger result for the stationary distribution as given in the following theorem.

Theorem 5.6 (Stationary Distributions). $\pi^* = [\pi_1^*, \ \pi_2^*, \ \ldots, \ \pi_N^*]$ *is a stationary distribution if and only if it satisfies*

$$\pi_j^* = \sum_{i=1}^{N} \pi_i^* p_{i,j}, \quad j \in S, \tag{5.43}$$

and

$$\sum_{j=1}^{N} \pi_j^* = 1. \tag{5.44}$$

Proof. First suppose π^* is a stationary distribution. This implies that if

$$P(X_0 = j) = \pi_j^*, \qquad j \in S,$$

then

$$P(X_1 = j) = \pi_j^*, \qquad j \in S.$$

But, using $n = 0$ in (5.38), we have

$$P(X_1 = j) = \sum_{i=1}^{N} P(X_0 = i) p_{i,j}.$$

Substituting the first two equations in the last we get (5.43). Equation (5.44) holds because π^* is a pmf.

Now suppose π^* satisfies (5.43) and (5.44). Suppose

$$P(X_0 = j) = \pi_j^*, \qquad j \in S.$$

Then, from (5.38), we have

$$P(X_1 = j) = \sum_{i=1}^{N} P(X_0 = i) p_{i,j}$$

$$= \sum_{i=1}^{N} \pi_i^* p_{i,j}$$

$$= \pi_j^* \qquad \text{due to (5.43).}$$

Thus the pmf of X_1 is π^*. Using (5.38) repeatedly we can show that the pmf of X_n is π^* for all $n \geq 0$. Hence π^* is a stationary distribution. □

Theorem 5.6 implies that, if there is a solution to (5.43) and (5.44), it is a stationary distribution. Also, note that the stationary distribution π^* satisfies the same balance equations, and the normalizing equation as the limiting distribution π. This yields the following corollary:

Corollary 5.3. *A limiting distribution, when it exists, is also a stationary distribution.*

Proof. Let π be a limiting distribution. Then, from Theorem 5.5, it satisfies (5.36) and (5.37). But these are the same as (5.43) and (5.44). Hence, from Theorem 5.6, $\pi^* = \pi$ is a stationary distribution. □

Example 5.19. Using Theorem 5.6 or Corollary 5.3 we see that $\pi^* = [.3704, .1111, .5185]$ is a (unique) limiting, as well as stationary, distribution to the DTMC of Example 5.17. Similarly, $\pi^* = [.05 \ .50 \ .45]$ is a stationary distribution for the DTMC in Example 5.18, but there is no limiting distribution for this DTMC. □

The next example shows that the limiting or stationary distributions need not be unique.

Example 5.20 (A DTMC with Multiple Limiting and Stationary Distributions). Consider a DTMC $\{X_n, n \geq 0\}$ with state space $\{1, 2, 3\}$ and transition matrix given below:

$$P = \begin{bmatrix} .20 & .80 & 0 \\ .10 & .90 & 0 \\ 0 & 0 & 1 \end{bmatrix}. \tag{5.45}$$

Computing the matrix powers P^n for increasing values of n we get

$$\lim_{n \to \infty} P^n = \begin{bmatrix} .1111 & .8889 & 0 \\ .1111 & .8889 & 0 \\ 0 & 0 & 1.0000 \end{bmatrix}.$$

This implies that the limiting distribution exists. Now let the initial distribution be $a = [a_1, a_2, a_3]$, and define

$$\pi_1 = .1111(a_1 + a_2),$$
$$\pi_2 = .8889(a_1 + a_2),$$
$$\pi_3 = a_3.$$

We see that π is a limiting distribution of $\{X_n, n \geq 0\}$. Thus the limiting distribution exists, but is not unique. It depends on the initial distribution. From Corollary 5.3 it follows that any of the limiting distributions is also a stationary distribution of this DTMC. $\square$

It should be clear by now that we need to find a normalized solution (i.e., a solution satisfying the normalizing equation) to the balance equations in order to study the limiting behavior of the DTMC. There is another important interpretation of the normalized solution to the balance equations as discussed below.

Let $N_j(n)$ be the number of times the DTMC visits state j over the time span $\{0, 1, \ldots, n\}$. We have studied the expected value of this quantity in Section 5.5. The *occupancy* of state j is defined as

$$\hat{\pi}_j = \lim_{n \to \infty} \frac{E(N_j(n))}{n + 1}. \tag{5.46}$$

Thus occupancy of state j is the same as the long-run fraction of the time the DTMC spends in state j. The next theorem shows that the *occupancy distribution*

$$\hat{\pi} = [\hat{\pi}_1, \hat{\pi}_2, \ldots, \hat{\pi}_N],$$

if it exists, satisfies the same balance and normalizing equations.

Theorem 5.7 (Occupancy Distribution). *If the occupancy distribution $\hat{\pi}$ exists, it satisfies*

$$\hat{\pi}_j = \sum_{i=1}^{N} \hat{\pi}_i p_{i.j}, \quad j \in S, \tag{5.47}$$

and

$$\sum_{j=1}^{N} \hat{\pi}_j = 1. \tag{5.48}$$

Proof. From Theorem 5.4, we have

$$\mathsf{E}(N_j(n)|X_0 = i) = m_{i,j}(n) = \sum_{r=0}^{n} p_{i,j}^{(r)}.$$

Hence,

$$\frac{\mathsf{E}(N_j(n)|X_0 = i)}{n+1} = \frac{1}{n+1} \sum_{r=0}^{n} p_{i,j}^{(r)}$$

$$= \frac{1}{n+1} \left(p_{i,j}^{(0)} + \sum_{r=1}^{n} p_{i,j}^{(r)} \right)$$

$$= \frac{1}{n+1} \left(p_{i,j}^{(0)} + \sum_{r=1}^{n} \sum_{k=1}^{N} p_{i,k}^{(r-1)} p_{k,j} \right)$$

(using Chapman–Kolmogorov equations)

$$= \frac{1}{n+1} \left(p_{i,j}^{(0)} + \sum_{k=1}^{N} \sum_{r=1}^{n} p_{i,k}^{(r-1)} p_{k,j} \right)$$

$$= \frac{1}{n+1} (p_{i,j}^{(0)}) + \frac{n}{n+1} \sum_{k=1}^{N} \frac{1}{n} \left(\sum_{r=0}^{n-1} p_{i,k}^{(r)} p_{k,j} \right).$$

Now let n tend to ∞,

$$\lim_{n\to\infty} \frac{\mathsf{E}(N_j(n)|X_0 = i)}{n+1} = \lim_{n\to\infty} \frac{1}{n+1}(p_{i,j}^{(0)}) + \lim_{n\to\infty} \left(\frac{n}{n+1} \right) \sum_{k=1}^{N} \lim_{n\to\infty} \frac{1}{n} \left(\sum_{r=0}^{n-1} p_{i,k}^{(r)} \right) p_{k,j}.$$

Assuming the limits exist, and using (5.46) and (5.33) we get

$$\hat{\pi}_j = \sum_{k=1}^{N} \hat{\pi}_k p_{k,j}$$

which is (5.47). The normalization equation (5.48) follows because

$$\sum_{j=1}^{N} N_j(n) = n+1.$$

$\square$

Thus the normalized solution of the balance equations can have as many as three interpretations: limiting distribution, stationary distribution, or occupancy distribution. The question is: Will there always be a solution? Will it be unique? When can this solution be interpreted as a limiting distribution, or stationary distribution, or occupancy distribution? Although these questions can be answered strictly in terms of solutions of linear systems of equations, it is more useful to develop the answers in terms of the DTMC framework. That is what we do below.

Definition 5.3 (Irreducible DTMC). A DTMC $\{X_n, n \geq 0\}$ on state space $S = \{1, 2, \ldots, N\}$ is said to be irreducible if, for every i and j in S, there is a $k > 0$ such that

$$P(X_k = j | X_0 = i) > 0. \tag{5.49}$$

A DTMC that is not irreducible is called reducible.

Note that the condition in (5.49) holds if and only if it is possible to go from any state i to any state j in the DTMC in one or more steps, or alternatively, there is a directed path from any node i to any node j in the transition diagram of the DTMC. It is in general easy to check if the DTMC is irreducible.

Example 5.21 (Irreducible DTMCs).
(a) The DTMC of Example 5.17 is irreducible, since the DTMC can visit any state from any other state in two or less number of steps.
(b) The DTMC of Example 5.5 is irreducible, since it can go from any state to any other state in one step.
(c) The 5-state DTMC of Example 5.7 is irreducible, since it can go from any state to any other state in four steps or less.
(d) The 9-state DTMC of Example 5.9 is irreducible, since it can go from any state to any other state in seven steps or less.
(e) The $(K + 1)$-state DTMC of Example 5.10 is irreducible, since it can go from any state to any other state in K steps or less. □

Example 5.22 (Reducible DTMCs). The DTMC of Example 5.20 is reducible, since this DTMC cannot visit state 3 from state 1 or 2. □

The usefulness of the concept of irreducibility arises from the following two theorems, whose proofs are beyond the scope of this book:

Theorem 5.8 (Unique Stationary Distribution). *A finite state irreducible DTMC has a unique stationary distribution, i.e., there is a unique normalized solution to the balance equation.*

Theorem 5.9 (Unique Occupancy Distribution). *A finite state irreducible DTMC has a unique occupancy distribution, and is equal to the stationary distribution.*

Next we introduce the concept of periodicity. This will help us decide when the limiting distribution exists.

Definition 5.4 (Periodicity). Let $\{X_n, n \geq 0\}$ be an irreducible DTMC on state space $S = \{1, 2, \ldots, N\}$, and let d be the largest integer such that

$$P(X_n = i | X_0 = i) > 0 \quad \Rightarrow \quad n \text{ is an integer multiple } d, \tag{5.50}$$

for all $i \in S$. The DTMC is said to be periodic with period d if $d > 1$ and aperiodic if $d = 1$.

A DTMC with period d can return to its starting state only at times $d, 2d, 3d, \ldots$. It is an interesting fact of irreducible DTMCs that it is sufficient to find the largest d satisfying (5.50) for any one state $i \in S$. All other states are guaranteed to produce the same d. This makes it easy to establish the periodicity of an irreducible DTMC. In particular, if $p_{i,i} > 0$ for any $i \in S$ for an irreducible DTMC, then d must be 1, and the DTMC must be aperiodic!

Periodicity is also easy to spot from the transition diagrams. First, define a directed cycle in the transition diagram as a directed path from any node to itself. If all the directed cycles in the transition diagram of the DTMC are multiples of some integer d, and this is the largest such integer, then this is the d of the above definition.

Example 5.23 (Aperiodic DTMCs). All the irreducible DTMCs mentioned in Example 5.21 are aperiodic, since each one of them has at least one state i with $p_{i,i} > 0$. □

Example 5.24 (Periodic DTMCs).

(a) Consider a DTMC on state space $\{1, 2\}$ with the following transition matrix:

$$P = \begin{bmatrix} 0 & 1 \\ 1 & 0 \end{bmatrix}. \tag{5.51}$$

This DTMC is periodic with period 2.

(b) Consider a DTMC on state space $\{1, 2, 3\}$ with the following transition matrix:

$$P = \begin{bmatrix} 0 & 1 & 0 \\ 0 & 0 & 1 \\ 1 & 0 & 0 \end{bmatrix}. \tag{5.52}$$

This DTMC is periodic with period 3.

(c) Consider a DTMC on state space $\{1, 2, 3\}$ with the following transition matrix:

$$P = \begin{bmatrix} 0 & 1 & 0 \\ .5 & 0 & .5 \\ 0 & 1 & 0 \end{bmatrix}. \tag{5.53}$$

This DTMC is periodic with period 2. □

The usefulness of the concept of irreducibility arises from the following main theorem, whose proof is beyond the scope of this book:

Theorem 5.10 (Unique Limiting Distribution). *A finite state irreducible aperiodic DTMC has a unique limiting distribution.*

The above theorem, along with Theorem 5.5, shows that the (unique) limiting distribution of an irreducible aperiodic DTMC is given by the solution to (5.36) and (5.37). From

Corollary 5.3, this is also the stationary distribution of the DTMC, and from Theorem 5.9 this is also the occupancy distribution of the DTMC.

We shall restrict ourselves to irreducible and aperiodic DTMCs in our study of limiting behavior. The limiting behavior of periodic and/or reducible DTMCs is more involved. For example, the pmf of X_n eventually cycles with period d if $\{X_n, n \geq 0\}$ is an irreducible periodic DTMC with period d. The stationary/limiting distribution of a reducible DTMC is not unique, and depends on the initial state of the DTMC. We refer the reader to an advanced text for a more complete discussion of these cases.

We end this section with several examples.

Example 5.25 (Three-State DTMC). Consider the DTMC of Example 5.17. This DTMC is irreducible and aperiodic. Hence the liming distribution, the stationary distribution, and the occupancy distribution all exist and are given by the unique solution to

$$[\pi_1 \ \pi_2 \ \pi_3] = [\pi_1 \ \pi_2 \ \pi_3] * \begin{bmatrix} .20 & .30 & .50 \\ .10 & .00 & .90 \\ .55 & .00 & .45 \end{bmatrix}$$

$\pi_0 = .2\pi_0 + .1\pi_1 + .55\pi_2$

$\pi_1 = .3\pi_0$

and the normalizing equation

$$\pi_1 + \pi_2 + \pi_3 = 1.$$

Note that although we have four equations in three unknowns, one of the balance equations is redundant. Solving the above equations simultaneously yields

$$\pi_1 = .3704, \qquad \pi_2 = .1111, \qquad \pi_3 = .5185.$$

This matches with our answer in Example 5.17. Thus we have

$$\pi = \pi^* = \hat{\pi} = [.3704 \ \ .1111 \ \ .5185].$$

Now consider the three-state DTMC from Example 5.18. This DTMC is irreducible but periodic. Hence there is no limiting distribution. However, the stationary distribution exists and is given by the solution to

$$[\pi_1^* \ \pi_2^* \ \pi_3^*] = [\pi_1^* \ \pi_2^* \ \pi_3^*] * \begin{bmatrix} 0 & 1 & 0 \\ .10 & 0 & .90 \\ 0 & 1 & 0 \end{bmatrix}$$

and

$$\pi_1^* + \pi_2^* + \pi_3^* = 1.$$

The solution is

$$[\pi_1^* \ \pi_2^* \ \pi_3^*] = [.0500 \ \ .5000 \ \ .4500].$$

This matches with the numerical analysis presented in Example 5.18. Since the DTMC is irreducible, the occupancy distribution is also given by $\hat{\pi} = \pi^*$. Thus the DTMC spends 45% of the time in state 3, in the long run. □

Example 5.26 (Telecommunications). Consider the DTMC model of the Tel-All data switch described in Example 5.14.

(a) Compute the long-run fraction of the time that the buffer is full.

Let X_n be the number of packets in the buffer at the beginning of the nth time slot. Then $\{X_n, n \geq 0\}$ is a DTMC on state space $\{0, 1, \ldots, 7\}$ with transition probability matrix P given in (5.30). We want to compute the long-run fraction of the time the buffer is full, i.e., the occupancy of state 7. Since this is an irreducible aperiodic DTMC, the occupancy distribution exists and is given by the solution to

$$\hat{\pi} = \hat{\pi} * P$$

and

$$\sum_{i=0}^{7} \hat{\pi}_i = 1.$$

The solution is given by

$$\hat{\pi} = [.0681 \quad .1171 \quad .1331 \quad .1361 \quad .1364 \quad .1364 \quad .1364 \quad .1364].$$

The occupancy of state 7 is .1364. Hence the buffer is full 13.64% of the time.

(b) Compute the expected number of packets waiting in the buffer in steady state.

Note that the DTMC has a limiting distribution, and is given by $\pi = \hat{\pi}$. Hence the expected number of packets in the buffer in steady state is given by

$$\lim_{n \to \infty} \mathsf{E}(X_n) = \sum_{i=0}^{7} i\pi_i = 3.7924.$$

Thus the buffer is a little more than half full on the average in steady state.   □

Example 5.27 (Manufacturing). Consider the manufacturing operation of the Gadgets-R-Us company as described in Example 5.13. Compute the long-run fraction of the time that both the machines are operating.

Let $\{X_n, n \geq 0\}$ be the DTMC described in Example 5.7 with state space $\{-2, -1, 0, 1, 2\}$ and the transition probability matrix given in (5.28). We are interested in computing the long-run fraction of the time that the DTMC spends in states $-1, 0, 1$. This is an irreducible and aperiodic DTMC. Hence the occupancy distribution exists and is given by the solution to

$$\hat{\pi} = \hat{\pi} * \begin{bmatrix} .0100 & .9900 & 0 & 0 & 0 \\ .00995 & .9851 & .00495 & 0 & 0 \\ 0 & .00995 & .9851 & .00495 & 0 \\ 0 & 0 & .00995 & .9851 & .00495 \\ 0 & 0 & 0 & .9950 & .0050 \end{bmatrix}$$

and

$$\sum_{i=-2}^{2} \hat{\pi}_i = 1.$$

Solving, we get

$$\hat{\pi} = [.0057 \quad .5694 \quad .2833 \quad .1409 \quad .0007].$$

Hence the long-run fraction of the time that both machines are working is given by

$$\hat{\pi}_{-1} + \hat{\pi}_0 + \hat{\pi}_1 = 0.9936. \qquad \qquad \square$$

5.7. Cost Models

Recall the inventory model of Example 5.2(c) where we were interested in computing the total cost of carrying the inventory over 10 weeks. In this section we develop methods of computing such costs. We start with a simple cost model first.

Let X_n be the state of a system at time n. Assume that $\{X_n, n \geq 0\}$ is a DTMC on state space $\{1, 2, \ldots, N\}$ with transition probability matrix P. Suppose the system incurs a random cost of $C(i)$ dollars every time it visits state i. Let $c(i) = \mathsf{E}(C(i))$ be the expected cost incurred at every visit to state i. Although we think of $c(i)$ as a cost per visit, it need not be so. It may be any other quantity, like reward per visit, or loss per visit, or profit per visit, etc. We shall consider two cost performance measures in the two subsections below.

5.7.1. Expected Total Cost over a Finite Horizon

In this subsection we shall develop methods of computing *expected total cost* (ETC) up to a given finite time n, called the horizon. The actual cost incurred at time r is $C(X_r)$. Hence the actual total cost up to time n is given by

$$\sum_{r=0}^{n} C(X_r),$$

and the ETC is given by

$$\mathsf{E}\left(\sum_{r=0}^{n} C(X_r)\right).$$

For $1 \leq i \leq N$, define

$$g(i, n) = \mathsf{E}\left(\sum_{r=0}^{n} C(X_r)|X_0 = i\right) \qquad (5.54)$$

as the ETC up to time n starting from state i. Next, let

$$c = \begin{bmatrix} c(1) \\ c(2) \\ \vdots \\ c(N) \end{bmatrix},$$

and

$$g(n) = \begin{bmatrix} g(1, n) \\ g(2, n) \\ \vdots \\ g(N, n) \end{bmatrix}.$$

Let $M(n)$ be the occupancy time matrix of the DTMC as defined in (5.31). The next theorem gives a method of computing $g(n)$ in terms of $M(n)$.

Theorem 5.11 (ETC: Finite Horizon).

$$g(n) = M(n) * c. \tag{5.55}$$

Proof. We have

$$g(i, n) = \mathsf{E}\left(\sum_{r=0}^{n} C(X_r)|X_0 = i\right)$$

$$= \sum_{r=0}^{n}\sum_{j=1}^{N} \mathsf{E}(C(X_r)|X_r = j)\mathsf{P}(X_r = j|X_0 = i)$$

$$= \sum_{r=0}^{n}\sum_{j=1}^{N} c(j)p_{i,j}^{(r)}$$

$$= \sum_{j=1}^{N}\left[\sum_{r=0}^{n} p_{i,j}^{(r)}\right]c(j)$$

$$= \sum_{j=1}^{N} m_{i,j}(n)c(j), \tag{5.56}$$

where the last equation follows from (5.33). This yields (5.55) in matrix form.   □

We illustrate with several examples.

Example 5.28 (Manufacturing). Consider the manufacturing model of Example 5.13. Assume that both bins are empty at the beginning of a shift. Compute the expected total number of assembled units produced during an 8-hour shift.

Let $\{X_n, n \geq 0\}$ be the DTMC described in Example 5.7. The transition probability matrix is given by (see (5.28))

$$P = \begin{bmatrix} .0100 & .9900 & 0 & 0 & 0 \\ .00995 & .9851 & .00495 & 0 & 0 \\ 0 & .00995 & .9851 & .00495 & 0 \\ 0 & 0 & .00995 & .9851 & .00495 \\ 0 & 0 & 0 & .9950 & .0050 \end{bmatrix}. \tag{5.57}$$

Recall that a_i is the probability that a component produced by machine i is nondefective, $i = 1, 2$. Let $c(i)$ be the expected number of assembled units produced in 1 hour if the DTMC is in state i at the beginning of the hour. (Note that $c(i)$ as defined here is not a cost, but can be treated as such!) Thus if $i = 0$, both the bins are empty, and a unit is assembled in the next hour if both machines produce nondefective components. Hence the expected number of assembled units produced per visit to state 0 is $a_1 a_2 = .99 * .995 = .98505$. Similar analysis for other states yields the following:

$$c(-2) = .99, \qquad c(-1) = .99, \qquad c(1) = .995, \qquad c(2) = .995.$$

We want to compute $g(0, 7)$. (Note that the production during the eighth hour is counted as production at time 7.) Using Theorem 5.4 and (5.55), we get

$$g(7) = \begin{bmatrix} 7.9195 \\ 7.9194 \\ 7.8830 \\ 7.9573 \\ 7.9580 \end{bmatrix}.$$

Hence the expected production during an 8-hour shift starting with both bins empty is 7.8830 units. If there were no defectives, the production would be 8 units. Thus the loss due to defective production is .1170 units in the shift!   □

Example 5.29 (Inventory Systems). Consider the DTMC model of the inventory system as described in Example 5.6. Suppose the store buys the PCs for \$1500 and sells them for \$1750. The weekly storage cost is \$50 per Pentium PC that is in the store at the beginning of the week. Compute the net revenue the store expects to get over the next 10 weeks, assuming that it begins with five PCs in stock at the beginning of the week.

Following Example 5.6, let X_n be the number of PCs in the store at the beginning of the nth week. $\{X_n, n \geq 0\}$ is a DTMC on state space $\{2, 3, 4, 5\}$ with the transition probability matrix given in (5.10). We are given $X_0 = 5$. If there are i PCs at the beginning of the nth week, the expected storage cost during that week is $50i$. Let D_n be the demand during the nth week. Then the expected number of PCs sold during the nth week is $\mathsf{E}(\min(i, D_n))$. Hence the expected net revenue is given as follows:

$$c(i) = -50i + (1750 - 1500)\mathsf{E}(\min(i, D_n)), \qquad 2 \leq i \leq 5.$$

Computing the above expectations we get

$$c = \begin{bmatrix} 337.7662 \\ 431.9686 \\ 470.1607 \\ 466.3449 \end{bmatrix}.$$

Note that the expected total net revenue over the next n weeks, starting in state i, is given by $g(i, n-1)$. Hence we need to compute $g(5, 9)$. Using Theorem 5.4 and (5.55), we get

$$g(9) = \begin{bmatrix} 4298.65 \\ 4381.17 \\ 4409.41 \\ 4404.37 \end{bmatrix}.$$

Hence the expected total net revenue over the next 10 weeks, starting with five PCs, is $4404.37. Note that the figure is higher if the initial inventory is 4! This is the result of storage costs. □

5.7.2. Long-Run Expected Cost Per Unit Time

The ETC $g(i, n)$ computed in the previous subsection tends to ∞ as n tends to ∞ in many examples. In such cases it makes more sense to compute the expected long-run cost rate, defined as

$$g(i) = \lim_{n \to \infty} \frac{g(i, n)}{n + 1}.$$

The following theorem shows that this long-run cost rate is independent of i in case the DTMC is irreducible, and gives an easy method of computing it:

Theorem 5.12 (Long-Run Cost Rate). *Suppose $\{X_n, n \geq 0\}$ is an irreducible DTMC with occupancy distribution $\hat{\pi}$. Then*

$$g = g(i) = \sum_{j=1}^{N} \hat{\pi}_j c(j). \tag{5.58}$$

Proof. From Theorem 5.7, we have

$$\lim_{n \to \infty} \frac{m_{i.j}(n)}{n + 1} = \hat{\pi}_j.$$

Using this, and Theorem 5.11, we get

$$g(i) = \lim_{n \to \infty} \frac{g(i, n)}{n + 1}$$

$$= \lim_{n \to \infty} \frac{1}{n + 1} \sum_{j=1}^{N} m_{i.j}(n) c(j)$$

$$= \sum_{j=1}^{N} \left[\lim_{n \to \infty} \frac{m_{i,j}(n)}{n+1} \right] c(j)$$

$$= \sum_{j=1}^{N} \hat{\pi}_j c(j).$$

This yields the desired result. □

The theorem is intuitive: In the long run, among all the visits to all the states, $\hat{\pi}_j$ is the fraction of the visits made by the DTMC to state j. The DTMC incurs a cost of $c(j)$ dollars for every visit to state j. Hence the expected cost per visit in the long run must be $\sum c(j)\hat{\pi}_j$. We can use Theorem 5.7 to compute the occupancy distribution $\hat{\pi}$. We illustrate with two examples below.

Example 5.30 (Manpower Planning). Consider the manpower planning model of Paper-Pushers Insurance Company, Inc. as described in Example 5.8. Suppose the company has 70 employees and this level does not change with time. Suppose the per person weekly payroll expenses are $400 for grade 1, $600 for grade 2, $800 for grade 3, and $1000 for grade 4. Compute the long-run weekly payroll expenses for the company.

We shall compute the long-run weekly payroll expenses for each employee slot and multiply that figure by 70 to get the final answer, since all employees behave identically. The grade of an employee evolves according to a DTMC with state space $\{1, 2, 3, 4\}$ and transition probability matrix given in (5.12). Since this is an irreducible DTMC the unique occupancy distribution is obtained by using (5.47) and (5.48) as

$$\hat{\pi} = [.2715 \quad .4546 \quad .1826 \quad .0913].$$

The cost vector is

$$c = [400 \quad 600 \quad 800 \quad 1000]'.$$

Hence the long-run weekly payroll expense for a single employee is

$$\sum_{j=1}^{4} \hat{\pi}_j c(j) = 618.7185.$$

For the 70 employees we get the total weekly payroll expense to be $70 * 618.7185 = \$43,310.29$. □

Example 5.31 (Telecommunications). Consider the model of the data switch as described in Examples 5.10 and 5.14. Compute the long-run packet loss rate if the parameters of the problem are as in Example 5.14.

Let $c(i)$ be the expected number of packets lost during the $(n+1)$ st slot if there were i packets in the buffer at the end of the nth slot. Following the analysis of Example 5.14

we get

$$c(i) = \sum_{r=K}^{\infty}(r - K)a_r \qquad \text{if} \quad i = 0,$$

$$= \sum_{r=K+1-i}^{\infty}(r - K - 1 + i)a_r \qquad \text{if} \quad 0 < i \le K,$$

where a_r is the probability that a Poisson random variable with parameter 1 takes a value r. Evaluating these sums we get

$$c = [.0000 \quad .0000 \quad .0001 \quad .0007 \quad .0043 \quad .0233 \quad .1036 \quad .3679].$$

The occupancy distribution of this DTMC is already computed in Example 5.26. It is given by

$$\hat{\pi} = [.0681 \quad .1171 \quad .1331 \quad .1361 \quad .1364 \quad .1364 \quad .1364 \quad .1364].$$

Hence the long-run rate of packet loss per slot is

$$\sum_{j=0}^{7}\hat{\pi}_j c(j) = .0682.$$

Since the arrival rate of packets is one packet per slot, this implies that the loss fraction is 6.82%. This is too high in practical applications. This loss can be reduced by either increasing the buffer size or reducing the input packet rate. Note that the expected number of packets lost during the nth slot, as computed in Example 5.14, was .0681 for $n = 80$. This agrees quite well with the long-run loss rate computed in this example. □

5.8. First Passage Times

We saw in Example 5.2(a) that one of the questions of interest in weather prediction was "How long will the current heat wave last?". If the heat wave is declared to be over when the temperature falls below 90°F, the problem can be formulated as follows: When will the stochastic process representing the temperature visit a state below 90°F? Questions of this sort lead us to study the *first passage time*, i.e., the random time at which a stochastic process "first passes into" a given subset of the state space. In this section we study the first passage times in DTMCs.

Let $\{X_n, n \ge 0\}$ be a DTMC on state space $S = \{1, 2, \ldots, N\}$ with transition probability matrix P. We shall first study a simple case: first passage time into state N, defined as follows.

$$T = \min\{n \ge 0 : X_n = N\}. \tag{5.59}$$

Note that T is *not* the minimum number of steps in which the DTMC can reach state N. It is the (random) number of steps that it takes to actually visit state N. Typically T can take values in $\{0, 1, 2, 3, \ldots\}$. We shall study the expected value of this random variable in detail below.

Let

$$m_i = \mathsf{E}(T|X_0 = i). \tag{5.60}$$

Clearly, $m_N = 0$. The next theorem gives a method of computing m_i, $1 \le i \le N - 1$.

Theorem 5.13 (Expected First Passage Times). $\{m_i, 1 \le i \le N - 1\}$ *satisfy the following:*

$$m_i = 1 + \sum_{j=1}^{N-1} p_{i,j} m_j, \qquad 1 \le i \le N - 1. \tag{5.61}$$

Proof. We condition on X_1. Suppose $X_0 = i$ and $X_1 = j$. If $j = N$ then $T = 1$, and if $j \neq N$ then the DTMC has already spent one time unit to go to state j, and the expected time from then on to reach state N is now given by m_j. Hence we get

$$\mathsf{E}(T|X_0 = i, X_1 = j) = \begin{cases} 1 & \text{if } j = N, \\ 1 + m_j & \text{if } j \neq N. \end{cases}$$

Unconditioning with respect to X_1 yields

$$\begin{aligned} m_i &= \mathsf{E}(T|X_0 = i) \\ &= \sum_{j=1}^{N} \mathsf{E}(T|X_0 = i, X_1 = j)\mathsf{P}(X_1 = j|X_0 = i) \\ &= \sum_{j=1}^{N-1}(1 + m_j)p_{i,j} + (1)(p_{i,N}) \\ &= \sum_{j=1}^{N}(1)p_{i,j} + \sum_{j=1}^{N-1} p_{i,j} m_j \\ &= 1 + \sum_{j=1}^{N-1} p_{i,j} m_j \end{aligned}$$

as desired.   □

The following examples illustrate the above theorem:

Example 5.32 (Machine Reliability). Consider the machine shop with two independent machines as described by the three-state DTMC $\{Y_n, n \ge 0\}$ in Example 5.4. Suppose both machines are up at time 0. Compute the expected time until both machines are down for the first time.

Let Y_n be the number of machines in the "up" state at the beginning of day n. From Example 5.4 we see that $\{Y_n, n \ge 0\}$ is a DTMC with state space $\{0, 1, 2\}$ and transition probability matrix given by

$$P = \begin{bmatrix} .0009 & .0582 & .9409 \\ .0006 & .0488 & .9506 \\ .0004 & .0392 & .9604 \end{bmatrix}.$$ (5.62)

Let T be the first passage time into state 0 (both machines down). We are interested in $m_2 = E(T|Y_0 = 2)$. Equations (5.61) become

$$m_2 = 1 + .9604m_2 + .0392m_1,$$
$$m_1 = 1 + .9506m_2 + .0488m_1.$$

Solving simultaneously, we get

$$m_1 = 2451 \text{ days}, \qquad m_2 = 2451.5 \text{ days}.$$

Thus the expected time until both machines are down is $2451.5/365 = 6.71$ years!   □

Example 5.33 (Manpower Planning). Consider the manpower model of Example 5.8. Compute the expected amount of time a new recruit spends with the company.

Note that the new recruit starts in grade 1. Let X_n be the grade of the new recruit at the beginning of the nth week. If the new recruit has left the company by the beginning of the nth week we set $X_n = 0$. Then, using the data in Example 5.8, we see that $\{X_n, n \geq 0\}$ is a DTMC on state space $\{0, 1, 2, 3, 4\}$ with the following transition probability matrix:

$$P = \begin{bmatrix} 1 & 0 & 0 & 0 & 0 \\ .020 & .950 & .030 & 0 & 0 \\ .008 & 0 & .982 & .010 & 0 \\ .020 & 0 & 0 & .975 & .005 \\ .010 & 0 & 0 & 0 & .990 \end{bmatrix}.$$ (5.63)

Note that state 0 is absorbing, since once the new recruit leaves the company the problem is finished. Let T be the first passage time into state 0. We are interested in $m_1 = E(T|X_0 = 1)$. Equations (5.61) can be written as follows:

$$m_1 = 1 + .950m_1 + .030m_2,$$
$$m_2 = 1 + .982m_2 + .010m_3,$$
$$m_3 = 1 + .975m_3 + .005m_4,$$
$$m_4 = 1 + .990m_4.$$

Solving simultaneously, we get

$$m_1 = 73.33, \qquad m_2 = 88.89, \qquad m_3 = 60, \qquad m_4 = 100.$$

Thus the new recruit stays with the company for 73.33 weeks or about 1.4 years.   □

So far we have dealt with a first passage time into a single state. What if we are interested in a first passage time into a set of states? We consider such a case next.

Let A be a subset of states in the state space, and define

$$T = \min\{n \geq 0 : X_n \in A\}. \tag{5.64}$$

Theorem 5.13 can be easily extended to the case of the first passage time defined above. Let $m_i(A)$ be the expected time to reach the set A starting from state i. Clearly, $m_i(A) = 0$ if $i \in A$. Following the same argument as in the proof of Theorem 5.13 we can show that

$$m_i(A) = 1 + \sum_{j \notin A} p_{i,j} m_j(A), \qquad i \notin A. \tag{5.65}$$

In matrix form the above equations can be written as

$$m(A) = e + P(A)m(A), \tag{5.66}$$

where $m(A)$ is a column vector $[m_i(A)]_{i \notin A}$, e is a column vector of ones, and $P(A) = [p_{i,j}]_{i,j \notin A}$ is a submatrix of P. A matrix language package can be used to solve this equation easily. We illustrate with an example.

Example 5.34 (Stock Market). Consider the model of the stock movement as described in Example 5.9. Suppose Mr Jones buys the stock when it is trading for \$5, and decides to sell it as soon as it trades at or above \$8. What is the expected amount of time that Mr Jones will end up holding the stock?

Let X_n be the value of the stock at the end of the nth day. From Example 5.9, $\{X_n, n \geq 0\}$ is a DTMC on state space $\{2, 3, \ldots, 9, 10\}$. We are given that $X_0 = 5$. Mr Jones will sell the stock as soon as X_n is 8 or 9 or 10. Thus we are interested in the first passage time T into the set $A = \{8, 9, 10\}$, in particular in $m_5(A)$. Equations (5.66) are

$$\begin{bmatrix} m_2(A) \\ m_3(A) \\ m_4(A) \\ m_5(A) \\ m_6(A) \\ m_7(A) \end{bmatrix} = \begin{bmatrix} 1 \\ 1 \\ 1 \\ 1 \\ 1 \\ 1 \end{bmatrix} + \begin{bmatrix} .6 & .2 & .2 & 0 & 0 & 0 \\ .4 & .2 & .2 & .2 & 0 & 0 \\ .2 & .2 & .2 & .2 & .2 & 0 \\ 0 & .2 & .2 & .2 & .2 & .2 \\ 0 & 0 & .2 & .2 & .2 & .2 \\ 0 & 0 & 0 & .2 & .2 & .2 \end{bmatrix} \begin{bmatrix} m_2(A) \\ m_3(A) \\ m_4(A) \\ m_5(A) \\ m_6(A) \\ m_7(A) \end{bmatrix}.$$

Solving the above equation we get

$$\begin{bmatrix} m_2(A) \\ m_3(A) \\ m_4(A) \\ m_5(A) \\ m_6(A) \\ m_7(A) \end{bmatrix} = \begin{bmatrix} 24.7070 \\ 23.3516 \\ 21.0623 \\ 17.9304 \\ 13.2601 \\ 9.0476 \end{bmatrix}.$$

Thus the expected time for the stock to reach 8 or more, starting from 5, is about 18 days.

□

Example 5.35 (Gambler's Ruin). Two gamblers, A and B, bet on successive independent tosses of a coin that lands heads up with probability p. If the coin turns up heads, gambler A wins a dollar from gambler B, and if the coin turns up tails, gambler B wins a dollar from gambler A. Thus the total number of dollars among the two gamblers stays fixed, say N. The game stops as soon as either gambler is ruined, i.e., is left with no money! Compute the expected duration of the game, assuming that the game stops as soon as one of the two gamblers is ruined. Assume the initial fortune of Gambler A is i.

Let X_n be the amount of money gambler A has after the nth toss. If $X_n = 0$, then gambler A is ruined and the game stops. If $X_n = N$, then gambler B is ruined and the game stops. Otherwise the game continues. We have

$$
X_{n+1} = \begin{cases} X_n & \text{if } X_n \text{ is 0 or } N, \\ X_n + 1 & \text{if } 0 < X_n < N \text{ and the coin turns up heads,} \\ X_n - 1 & \text{if } 0 < X_n < N \text{ and the coin turns up tails.} \end{cases}
$$

Since the successive coin tosses are independent, we see that $\{X_n, n \geq 0\}$ is a DTMC on state space $\{0, 1, \ldots, N - 1, N\}$ with the following transition probability matrix (with $q = 1 - p$):

$$
P = \begin{bmatrix}
1 & 0 & 0 & \cdots & 0 & 0 & 0 \\
q & 0 & p & \cdots & 0 & 0 & 0 \\
0 & q & 0 & \cdots & 0 & 0 & 0 \\
\vdots & \vdots & \vdots & \ddots & \vdots & \vdots & \vdots \\
0 & 0 & 0 & \cdots & 0 & p & 0 \\
0 & 0 & 0 & \cdots & q & 0 & p \\
0 & 0 & 0 & \cdots & 0 & 0 & 1
\end{bmatrix}. \tag{5.67}
$$

The game ends when the DTMC visits state 0 or N. Thus we are interested $m_i(A)$, where $A = \{0, N\}$. Equations (5.65) are

$$
m_0(A) = 0,
$$
$$
m_i(A) = 1 + q m_{i-1}(A) + p m_{i+1}(A), \qquad 1 \leq i \leq N - 1,
$$
$$
m_N(A) = 0.
$$

We leave it to the reader to verify that the solution, whose derivation is rather tedious, is given by

$$
m_i(A) = \begin{cases} \dfrac{i}{q - p} - \dfrac{N}{q - p} \cdot \dfrac{1 - (q/p)^i}{1 - (q/p)^N} & \text{if } q \neq p, \\ (N - i)(i) & \text{if } q = p. \end{cases} \tag{5.68}
$$

$\square$

5.9. Problems

CONCEPTUAL PROBLEMS

5.1. Let $\{X_n, n \geq 0\}$ be a time-homogeneous DTMC on state space $S = \{1, 2, \ldots, N\}$, with transition probability matrix P. Then, for $i_0, i_1, \ldots, i_{k-1}, i_k \in S$, show that

$$P(X_1 = i_1, \ldots, X_{k-1} = i_{k-1}, X_k = i_k | X_0 = i_0) = p_{i_0.i_1} \cdots p_{i_{k-1}.i_k}.$$

5.2. Let $\{X_n, n \geq 0\}$ be a time-homogeneous DTMC on state space $S = \{1, 2, \ldots, N\}$, with transition probability matrix P. Prove or disprove by counterexample:

$$P(X_1 = i, X_2 = j, X_3 = k) = P(X_2 = i, X_3 = j, X_4 = k).$$

5.3. Consider the machine reliability model of Example 5.4. Now suppose that there are three independent and identically behaving machines in the shop. If a machine is up at the beginning of a day, it stays up at the beginning of the next day with probability p, and if it is down at the beginning of a day, it stays down at the beginning of the next day with probability q, where $0 \leq p, q \leq 1$ are fixed numbers. Let X_n be the number of working machines at the beginning of the nth day. Show that $\{X_n, n \geq 0\}$ is a DTMC and display its transition probability matrix.

5.4. Let P be an $N \times N$ stochastic matrix. Using the probabilistic interpretation, show that P^n is also a stochastic matrix.

5.5. Prove Corollaries 5.1 and 5.2.

5.6. Let $\{X_n, n \geq 0\}$ be a DTMC on state space $S = \{1, 2, \ldots, N\}$ with transition probability matrix P. Let $Y_n = X_{2n}, n \geq 0$. Show that $\{Y_n, n \geq 0\}$ is a DTMC on S with transition matrix P^2.

5.7. Let $\{X_n, n \geq 0\}$ be a DTMC on state space $S = \{1, 2, \ldots, N\}$ with transition probability matrix P. Suppose $X_0 = i$ with probability 1. The sojourn time T_i of the DTMC in state i is said to be k if $\{X_0 = X_1 = \cdots = X_{k-1} = i, X_k \neq i\}$. Show that T_i is a $G(1 - p_{i.i})$ random variable.

5.8. Consider a machine that works as follows. If it is up at the beginning of a day, it stays up at the beginning of the next day with probability p, and fails with probability $1 - p$. It takes exactly 2 days for the repairs, at the end of which the machine is as good as new. Let X_n be the state of the machine at the beginning of day n, where the state is 0 if the machine has just failed, 1 if 1 day's worth of repair work is done on it, and 2 if it is up. Show that $\{X_n, n \geq 0\}$ is a DTMC and display its transition probability matrix.

5.9. Items arrive at a machine shop in a deterministic fashion at a rate of one per minute. Each item is tested before it is loaded onto the machine. An item is found to be nondefective with probability p, and defective with probability $1 - p$. If an item is found defective, it is discarded. Otherwise, it is loaded onto the machine. The machine takes exactly 1 minute to process the item, after which it is ready to process the next one. Let X_n be 0 if the machine is idle at the beginning of the nth minute, and 1 if it is starting the processing of an item. Show that $\{X_n, n \geq 0\}$ is a DTMC and display its transition probability matrix.

5.10. Consider the system of Conceptual Problem 5.9. Now suppose the machine can process two items simultaneously. However, it takes 2 minutes to complete the processing. There is a bin in front of the machine where there is room to store two nondefective items. As soon as there are two items in the bin they are loaded onto the machine and the machine starts processing them. Model this system as a DTMC.

5.11. Consider the system of Conceptual Problem 5.10. However, now suppose that the machine starts working on whatever items are waiting in the bin when it becomes idle. It takes 2 minutes to complete the processing whether the machine is processing one or two items. Processing on a new item cannot start unless the machine is idle. Model this as a DTMC.

5.12. The weather at a resort city is either sunny or rainy. The weather tomorrow depends on the weather today and yesterday as follows: If it was sunny yesterday and today, it will be sunny tomorrow with probability .9. If it was rainy yesterday, but sunny today, it will be sunny tomorrow with probability .8. If it was sunny yesterday, but rainy today, it will be sunny tomorrow with probability .7. If it was rainy yesterday and today, it will be sunny tomorrow with probability .6. Define today's state of the system as the pair (weather yesterday, weather today). Model this system as a DTMC, making appropriate independence assumptions.

5.13. Consider the following weather model: the weather normally behaves as in Example 5.5. However, when the cloudy spell lasts for 2 or more days, it continues to be cloudy for another day with probability .8 or turns rainy with probability .2. Develop a four-state DTMC model to describe this behavior.

5.14. N points, labeled $1, 2, \ldots, N$ are arranged clockwise on a circle and a particle moves on it as follows: If the particle is on point i at time n, it moves one step in clockwise fashion with probability p, or one step in counterclockwise fashion with probability $1 - p$, to move to a new point at time $n + 1$. Let X_n be the position of the particle at time n. Show that $\{X_n, n \geq 0\}$ is a DTMC and display its transition probability matrix.

5.15. Let $\{X_n, n \geq 0\}$ be an irreducible DTMC on state space $\{1, 2, \ldots, N\}$. Let u_i be the probability that the DTMC visits state 1 before it visits state N, starting from state i. Using the first step analysis, show that

$$u_1 = 1,$$

$$u_i = \sum_{j=1}^{N} p_{i,j} u_j, \qquad 2 \leq i \leq N - 1,$$

$$u_N = 0.$$

5.16. A total of N balls are put in two urns, so that initially urn A has i balls and urn B has $N - i$ balls. At each step, one ball is chosen at random from the N balls. If it is from urn A it is moved to urn B, and vice versa. Let X_n be the number of balls in urn A after n steps. Show that $\{X_n, n \geq 0\}$ is a DTMC assuming that the successive random drawings of the balls are independent. Display the transition probability matrix of the DTMC.

5.17. The following selection procedure is used to select one of two brands, say A and B, of infrared light bulbs. Suppose the brand A light bulbs life-times are iid $\exp(\lambda)$ random variables, and brand B light bulbs lifetimes are iid $\exp(\mu)$ random variables. One light bulb from each brand is turned on simultaneously, and the experiment ends when one of the two light bulbs fails. Brand A wins a point if the brand A light bulb outlasts brand B, and vice versa. (The probability that the bulbs fail simultaneously is zero.) The experiment is repeated with new light bulbs until one of the brands accumulates five points more than the other, and that brand is selected as the better brand. Let X_n be the number of points for brand A minus the number of points accumulated by brand B after n experiments. Show that $\{X_n, n \geq 0\}$ is a DTMC and display its transition probability matrix. (*Hint*: Once X_n takes a value 5 or -5, it stays there forever.)

5.18. Let $\{X_n, n \geq 0\}$ be a DTMC on state space $\{1, 2, \ldots, N\}$. Suppose it incurs a cost of $c(i)$ dollars every time it visits state i. Let $g(i)$ be the total expected cost incurred by the DTMC until it visits state N starting from state i. Derive the following equations:

$$g(N) = 0,$$

$$g(i) = c(i) + \sum_{j=1}^{N} p_{i,j} g(j), \qquad 1 \leq j \leq N - 1.$$

5.19. Here is another useful cost model: the system incurs a random cost of $C(i, j)$ dollars whenever it undergoes a transition from state i to j in one step. This model is called the *cost per transition model*. Define

$$c(i) = \sum_{j=1}^{N} \mathsf{E}(C(i, j)) p_{i,j}, \qquad 1 \leq i \leq N.$$

Show that $g(i, T)$, the total cost over a finite horizon T, under this cost model satisfies Theorem 5.11 with $c(i)$ as defined above. Also show that, $g(i)$, the long-run cost rate satisfies Theorem 5.12.

COMPUTATIONAL PROBLEMS

5.1. Consider the Telecommunication model of Example 5.14. Suppose the buffer is full at the end of the third time slot. Compute the expected number of packets in the buffer at the end of the fifth time slot.

5.2. Consider the DTMC in Conceptual Problem 5.14 with $N = 5$. Suppose the particle starts on point 1. Compute the probability distribution of its position at time 3.

5.3. Consider the Stock Market model of Example 5.9. Suppose Mr BigShot has bought 100 stocks at $5 per share. Compute the expected net change in the value of his investment in 5 days.

5.4. Mr Smith is a coffee addict. He keeps switching between three brands of coffee, say A, B, and C, from week to week according to a DTMC with the following transition probability matrix:

$$P = \begin{bmatrix} .10 & .30 & .60 \\ .10 & .50 & .40 \\ .30 & .20 & .50 \end{bmatrix}. \tag{5.69}$$

If he is using brand A this week (i.e., week 1), what is the probability distribution of the brand he will be using in week 10?

5.5. Consider the Telecommunication model of Example 5.10. Suppose the buffer is full at the beginning. Compute the expected number of packets in the buffer at time n for $n = 1, 2, 5,$ and 10, assuming that the buffer size is 10 and that the number of packets arriving during one time slot is a binomial random variable with parameters $(5, .2)$.

5.6. Consider the machine described in Conceptual Problem 5.8. Suppose the machine is initially up. Compute the probability that the machine is up at time $n = 5, 10, 15,$ and 20. (Assume $p = .95$.)

5.7. Consider the Paper-Pushers Insurance Company, Inc. of Example 5.8. Suppose it has 100 employees at the beginning of week 1, distributed as follows: 50 in grade 1, 25 in grade 2, 15 in grade 3, and 10 in grade 4. If employees behaves independently of each other, compute the expected number of employees in each grade at the beginning of week 4.

5.8. Consider the Machine Reliability model in Example 5.4 with one machine. Suppose the machine is up at the beginning of day 0. Compute the probability that the state of the machine at the beginning of the next 3 days is up, down, down, in that order.

5.9. Consider the Machine Reliability model in Example 5.4 with two machines. Suppose both machines are up at the beginning of day 0. Compute the probability that the number of working machines at the beginning of the next 3 days is 2, 1, and 2, in that order.

5.10. Consider the weather model of Example 5.5. Compute the probability that once the weather becomes sunny, the sunny spell lasts for at least 3 days.

5.11. Compute the expected length of a rainy spell in the weather model of Example 5.5.

5.12. Consider the inventory system of Example 5.6 with a starting inventory of five PCs on a Monday. Compute the probability that the inventory trajectory over the next four Mondays is as follows: 4, 2, 5, and 3.

5.13. Consider the inventory system of Example 5.6 with a starting inventory of five PCs on a Monday. Compute the probability that an order is placed at the end of the first week for more PCs.

5.14. Consider the Manufacturing Model of Example 5.7. Suppose both bins are empty at time 0. Compute the probability that both bins stay empty at times $n = 1, 2$, and then machine 1 is shut down at time $n = 4$.

5.15. Compute the occupancy matrix $M(10)$ for the DTMCs with transition matrices as given below:

(a)
$$P = \begin{bmatrix} .10 & .30 & .20 & .40 \\ .10 & .30 & .40 & .20 \\ .30 & .30 & .10 & .30 \\ .15 & .25 & .35 & .25 \end{bmatrix}.$$

(b)
$$P = \begin{bmatrix} 0 & 1 & 0 & 0 \\ 0 & 0 & 1 & 0 \\ 0 & 0 & 0 & 1 \\ 1 & 0 & 0 & 0 \end{bmatrix}.$$

(c)
$$P = \begin{bmatrix} .10 & 0 & .90 & 0 \\ 0 & .30 & 0 & .70 \\ .30 & 0 & .70 & 0 \\ 0 & .25 & 0 & .75 \end{bmatrix}.$$

(d)
$$P = \begin{bmatrix} .10 & .30 & 0 & .60 \\ .10 & .30 & 0 & .60 \\ .30 & .10 & .10 & .50 \\ .5 & .25 & 0 & .25 \end{bmatrix}.$$

5.16. Consider the inventory system of Example 5.6. Compute the occupancy matrix $M(52)$. Using this compute the expected number of weeks that Computers-R-Us starts with a full inventory (i.e., five PCs) during a year, given that it started the first week of the year with an inventory of five PCs.

5.17. Consider the Manufacturing Model of Example 5.13. Suppose at time 0 there is one item in bin 1, and bin 2 is empty. Compute the expected amount of time that machine 1 is turned off during an 8-hour shift.

5.18. Consider the Telecommunications Model of Example 5.14. Suppose the buffer is empty at time 0. Compute the expected number of slots that have an empty buffer at the end during the next 50 slots.

5.19. Classify the DTMCs with the transition matrices given in Computational Problem 5.15 as irreducible or reducible.

5.20. Classify the irreducible DTMCs with the transition matrices given below as periodic or aperiodic:

(a)
$$P = \begin{array}{c} \\ 0 \\ 1 \\ 2 \\ 3 \end{array} \begin{array}{cccc} 0 & 1 & 2 & 3 \\ \left[\begin{array}{cccc} .10 & .30 & .20 & .40 \\ .10 & .30 & .40 & .20 \\ .30 & .10 & .10 & .50 \\ .15 & .25 & .35 & .25 \end{array}\right] \end{array}.$$

(b)
$$P = \begin{bmatrix} 0 & 1 & 0 & 0 \\ 0 & 0 & 1 & 0 \\ 0 & 0 & 0 & 1 \\ 1 & 0 & 0 & 0 \end{bmatrix}.$$

(c)
$$P = \begin{array}{c} \\ 0 \\ 1 \\ 2 \\ 3 \end{array} \begin{array}{cccc} 0 & 1 & 2 & 3 \\ \left[\begin{array}{cccc} 0 & .20 & .30 & .50 \\ 1 & 0 & 0 & 0 \\ 1 & 0 & 0 & 0 \\ 1 & 0 & 0 & 0 \end{array}\right] \end{array}.$$

(d)
$$P = \begin{array}{c} \\ 0 \\ 1 \\ 2 \\ 3 \end{array} \begin{array}{cccc} 0 & 1 & 2 & 3 \\ \left[\begin{array}{cccc} 0 & 0 & .40 & .60 \\ 1 & 0 & 0 & 0 \\ 0 & 1 & 0 & 0 \\ 0 & 1 & 0 & 0 \end{array}\right] \end{array}.$$

5.21. Compute a normalized solution to the balance equations for the DTMC in Computational Problem 5.20(a). When possible, compute:

(1) the limiting distribution;

(2) the stationary distribution;

(3) the occupancy distribution.

5.22. Do Computational Problem 5.21 for Computational Problem 5.20(b).

5.23. Do Computational Problem 5.21 for Computational Problem 5.20(c).

5.24. Do Computational Problem 5.21 for Computational Problem 5.20(d).

5.25. Consider the DTMC of Computational Problem 5.5. Compute:

(1) the long-run fraction of the time that the buffer is full;

(2) the expected number of packets in the buffer in the long run.

5.26. Consider Computational Problem 5.7. Compute the expected number of employees in each grade in steady state.

5.27. Consider the weather model of Conceptual Problem 5.12. Compute the long-run fraction of the days that are rainy.

5.28. Consider the weather model of Conceptual Problem 5.13. Compute the long-run fraction of the days that are sunny.

5.29. What fraction of the time does the coffee addict of Computational Problem 5.4 consume brand A coffee?

5.30. Consider the machine as described in Conceptual Problem 5.8. What is the long-run fraction of the time that this machine is up? (Assume $p = .90$.)

5.31. Consider the Manufacturing Model of Example 5.13. Compute the expected number of components in bins A and B in steady state.

5.32. Consider the Stock Market Model of Example 5.9. What fraction of the time does the chief financial officer have to interfere in the stock market to control the price of the stock?

5.33. Consider the single machine production system of Conceptual Problem 5.10. Compute the expected number of items processed by the machine in 10 minutes, assuming that the bin is empty and the machine is idle to begin with. (Assume $p = .95$.)

5.34. Do Computational Problem 5.33 for the production system of Conceptual Problem 5.11. (Assume $p = .95$.)

5.35. Which one of the two production systems described in Conceptual Problems 5.10 and 5.11 has a higher per minute rate of production in steady state?

5.36. Consider the three-machine workshop described in Conceptual Problem 5.3. Suppose each working machine produces a revenue of $500 per day, while repairs cost $300 per day per machine. What is the net rate of revenue per day in steady state? (*Hint*: Can we consider the problem with one machine to obtain the answer for three machines?)

5.37. Consider the inventory system of Example 5.29. Compute the long-run expected cost per day of operating this system.

5.38. Consider the Manufacturing System of Example 5.13. Compute the expected number of assemblies produced per hour in steady state.

5.39. (Computational Problem 5.38 continued.) What will be the increase in the production rate (in number of assemblies per hour) if we provide bins of capacity 3 to the two machines in Example 5.13?

5.40. Compute the expected number of packets transmitted per unit time by the data switch of Example 5.14. How is this connected to the packet loss rate computed in Example 5.31?

5.41. Consider the brand switching model of Computational Problem 5.4. Suppose the per pound cost of coffee is $6, $8, and $15, for brands A, B, and C, respectively. Assuming

Mr Smith consumes one pound of coffee per week, what is his long-run expected coffee expense per week?

5.42. Compute the expected time to go from state 1 to 4 in the DTMCs of Computational Problem 5.20(a) and (c).

5.43. Compute the expected time to go from state 1 to 4 in the DTMCs of Computational Problem 5.20(b) and (d).

5.44. Consider the selection procedure of Conceptual Problem 5.17. Suppose the mean lifetime of Brand A light bulbs is 1, while that of Brand B light bulbs is 1.25. Compute the expected number of experiments done before the selection procedure ends. (*Hint*: Use the Gambler's Ruin model of Example 5.35.)

5.45. Consider the DTMC model of the data switch as described in Example 5.14. Suppose the buffer is full to begin with. Compute the expected amount of time (counted in number of time slots) before the buffer becomes empty.

5.46. Do Computational Problem 5.45 for the data buffer described in Computational Problem 5.5.

5.47. Consider the manufacturing model of Example 5.13. Compute the expected time (in hours) before one of the two machines is shut down, assuming that both bins are empty at time 0.

6

Continuous-Time Markov Models

6.1. Continuous-Time Stochastic Processes

In the previous chapter we studied stochastic models of randomly evolving systems that are observed at discrete-times $n = 0, 1, 2, \ldots$, etc., with X_n being the state of the system at time n. In this chapter we shall consider randomly evolving systems that are observed continuously at all times $t \geq 0$, with $X(t)$ being the state of the system at time t. The set of states that the system can be in at any given time is called the *state space* of the system and is denoted by S. The process $\{X(t), t \geq 0\}$ is called a *continuous-time stochastic process* with state space S. We illustrate with a few examples.

Example 6.1.
(a) Suppose a machine can be in two states: up or down. Let $X(t)$ be the state of the machine at time t. Then $\{X(t), t \geq 0\}$ is a continuous-time stochastic process with state space {up, down}.
(b) A multitasking computer can execute many computer programs simultaneously. Let $X(t)$ be the number of jobs running on such a multitasking computer at time t. Then $\{X(t), t \geq 0\}$ is a continuous-time stochastic process with state space $\{0, 1, 2, \ldots, K\}$, where K is the maximum number of jobs that can be handled by the computer at one time.
(c) Let $X(t)$ be the number of customers that enter a bookstore during time $[0, t]$. Then $\{X(t), t \geq 0\}$ is a continuous-time stochastic process with state space $\{0, 1, 2, \ldots\}$.
(d) Let $X(t)$ be the number of customers that are in the bookstore at time t. Then $\{X(t), t \geq 0\}$ is a continuous-time stochastic process with state space $\{0, 1, 2, \ldots\}$.
(e) Let $X(t)$ be the temperature at a given location at time t. Then $\{X(t), t \geq 0\}$ is a continuous-time stochastic process with state space $(-150, 130)$. Note that this implies that the temperature never goes below $-150°$ or above $130°$ Fahrenheit. □

As in Chapter 5, here we are interested in "studying" the systems modeled by a continuous-time stochastic process $\{X(t), t \geq 0\}$. The following example illustrates the various aspects of such a "study."

Example 6.2.
(a) Consider the two state model of a machine as described in Example 6.1(a). The typical quantities of interest are:

(1) The probability that the machine is up at time t, which can be mathematically expressed as $\mathsf{P}(X(t) = \text{up})$.

(2) Let $W(t)$ be the total amount of time the machine is up during the interval $[0, t]$. We are interested in the expected duration of the time the machine is up, up to time t, namely, $\mathsf{E}(W(t))$.

(3) The long-run fraction of the time the machine is up, is defined as

$$\lim_{t \to \infty} \frac{\mathsf{E}(W(t))}{t}.$$

(4) Suppose it costs $\$c$ per unit time to repair the machine. What is the expected total repair cost up to time T? What is the long-run repair cost per unit time?

(b) Consider the computer system of Example 6.1(b). The typical quantities of interest are:

(1) the probability that the computer system is idle at time t;

(2) the expected number of jobs in the system at time t, viz., $\mathsf{E}(X(t))$;

(3) the expected time it takes to go from idle to full.   □

As in the case of discrete time stochastic processes, the answers to the questions raised in the above example are easy to obtain if the continuous-time stochastic process has some special structure. In the discrete case we discovered that the Markov property provided just such a structure. Hence we search for continuous-time stochastic processes with the Markov property.

6.2. Continuous-Time Markov Chains

We consider a system that is observed continuously, with $X(t)$ being the state at time t, $t \geq 0$. Following the definition of DTMCs we define *Continuous-Time Markov Chains* (CTMCs) below.

Definition 6.1 (Continuous-Time Markov Chain (CTMC)). A stochastic process $\{X(t), t \geq 0\}$ on state space S is called a CTMC if, for all i and j in S,

$$\mathsf{P}(X(s + t) = j | X(s) = i, X(u), 0 \leq u \leq s) = \mathsf{P}(X(s + t) = j | X(s) = i),$$

$$t, s \geq 0. \quad (6.1)$$

The CTMC $\{X(t), t \geq 0\}$ is said to be time homogeneous if, for $t, s \geq 0$,

$$P(X(s + t) = j | X(s) = i) = P(X(t) = j | X(0) = i). \tag{6.2}$$

Equation (6.1) implies that the evolution of a CTMC from a fixed time s onward depends on $X(u)$, $0 \leq u \leq s$ (the history of the CTMC up to time s) only via $X(s)$ (the state of the CTMC at time s). Furthermore, for time-homogeneous CTMCs, it is also independent of s! In this chapter we shall assume, unless otherwise mentioned, that all CTMCs are time homogeneous, and have a finite state space $\{1, 2, \ldots, N\}$. For such CTMCs we define

$$p_{i,j}(t) = P(X(t) = j | X(0) = i), \qquad 1 \leq i, j \leq N. \tag{6.3}$$

The N^2 entities $p_{i,j}(t)$ are arranged in a matrix form as follows:

$$P(t) = \begin{bmatrix} p_{1,1}(t) & p_{1,2}(t) & p_{1,3}(t) & \cdots & p_{1,N}(t) \\ p_{2,1}(t) & p_{2,2}(t) & p_{2,3}(t) & \cdots & p_{2,N}(t) \\ p_{3,1}(t) & p_{3,2}(t) & p_{3,3}(t) & \cdots & p_{3,N}(t) \\ \vdots & \vdots & \vdots & \ddots & \vdots \\ p_{N,1}(t) & p_{N,2}(t) & p_{N,3}(t) & \cdots & p_{N,N}(t) \end{bmatrix}. \tag{6.4}$$

The above matrix is called the *transition probability matrix* of the CTMC $\{X(t), t \geq 0\}$. Note that $P(t)$ is a matrix of functions of t, i.e., it has to be specified for each t. The following theorem gives the important properties of the transition probability matrix $P(t)$:

Theorem 6.1 (Properties of $P(t)$). *A transition probability matrix $P(t) = [p_{i,j}(t)]$ of a time-homogeneous CTMC on state space $S = \{1, 2, \ldots, N\}$ satisfies the following:*

(1) $p_{i,j}(t) \geq 0, \qquad 1 \leq i, j \leq N; \quad t \geq 0,$ \hfill (6.5)

(2) $\displaystyle\sum_{j=1}^{N} p_{i,j}(t) = 1, \qquad 1 \leq i \leq N; \quad t \geq 0,$ \hfill (6.6)

(3) $\displaystyle p_{i,j}(s + t) = \sum_{k=1}^{N} p_{i,k}(s) p_{k,j}(t) = \sum_{k=1}^{N} p_{i,k}(t) p_{k,j}(s), \qquad 1 \leq i \leq N; \quad t, s \geq 0.$ \hfill (6.7)

Proof. (1) The nonnegativity of $p_{i,j}(t)$ follows because it is a (conditional) probability.

(2) To prove the second assertion, we have

$$\sum_{j=1}^{N} p_{i,j}(t) = \sum_{j=1}^{N} P(X(t) = j | X(0) = i)$$
$$= P(X(t) \in S | X(0) = i).$$

Since $X(t)$ must take some value in the state space S, regardless of the value of $X(0)$, it follows that the last quantity is 1.

(3) To prove (6.7), we have

$$p_{i,j}(s+t) = P(X(s+t) = j | X(0) = i)$$

$$= \sum_{k=1}^{N} P(X(s+t) = j | X(s) = k, X(0) = i) P(X(s) = k | X(0) = i)$$

$$= \sum_{k=1}^{N} P(X(s+t) = j | X(s) = k) p_{i,k}(s)$$

(due to the Markov property at s)

$$= \sum_{k=1}^{N} P(X(t) = j | X(0) = k) p_{i,k}(s)$$

(due to time homogeneity)

$$= \sum_{k=1}^{N} p_{i,k}(s) p_{k,j}(t).$$

The other part of (6.7) follows by interchanging s and t. This proves the theorem.   □

Equation (6.7) states the *Chapman–Kolmogorov equations* for CTMCs, and can be written in matrix form as

$$P(t+s) = P(s)P(t) = P(t)P(s). \tag{6.8}$$

Thus the transition probability matrices $P(s)$ and $P(t)$ commute! Since AB is generally not the same as BA for square matrices A and B, the above property implies that the transition probability matrices are very special indeed!

Example 6.3 (Two-State CTMC). Suppose the lifetime of a high altitude satellite is an $\mathrm{Exp}(\mu)$ random variable. Once it fails, it stays failed forever since no repair is possible. Let $X(t)$ be 1 if the satellite is operational at time t, and 0 otherwise.

We first show that $\{X(t), t \geq 0\}$ is a CTMC. The state space is $\{0, 1\}$. We directly check (6.1) and (6.2). Consider the case $i = j = 1$. Let the lifetime of the satellite be T. Then $X(s) = 1$ if and only if $T > s$, and if $X(s) = 1$ then $X(u)$ must be 1 for $0 \leq u \leq s$. Hence

$$P(X(s+t) = 1 | X(s) = 1, X(u) : 0 \leq u \leq s) = P(T > s+t | T > s)$$

$$= \frac{P(T > s+t; T > s)}{P(T > s)}$$

$$= \frac{e^{-\mu(s+t)}}{e^{-\mu s}}$$

$$= e^{-\mu t}$$

$$= P(X(t) = 1 | X(0) = 1).$$

Performing similar calculations for all pairs i, j we establish that $\{X(t), t \geq 0\}$ is a CTMC. Next we compute its transition probability matrix. We have

$$p_{0,0}(t) = \mathsf{P}(\text{satellite is down at } t | \text{satellite is down at } 0) = 1,$$

$$p_{1,1}(t) = \mathsf{P}(\text{satellite is up at } t | \text{satellite is up at } 0) = \mathsf{P}(T > t) = e^{-\mu t}.$$

Hence the transition probability matrix is given by

$$P(t) = \begin{bmatrix} 1 & 0 \\ 1 - e^{-\mu t} & e^{-\mu t} \end{bmatrix}. \tag{6.9}$$

We can check by direct calculations that (6.8) are satisfied. $\qquad\square$

In Chapter 5 we described a DTMC by giving its one-step transition probability matrix. How do we describe a CTMC? Giving $P(t)$ for each t is too complicated for most CTMCs. We need a simpler method of describing a CTMC. We develop such a method below, based on the insight provided by the above example.

The proof that the $\{X(t), t \geq 0\}$ in Example 6.3 is a CTMC is based on the following property of an exponential random variable. Suppose T is an $\text{Exp}(\mu)$ random variable. Then, for all $s, t > 0$,

$$\begin{aligned} \mathsf{P}(T > s + t | T > s) &= \frac{\mathsf{P}(T > s + t; T > s)}{\mathsf{P}(T > s)} \\ &= \frac{e^{-\mu(s+t)}}{e^{-\mu s}} \\ &= e^{-\mu t} \\ &= \mathsf{P}(T > t). \end{aligned}$$

Interpreting T as the lifetime (in years, say) of a machine, we see that the probability that an s-year-old machine will last another t years is the same as the probability that a new machine will last t years. It is as if the machine has no memory of how old it is. This property of an exponential random variable is called the *memoryless property*. We have encountered it earlier in Example 2.22. What is more important is that the exponential is the *only* random variable that has this property for all positive values of s and t. (A geometric random variable has this property for all nonnegative integer values of s and t.) We refer the reader to a more advanced book for a proof of this fact.

The above discussion implies that the stochastic process $\{X(t), t \geq 0\}$ of Example 6.3 can be a CTMC if and only if the lifetime of the satellite is an exponential random variable. This suggests that we build CTMCs using exponential random variables. This is what we do next.

Let $X(t)$ be the state of a system at time t. Suppose the state space of the stochastic process $\{X(t), t \geq 0\}$ is $\{1, 2, \ldots, N\}$. The random evolution of the system occurs as follows: suppose the system starts in state i. It stays there for an $\text{Exp}(r_i)$ amount of time, called the *sojourn time in state i*. At the end of the sojourn time in state i, the system makes a

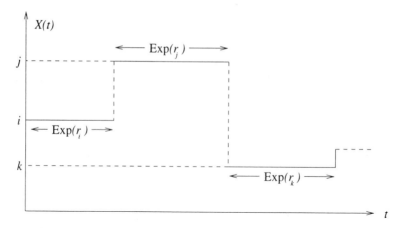

FIGURE 6.1. A typical sample path of a CTMC.

sudden transition to state $j \neq i$ with probability $p_{i,j}$, independent of how long the system has been in state i. Once in state j, it stays there for an $\text{Exp}(r_j)$ amount of time and then moves to a new state $k \neq j$ with probability $p_{j,k}$, independently of the history of the system so far. And it continues this way forever. A typical sample path of the system is shown in Figure 6.1.

Three remarks are in order at this time. First, the jump probabilities $p_{i,j}$ are not to be confused with the transition probabilities $p_{i,j}(t)$ of (6.3). Think of $p_{i,j}$ as the probability that the system moves to state j when it moves out of state i. Second, $p_{i,i} = 0$. This is because, by definition, the sojourn time in state i is the time the system spends in state i until it moves out of it. Hence, a transition from i to i is not allowed. (Contrast this with the DTMC.) Third, in case state i is absorbing (i.e., the system stays in state i forever once it gets there), we set $r_i = 0$, interpreting an $\text{Exp}(0)$ random variable as taking a value equal to ∞ with probability 1. In such a case, $p_{i,j}$'s have no meaning and can be left undefined.

Theorem 6.2. *The stochastic process $\{X(t), t \geq 0\}$ with parameters r_i, $1 \leq i \leq N$, and $p_{i,j}$, $1 \leq i, j \leq N$, as described above is a CTMC.*

Idea of Proof. The rigorous proof of this theorem is tedious, but the idea is simple. Let $s, t \geq 0$ be fixed. Suppose we are given the history of the system up to s, i.e., we are given $X(u), 0 \leq u \leq s$. Also, suppose $X(s) = i$. Then using the memoryless property of the exponential random variable, we see that the remaining sojourn time of the system in state i from time s onward is again an $\text{Exp}(r_i)$ random variable, independent of the past. At the end of the sojourn time the system will move to state j with probability $p_{i,j}$. The subsequent sojourn times and the transitions are independent of the history by construction. Hence the evolution of the system from time s onward depends only on i, the state of the system at time s. Hence $\{X(t), t \geq 0\}$ is a CTMC. $\qquad \square$

What is even more important is that all CTMCs on finite state spaces that have nonzero sojourn times in each state can be described this way. We shall not prove this fact here. In all our applications this condition will be met and hence we can describe a CTMC by giving the parameters $\{r_i, 1 \leq i \leq N\}$ and $\{p_{i,j}, 1 \leq i, j \leq N\}$. This is a much simpler description than giving the transition probability matrix $P(t)$ for all $t \geq 0$.

Analogous to DTMCs, a CTMC can also be represented graphically by means of a directed graph as follows. The directed graph has one node (or vertex) for each state. There is a directed arc from node i to node j if $p_{i,j} > 0$. The quantity $r_{i,j} = r_i p_{i,j}$, called the *transition rate from i to j*, is written next to this arc. Note that there are no self-loops (arcs from i to i) in this representation. This graphical representation is called the *rate diagram* of the CTMC.

Rate diagrams are useful visual representations. We can understand the dynamics of the CTMC by visualizing a particle that moves from node to node in the rate diagram as follows: It stays on node i for an $\text{Exp}(r_i)$ amount of time and then chooses one of the outgoing arcs from node i with probabilities proportional to the rates on the arcs, and moves to the node at the other end of the arc. This motion continues forever. The node occupied by the particle at time t is the state of the CTMC at time t.

Note that we can recover the original parameters $\{r_i, 1 \leq i \leq N\}$ and $\{p_{i,j}, 1 \leq i, j \leq N\}$ from the rates $\{r_{i,j}, 1 \leq i, j \leq N\}$ shown in the rate diagram as follows:

$$r_i = \sum_{j=1}^{N} r_{i,j},$$

$$p_{i,j} = \frac{r_{i,j}}{r_i} \quad \text{if} \quad r_i \neq 0.$$

It is convenient to put all the rates $r_{i,j}$ in a matrix form as follows:

$$R = \begin{bmatrix} r_{1,1} & r_{1,2} & r_{1,3} & \cdots & r_{1,N} \\ r_{2,1} & r_{2,2} & r_{2,3} & \cdots & r_{2,N} \\ r_{3,1} & r_{3,2} & r_{3,3} & \cdots & r_{3,N} \\ \vdots & \vdots & \vdots & \ddots & \vdots \\ r_{N,1} & r_{N,2} & r_{N,3} & \cdots & r_{N,N} \end{bmatrix}. \tag{6.10}$$

Note that $r_{i,i} = 0$ for all $1 \leq i \leq N$, and hence the diagonal entries in R are always zero. R is called the *rate matrix* of the CTMC. It is closely related to $Q = [q_{i,j}]$, called the *generator matrix* of the CTMC, which is defined as follows:

$$q_{i,j} = \begin{cases} -r_i & \text{if } i = j, \\ r_{i,j} & \text{if } i \neq j. \end{cases} \tag{6.11}$$

Thus generator matrix Q is the same as the rate matrix R with the diagonal elements replaced by $-r_i$'s. It is common in the literature to describe a CTMC by the Q matrix. However we shall describe it by the R matrix. Clearly, one can be obtained from the other.

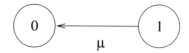

FIGURE 6.2. Rate diagram of the satellite system.

Example 6.4 (Two-State CTMC). Consider the CTMC of Example 6.3. The system stays in state 1 (satellite up) for an $\text{Exp}(\mu)$ amount of time and then moves to state 0. Hence, $r_1 = \mu$ and $p_{1,0} = 1$. Once the satellite fails (state 0) it stays failed forever. Thus state 0 is absorbing. Hence $r_0 = 0$, and $p_{0,1}$ is left undefined. The rate matrix is as follows:

$$R = \begin{bmatrix} 0 & 0 \\ \mu & 0 \end{bmatrix}. \tag{6.12}$$

The generator matrix is given by

$$Q = \begin{bmatrix} 0 & 0 \\ \mu & -\mu \end{bmatrix}. \tag{6.13}$$

The rate diagram is shown in Figure 6.2.   □

Example 6.5 (Two-State Machine). Consider a machine that operates for an $\text{Exp}(\mu)$ amount of time and then fails. Once it fails, it gets repaired. The repair time is an $\text{Exp}(\lambda)$ random variable, and is independent of the past. The machine is as good as new after the repair is complete. Let $X(t)$ be the state of the machine at time t, 1 if it is up and 0 if it is down. Model this as a CTMC.

The state space of $\{X(t), t \geq 0\}$ is $\{0, 1\}$. The sojourn time in state 0 is the repair time, which is given to be an $\text{Exp}(\lambda)$ random variable. Hence $r_0 = \lambda$. After the repair is complete, the machine is up with probability 1, hence $p_{0,1} = 1$. Similarly, $r_1 = \mu$ and $p_{1,0} = 1$. The independence assumptions are satisfied, thus making $\{X(t), t \geq 0\}$ a CTMC. The rate matrix is as follows:

$$R = \begin{bmatrix} 0 & \lambda \\ \mu & 0 \end{bmatrix}. \tag{6.14}$$

The generator matrix is given by

$$Q = \begin{bmatrix} -\lambda & \lambda \\ \mu & -\mu \end{bmatrix}. \tag{6.15}$$

The rate diagram is shown in Figure 6.3. Note that describing this CTMC by giving its $P(t)$

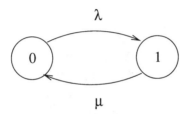

FIGURE 6.3. Rate diagram of the two-state machine.

matrix would be very difficult, since the machine can undergo many failures and repairs up to time t.   □

6.3. Exponential Random Variables

It is clear from the discussion in the previous section that the exponential random variables play an important role in CTMCs. We have already seen the most important property of exponential random variables earlier: the memoryless property. Here we study a few other properties of these random variables, that will be useful in building CTMC models.

Let T_i, $1 \le i \le k$, be an $\text{Exp}(\lambda_i)$ random variable, and suppose $T_1, T_2, \ldots, T_k$ are independent. Let

$$T = \min\{T_1, T_2, \ldots, T_k\}.$$

Think of T_i as the time when the ith event occurs. Then T is the time when the first of these k events occurs. Here we compute the complementary cdf of T. Using the argument in Example 3.10, and the independence of T_i, $1 \le i \le k$, we get

$$
\begin{aligned}
\mathsf{P}(T > t) &= \mathsf{P}(\min\{T_1, T_2, \ldots, T_k\} > t) \\
&= \mathsf{P}(T_1 > t, T_2 > t, \ldots, T_k > t) \\
&= \mathsf{P}(T_1 > t)\mathsf{P}(T_2 > t) \cdots \mathsf{P}(T_k > t) \\
&= e^{-\lambda_1 t} e^{-\lambda_2 t} \cdots e^{-\lambda_k t} \\
&= e^{-\lambda t},
\end{aligned}
\tag{6.16}
$$

where

$$\lambda = \sum_{i=1}^{k} \lambda_i. \tag{6.17}$$

This implies that T is an $\text{Exp}(\lambda)$ random variable.

Next define $Z = i$ if $T_i = T$, i.e., if event i is the first of the k events to occur. Note that T_i's are continuous random variables, and hence the probability that two or more of them will be equal is zero. Thus there is no ambiguity in defining Z. Following the steps as in Example 4.6 we get, for $i = 1$,

$$
\begin{aligned}
P(Z = 1) &= P(T_1 = T) \\
&= P(T_1 = \min\{T_1, T_2, \ldots, T_k\}) \\
&= \int_0^\infty P(T_1 = \min\{T_1, T_2, \ldots, T_k\} | T_1 = x)\lambda_1 e^{-\lambda_1 x}\, dx \\
&= \int_0^\infty P(T_2 > x, T_3 > x, \ldots, T_k > x | T_1 = x)\lambda_1 e^{-\lambda_1 x}\, dx \\
&= \int_0^\infty e^{-\lambda_2 x} e^{-\lambda_3 x} \cdots e^{-\lambda_k x} \lambda_1 e^{-\lambda_1 x}\, dx \\
&= \lambda_1 \int_0^\infty e^{-\lambda x}\, dx \\
&= \frac{\lambda_1}{\lambda}.
\end{aligned}
$$

Thus in general we get

$$
P(Z = i) = \frac{\lambda_i}{\lambda}, \qquad 1 \le i \le N. \tag{6.18}
$$

Thus the probability that event i is the first to occur among the k events is proportional to λ_i.

Next we study the joint distribution of (Z, T). We have

$$
\begin{aligned}
P(Z = 1; T > t) &= \int_t^\infty P(Z = 1; T > t | T_1 = x)\lambda_1 e^{-\lambda_1 x}\, dx \\
&= \int_t^\infty P(T_2 > x, T_3 > x, \ldots, T_k > x | T_1 = x)\lambda_1 e^{-\lambda_1 x}\, dx \\
&= \lambda_1 \int_t^\infty e^{-\lambda x}\, dx \\
&= \frac{\lambda_1}{\lambda} \lambda e^{-\lambda t} \\
&= P(Z = 1)P(T > t).
\end{aligned}
$$

In general we get

$$
P(Z = i; T > t) = P(Z = i)P(T > t), \qquad 1 \le i \le k, \tag{6.19}
$$

implying that Z and T are independent random variables! Thus the time until the occurrence of the first of the k events is independent of which of the k events occurs first! In yet other terms, the conditional distribution of T, given that $Z = i$, is $\text{Exp}(\lambda)$ and not $\text{Exp}(\lambda_i)$ as we might have guessed.

In the next section we use the above properties of exponential random variables in developing CTMC models of real-life systems. We give the rate matrix R for the CTMCs and the rate diagrams, but omit the generator matrix Q.

6.4. Examples of CTMCs: I

Example 6.6 (Two Machines Workshop). Consider a workshop with two independent machines, each with its own repair person, and each machine behaving as described in Example 6.5. Let $X(t)$ be the number of machines operating at time t. Model $\{X(t), t \geq 0\}$ as a CTMC.

The state space of $\{X(t), t \geq 0\}$ is $\{0, 1, 2\}$. Consider state 0. In this state both machines are down, and hence under repair. Due to the memoryless property, the remaining repair times of the two machines are iid $\text{Exp}(\lambda)$ random variables. The system moves from state 0 to state 1 as soon as one of the two machines finishes service. Hence the sojourn time in state 0 is a minimum of two iid $\text{Exp}(\lambda)$ random variables. Hence, by (6.16) the sojourn time is an $\text{Exp}(2\lambda)$ random variable. Hence $r_0 = 2\lambda$ and $p_{0.1} = 1$. The transition rate from state 0 to state 1 is thus $r_{0.1} = 2\lambda$.

Now consider state 1, when one machine is up and one is down. Let T_1 be the remaining repair time of the down machine, and let T_2 be the remaining repair time of the up machine. Again from the memoryless property of the exponentials, T_1 is $\text{Exp}(\lambda)$ and T_2 is $\text{Exp}(\mu)$, and they are independent. The sojourn time in state 1 is $\min(T_1, T_2)$, which is an $\text{Exp}(\lambda + \mu)$ random variable. Hence $r_1 = \lambda + \mu$. The system moves to state 2 if $T_1 < T_2$, which has probability $\lambda/(\lambda + \mu)$, and it moves to state 0 if $T_2 < T_1$, which has probability $\mu/(\lambda + \mu)$. Note that the next state is independent of the sojourn time from (6.19). Hence $p_{1.2} = \lambda/(\lambda + \mu)$, and $p_{1.0} = \mu/(\lambda + \mu)$. The transition rates are $r_{1.2} = \lambda$ and $r_{1.0} = \mu$.

A similar analysis of state 2 yields: $r_2 = 2\mu$, $p_{2.1} = 1$, and $r_{2.1} = 2\mu$. Thus $\{X(t), t \geq 0\}$ is a CTMC with rate matrix as follows:

$$R = \begin{bmatrix} 0 & 2\lambda & 0 \\ \mu & 0 & \lambda \\ 0 & 2\mu & 0 \end{bmatrix}. \tag{6.20}$$

The rate diagram is as shown in Figure 6.4.   □

The above example produces the following extremely useful insight: suppose the state of the system at time t is $X(t) = i$. Suppose there are r different events that can trigger a transition in the system state from state i to state j. Suppose the occurrence time of the kth event ($1 \leq k \leq r$) is an $\text{Exp}(\lambda_k)$ random variable. Then assuming the r events are independent of each other and of the history of the system up to time t, the transition rate from state i to state j is given by

$$r_{i.j} = \lambda_1 + \lambda_2 + \cdots + \lambda_k.$$

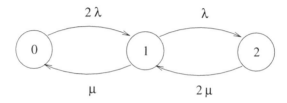

FIGURE 6.4. Rate diagram of the two-machine workshop.

If the transition rates can be computed this way for all i and j in the state space, then $\{X(t), t \geq 0\}$ is a CTMC with transition rates $r_{i,j}$. We use this observation in the remaining examples.

Example 6.7 (General Machine Shop). A machine shop consists of N machines and M repair persons. ($M \leq N$.) The machines are identical and the lifetimes of the machines are independent $\mathrm{Exp}(\mu)$ random variables. When the machines fail, they are served in the order of failure by the M repair persons. Each failed machine needs one and only one repair person, and the repair times are independent $\mathrm{Exp}(\lambda)$ random variables. A repaired machine behaves like a new machine. Let $X(t)$ be the number of machines that are functioning at time t. Model $\{X(t), t \geq 0\}$ as a CTMC.

We shall explain with a special case: four machines and two repair persons. The general case is similar. For the special case, the state space of $\{X(t), t \geq 0\}$ is $S = \{0, 1, 2, 3, 4\}$. We obtain the transition rates below: In state 0, all machines are down, two are under repair. (The other two are waiting for a repair person.) The repair times are iid $\mathrm{Exp}(\lambda)$, and either of the two repair completions will move the system to state 1. Hence $r_{0,1} = \lambda + \lambda = 2\lambda$. In state 1, one machine is up, three are down, two of which are under repair. Either repair completion will send the system to state 2. Hence $r_{1,2} = 2\lambda$. The failure of the functioning machine, which will happen after an $\mathrm{Exp}(\mu)$ amount of time, will trigger a transition to state 0. Hence $r_{1,0} = \mu$. Similarly, for state 2 we get $r_{2,3} = 2\lambda$ and $r_{2,1} = 2\mu$. In state 3, three machines are functioning, one is down and under repair. (One repair person is idle.) Thus there are three independent failure events that will trigger a transition to state 2, and a repair completion will trigger a transition to state 4. Hence $r_{3,2} = 3\mu$ and $r_{3,4} = \lambda$. Similarly, for state 4 we get $r_{4,3} = 4\mu$.

Thus $\{X(t), t \geq 0\}$ is a CTMC with a rate matrix as follows:

$$R = \begin{bmatrix} 0 & 2\lambda & 0 & 0 & 0 \\ \mu & 0 & 2\lambda & 0 & 0 \\ 0 & 2\mu & 0 & 2\lambda & 0 \\ 0 & 0 & 3\mu & 0 & \lambda \\ 0 & 0 & 0 & 4\mu & 0 \end{bmatrix}. \tag{6.21}$$

The rate diagram is as shown in Figure 6.5.   □

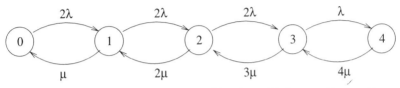

FIGURE 6.5. Rate diagram of the general machine shop with four machines and two repair persons.

Example 6.8 (Airplane Reliability). A commercial jet airplane has four engines, two on each wing. Each engine lasts for a random amount of time that is exponentially distributed with parameter λ, and then fails. If the failure takes place in flight, there can be no repair. The airplane needs at least one engine on each wing to function properly in order to fly safely. Model this system so that we can predict the probability of a trouble-free flight.

Let $X_L(t)$ be the number of functioning engines on the left wing at time t, and let $X_R(t)$ be the number of functioning engines on the right wing at time t. The state of the system at time t is given by $X(t) = (X_L(t), X_R(t))$. Assuming the engine failures to be independent of each other, we see that $\{X(t), t \geq 0\}$ is a CTMC on state space

$$S = \{(0, 0), (0, 1), (0, 2), (1, 0), (1, 1), (1, 2), (2, 0), (2, 1), (2, 2)\}.$$

Note that the flight continues in a trouble-free fashion as long as $X_L(t) \geq 1$ and $X_R(t) \geq 1$, i.e., as long as the CTMC is in states $(1, 1), (1, 2), (2, 1), (2, 2)$. The flight crashes as soon as the CTMC visits any state in $\{(0, 0), (0, 1), (0, 2), (1, 0), (2, 0)\}$. We assume that the CTMC continues to evolve even after the airplane crashes!

Label the states in S 1 through 9 in the order in which they are listed above. Performing the standard triggering event analysis (assuming that no repairs are possible), we get the following rate matrix:

$$R = \begin{bmatrix} 0 & 0 & 0 & 0 & 0 & 0 & 0 & 0 & 0 \\ \lambda & 0 & 0 & 0 & 0 & 0 & 0 & 0 & 0 \\ 0 & 2\lambda & 0 & 0 & 0 & 0 & 0 & 0 & 0 \\ \lambda & 0 & 0 & 0 & 0 & 0 & 0 & 0 & 0 \\ 0 & \lambda & 0 & \lambda & 0 & 0 & 0 & 0 & 0 \\ 0 & 0 & \lambda & 0 & 2\lambda & 0 & 0 & 0 & 0 \\ 0 & 0 & 0 & 2\lambda & 0 & 0 & 0 & 0 & 0 \\ 0 & 0 & 0 & 0 & 2\lambda & 0 & \lambda & 0 & 0 \\ 0 & 0 & 0 & 0 & 0 & 2\lambda & 0 & 2\lambda & 0 \end{bmatrix}. \tag{6.22}$$

The rate diagram is as shown in Figure 6.6.   □

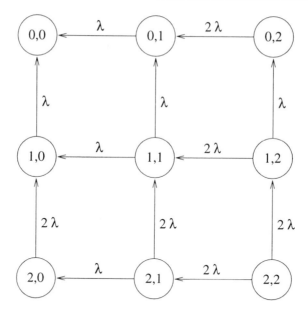

FIGURE 6.6. Rate diagram of the airplane reliability system.

6.5. Poisson Processes

Frequently we encounter systems whose state transitions are triggered by streams of events that occur one at a time, for example, arrivals to a queueing system, shocks to an engineering system, earthquakes in a geological system, biological stimuli in a neural system, accidents in a given city, claims on an insurance company, demands on an inventory system, failures in a manufacturing system, etc. The modeling of these systems becomes more tractable if we assume that the successive inter-event times are iid exponential random variables. Hence we give such a stream of events a special name: *Poisson Process*. We define it formally below.

Let S_n be the occurrence time of the nth event. Assume $S_0 = 0$ and define

$$T_n = S_n - S_{n-1}, \qquad n \geq 1.$$

Thus T_n is the time between the occurrence of the nth and the $(n-1)$st event. Let $N(t)$ be the total number of events that occur during the interval $(0, t]$. Thus an event at 0 is not counted, but an event at t, if any, is counted in $N(t)$.

Definition 6.2 (Poisson Process). The stochastic process $\{N(t), t \geq 0\}$, where $N(t)$ is as defined above, is called a Poisson Process with rate λ (denoted by PP(λ)) if $\{T_n, n \geq 1\}$ is a sequence of iid Exp(λ) random variables.

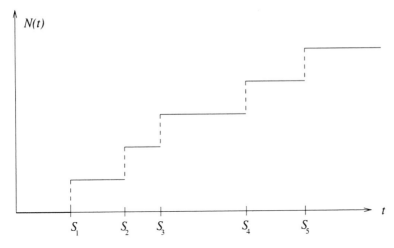

FIGURE 6.7. A typical sample path of a Poisson Process.

A typical sample path of a Poisson process is shown in Figure 6.7. The next theorem gives the pmf of $N(t)$ for a given t. It also justifies the name Poisson process for this stochastic process.

Theorem 6.3 (Poisson Process). *Let $\{N(t), t \geq 0\}$ be a PP(λ). For a given t, $N(t)$ is a Poisson random variable with parameter λt, i.e.,*

$$P(N(t) = k) = e^{-\lambda t} \frac{(\lambda t)^k}{k!}, \qquad k \geq 0. \tag{6.23}$$

Proof. We shall first compute $P(N(t) \geq k)$ and then obtain $P(N(t) = k)$ by using

$$P(N(t) = k) = P(N(t) \geq k) - P(N(t) \geq k + 1). \tag{6.24}$$

Suppose $N(t) \geq k$. Then the kth event in the Poisson process must have occurred at or before t, i.e., $S_k \leq t$. On the other hand, suppose $S_k \leq t$. Then the kth event has occurred at or before t. Hence there must be at least k events up to t, i.e., $N(t) \geq k$. Thus the events $\{N(t) \geq k\}$ and $\{S_k \leq t\}$ are identical, and must have the same probability. However, S_k is a sum of k iid Exp(λ) random variables. Hence, from Conceptual Problem 3.6, it is an Erl(k, λ) random variable. Using (2.20), we get

$$P(S_k \leq t) = 1 - \sum_{r=0}^{k-1} e^{-\lambda t} \frac{(\lambda t)^r}{r!}.$$

Hence

$$P(N(t) \geq k) = 1 - \sum_{r=0}^{k-1} e^{-\lambda t} \frac{(\lambda t)^r}{r!}.$$

Substituting in (6.24) we get

$$P(N(t) = k) = P(N(t) \geq k) - P(N(t) \geq k + 1)$$

$$= 1 - \sum_{r=0}^{k-1} e^{-\lambda t} \frac{(\lambda t)^r}{r!} - \left(1 - \sum_{r=0}^{k} e^{-\lambda t} \frac{(\lambda t)^r}{r!} \right)$$

$$= e^{-\lambda t} \frac{(\lambda t)^k}{k!}.$$

This completes the proof.   □

The usefulness of Poisson processes in building CTMC models derives from the following important property: Suppose $\{N(t), t \geq 0\}$ is PP(λ), and $N(t) = k$ is given. Then S_{k+1} is the time of occurrence of the first event after t. Using the memoryless property of the exponential random variables we can show that the time until occurrence of this next event, namely $S_{k+1} - t$, is exponentially distributed with parameter λ, and is independent of the history of the Poisson process up to t. Thus the time until the occurrence of the next event in a Poisson process is always Exp(λ), regardless of the occurrence of events so far.

Poisson processes and their variations and generalizations have been extensively studied, and we refer the reader to a higher level text for more material on them. The material presented here is sufficient for the purposes of modeling encountered in this book. We shall study one generalization (called the Compound Poisson Process) as an example of a cumulative process in Section 7.3.

Using the properties of the exponential distributions and the Poisson processes, we develop several examples of CTMCs in the next section. As before, we give the rate matrix and the rate diagram for each model, but omit the generator matrix.

6.6. Examples of CTMCs: II

Example 6.9 (Finite Capacity Single Server Queue). Customers arrive at an automatic teller machine (ATM) according to a PP(λ). The space in front of the ATM can accommodate at most K customers. Thus if there are K customers waiting at the ATM, and a new customer arrives, he or she simply walks away and is lost forever. The customers form a single line and use the ATM in a first-come first-served fashion. The processing times at the ATM for the customers are iid Exp(μ) random variables. Let $X(t)$ be the number of customers at the ATM at time t. Model $\{X(t), t \geq 0\}$ as a CTMC.

The state space of the system is $S = \{0, 1, 2, \ldots, K\}$. We shall obtain the transition rates below. In state 0, the system is empty, and the only event that can occur is an arrival, which occurs after an Exp(λ) amount of time (due to the PP(λ) arrival process) and triggers a transition to state 1. Hence $r_{0,1} = \lambda$. In state i, $1 \leq i \leq K - 1$, there are two possible events: an arrival and a departure. An arrival occurs after an Exp(λ) amount of time and takes the system to state $i + 1$; while the departure occurs at the next service completion time

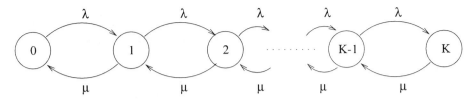

FIGURE 6.8. Rate diagram of a single server queue.

(which occurs after an $\text{Exp}(\mu)$ amount of time due to the memoryless property of the service times) and takes the system to state $i - 1$. Hence $r_{i,i+1} = \lambda$ and $r_{i,i-1} = \mu$, $1 \leq i \leq K - 1$. Finally, in state K, there is no arrival due to capacity limitation. Thus the only event that can occur is a service completion, taking the system to state $K - 1$. Hence $r_{K,K-1} = \mu$. Thus $\{X(t), t \geq 0\}$ is a CTMC with rate matrix as follows:

$$R = \begin{bmatrix} 0 & \lambda & 0 & \cdots & 0 & 0 \\ \mu & 0 & \lambda & \cdots & 0 & 0 \\ 0 & \mu & 0 & \cdots & 0 & 0 \\ \vdots & \vdots & \vdots & \ddots & \vdots & \vdots \\ 0 & 0 & 0 & \cdots & 0 & \lambda \\ 0 & 0 & 0 & \cdots & \mu & 0 \end{bmatrix}. \tag{6.25}$$

The rate diagram is as shown in Figure 6.8. We shall see in Chapter 8 that $\{X(t), t \geq 0\}$ is called an $M/M/1/K$ queue in queueing terminology.   □

The CTMCs in Examples 6.7 and 6.9 have a similar structure: both CTMCs move from state i to state $i - 1$ or to state $i + 1$. There are no other transitions. We describe a general class of such DTMCs below.

Example 6.10 (Finite Birth and Death Process). A CTMC on state space $\{0, 1, 2, \ldots, K\}$ with the following rate matrix

$$R = \begin{bmatrix} 0 & \lambda_0 & 0 & \cdots & 0 & 0 \\ \mu_1 & 0 & \lambda_1 & \cdots & 0 & 0 \\ 0 & \mu_2 & 0 & \cdots & 0 & 0 \\ \vdots & \vdots & \vdots & \ddots & \vdots & \vdots \\ 0 & 0 & 0 & \cdots & 0 & \lambda_{K-1} \\ 0 & 0 & 0 & \cdots & \mu_K & 0 \end{bmatrix}, \tag{6.26}$$

is called a *finite birth and death process*. The transitions from i to $i + 1$ are called the births, and the λ_i's are called the birth parameters. The transitions from i to $i - 1$ are called the deaths, and the μ_i's are called the death parameters. It is convenient to define $\lambda_K = 0$ and

$\mu_0 = 0$ signifying that there are no births in state K and no deaths in state 0. A birth and death process spends an $\text{Exp}(\lambda_i + \mu_i)$ amount of time in state i and then jumps to state $i + 1$ with probability $\lambda_i/(\lambda_i + \mu_i)$, or to state $i - 1$ with probability $\mu_i/(\lambda_i + \mu_i)$.   □

Birth and death processes occur quite often in applications and hence form an important class of CTMCs. For example, the CTMC model of the general machine shop in Example 6.7 is a birth and death process on $\{0, 1, \ldots, N\}$ with birth parameters

$$\lambda_i = \min(N - i, K)\lambda, \qquad 0 \le i \le N,$$

and death parameters

$$\mu_i = \min(i, M)\mu, \qquad 0 \le i \le N.$$

The CTMC model of the single server queue as described in Example 6.9 is a birth and death process on $\{0, 1, \ldots, K\}$ with birth parameters

$$\lambda_i = \lambda, \qquad 0 \le i \le K - 1,$$

and death parameters

$$\mu_i = \mu, \qquad 1 \le i \le K.$$

We illustrate with more examples.

Example 6.11 (Telephone Switch). A telephone switch can handle K calls at any one time. Calls arrive according to a Poisson process with rate λ. If the switch is already serving K calls when a new call arrives, then the new call is lost. If a call is accepted, it lasts for an $\text{Exp}(\mu)$ amount of time and then terminates. All call durations are independent of each other. Let $X(t)$ be the number of calls that are being handled by the switch at time t. Model $\{X(t), t \ge 0\}$ as a CTMC.

The state space of $\{X(t), t \ge 0\}$ is $\{0, 1, 2, \ldots, K\}$. In state i, $0 \le i \le K - 1$, an arrival triggers a transition to state $i + 1$ with rate λ. Hence $r_{i,i+1} = \lambda$. In state K there are no arrivals. In state i, $1 \le i \le K$, any of the i calls may complete, and trigger a transition to state $i - 1$. The transition rate is $r_{i,i-1} = i\mu$. In state 0 there are no departures. Thus $\{X(t), t \ge 0\}$ is a finite birth and death process with birth parameters

$$\lambda_i = \lambda, \qquad 0 \le i \le K - 1,$$

and death parameters

$$\mu_i = i\mu, \qquad 0 \le i \le K.$$

We shall see in Chapter 8 that $\{X(t), t \ge 0\}$ is called an $M/M/K/K$ queue in queueing terminology.   □

Example 6.12 (Call Center). An airline phone-reservation system is called a call center, and is staffed by M reservation clerks called agents. An incoming call for reservations is

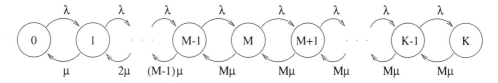

FIGURE 6.9. Rate diagram of the call center.

handled by an agent if one is available, otherwise the caller is put on hold. The system can put a maximum of H callers on hold. When an agent becomes available, the callers on hold are served in order of arrival. When all the agents are busy and there are H calls on hold, any additional callers get a busy signal and are permanently lost. Let $X(t)$ be the number of calls in the system, those handled by the agents plus any on hold, at time t. Assume the calls arrive according to a PP(λ) and the processing time of the calls are iid Exp(μ) random variables. Model $\{X(t), t \geq 0\}$ as a CTMC.

The state space is $\{0, 1, 2, \ldots, K\}$, where $K = M + H$. Suppose $X(t) = i$. For $0 \leq i \leq K - 1$ the transition to state $i + 1$ is caused by an arrival, which occurs at rate λ. In state K there are no arrivals. For $1 \leq i \leq K$, the transition to state $i - 1$ is caused by the completion of the processing of any of the $\min(M, i)$ calls that are under service. The completion rate of individual call is μ, hence the transition rate from state i to $i - 1$ is $\min(i, M)\mu$. Thus $\{X(t), t \geq 0\}$ is a finite birth and death process with birth parameters

$$\lambda_i = \lambda, \qquad 0 \leq i \leq K - 1,$$

and death parameters

$$\mu_i = \min(i, M)\mu, \qquad 0 \leq i \leq K.$$

The rate diagram is as shown in Figure 6.9. We shall see in Chapter 8 that $\{X(t), t \geq 0\}$ is called an $M/M/s/K$ queue in queueing terminology. □

Example 6.13 (Leaky Bucket). Leaky bucket is a traffic control mechanism proposed for high-speed digital telecommunication networks. It consists of two buffers: a token buffer of size M and a data buffer of size D. Tokens are generated according to a PP(μ) and are stored in the token buffer. Tokens generated while the token buffer is full are lost. Data packets arrive according to a PP(λ). If there is a token in the token buffer when a packet arrives, it immediately removes one token from the token buffer and enters the network. If there are no tokens in the token buffer, it waits in the data buffer, if the data buffer is not full. If the data buffer is full, the packet is lost. Let $Y(t)$ be the number of tokens in the token buffer at time t, and let $Z(t)$ be the number of data packets in the data buffer at time t. Model this system as a CTMC.

First note that if there are packets in the data buffer, the token buffer must be empty (otherwise the packets would have already entered the network), and vice versa. Consider a new process $\{X(t), t \geq 0\}$ defined by

$$X(t) = M - Y(t) + Z(t).$$

The state space of $\{X(t), t \geq 0\}$ is $\{0, 1, \ldots, K\}$ where $K = M + D$. Also, if $0 \leq X(t) \leq M$, we must have $Z(t) = 0$, and $Y(t) = M - X(t)$; and if $M \leq X(t) \leq K$, we must have $Y(t) = 0$, and $Z(t) = X(t) - M$. Thus we can recover $Y(t)$ and $Z(t)$ from the knowledge of $X(t)$.

Now, suppose $1 \leq X(t) \leq K - 1$. If a token arrives (this happens at rate μ) $X(t)$ decreases by 1, and if a data packet arrives (this happens at rate λ) it increases by 1. If $X(t) = 0$, then $Y(t) = M$, and tokens cannot enter the token buffer, while data packets continue to arrive. If $X(t) = K$, then $Z(t) = D$ and packets cannot enter the data buffer, however tokens continue to enter. Thus $\{X(t), t \geq 0\}$ is a finite birth and death process with birth parameters

$$\lambda_i = \lambda, \qquad 0 \leq i \leq K - 1,$$

and death parameters

$$\mu_i = \mu, \qquad 1 \leq i \leq K.$$

Thus $\{X(t), t \geq 0\}$ is the same process as in the single server queue of Example 6.9!   □

Example 6.14 (Inventory Management). A retail store manages the inventory of washing machines as follows. When the number of washing machines in stock decreases to a fixed number k, an order is placed with the manufacturer for r new washing machines. It takes a random amount of time for the order to be delivered to the retailer. If the inventory is at most k when an order is delivered (including the newly delivered order), another order for r items is placed immediately. Suppose the delivery times are iid Exp(λ), and that the demands for the washing machines occur according to a PP(μ). Demands that cannot be immediately satisfied are lost. Model this system as a CTMC.

Let $X(t)$ be the number of machines in stock at time t. Note that the maximum number of washing machines in stock is $K = k + r$, which happens if the order is delivered before the next demand occurs. The state space is thus $\{0, 1, \ldots, K\}$. In state 0, the demands are lost, and the stock jumps to r when the current outstanding order is delivered (this happens at rate λ). Hence we have $r_{0,r} = \lambda$. In state i, $1 \leq i \leq k$, one order is outstanding. The state changes to $i - 1$ if a demand occurs (this happens at rate μ), and to $i + r$ if the order is delivered. Thus we have $r_{i,i+r} = \lambda$ and $r_{i,i-1} = \mu$. Finally, if $X(t) = i, k+1 \leq i \leq K$, there are no outstanding orders. Hence, the only transition is from i to $i - 1$, and that happens when a demand occurs. Thus, $r_{i,i-1} = \mu$. Thus $\{X(t), t \geq 0\}$ is a CTMC. The rate matrix

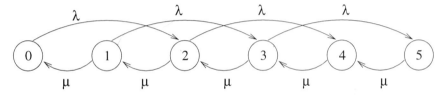

FIGURE 6.10. Rate diagram of the inventory management example with $k = 3, r = 2$.

is shown below for the case $k = 3, r = 2$,

$$R = \begin{bmatrix} 0 & 0 & \lambda & 0 & 0 & 0 \\ \mu & 0 & 0 & \lambda & 0 & 0 \\ 0 & \mu & 0 & 0 & \lambda & 0 \\ 0 & 0 & \mu & 0 & 0 & \lambda \\ 0 & 0 & 0 & \mu & 0 & 0 \\ 0 & 0 & 0 & 0 & \mu & 0 \end{bmatrix}. \tag{6.27}$$

The rate diagram is as shown in Figure 6.10. □

Example 6.15 (Manufacturing). A simple manufacturing facility consists of a single machine that can be turned on or off. If the machine is on it produces items according to a Poisson process with rate λ. Demands for the items arrive according to a PP(μ). The machine is controlled as follows: if the number of items in stock reaches a maximum number K (the storage capacity), the machine is turned off. It is turned on again when the number of items in stock decreases to k. Model this system as a CTMC.

Let $X(t)$ be the number of items in stock at time t. $\{X(t), t \geq 0\}$ is not a CTMC, since we don't know whether the machine is on or off if $k < X(t) < K$. Let $Y(t)$ be the state of the machine at time t, 1 if it is on, and 0 if it is off. Then $\{(X(t), Y(t)), t \geq 0\}$ is a CTMC with state space

$$S = \{(i, 1), 0 \leq i < K\} \cup \{(i, 0), k < i \leq K\}.$$

Note that the machine is always on if the number of items is k or less. Hence we do not need the states $(i, 0), 0 \leq i \leq k$. The usual triggering event analysis yields the following transition rates:

$$r_{(i,1),(i+1,1)} = \lambda, \quad 0 \leq i < K - 1,$$
$$r_{(K-1,1),(K,0)} = \lambda,$$
$$r_{(i,1),(i-1,1)} = \mu, \quad 1 \leq i \leq K - 1,$$
$$r_{(i,0),(i-1,0)} = \mu, \quad k + 1 < i \leq K,$$
$$r_{(k+1,0),(k,1)} = \mu.$$

The rate diagram is shown in Figure 6.11. □

Armed with these examples, we proceed with the analysis of CTMCs in the next section.

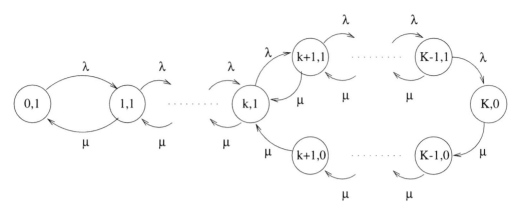

FIGURE 6.11. Rate diagram of the manufacturing system.

6.7.   Transient Analysis: Uniformization

In this section we develop methods of computing the transient distribution of $\{X(t), t \geq 0\}$, that is, we compute the pmf of $X(t)$ for a given t. We assume that the probability distribution of the initial state, $X(0)$, is known. Then we have

$$P(X(t) = j) = \sum_{i=1}^{N} P(X(t) = j | X(0) = i) P(X(0) = i), \qquad 1 \leq j \leq N. \qquad (6.28)$$

Thus we need to compute $P(X(t) = j | X(0) = i) = p_{i.j}(t)$ in order to obtain the pmf of $X(t)$. Hence we concentrate on computing the transition matrix $P(t) = [p_{i.j}(t)]$ in this section. We first introduce the notation needed before stating the main result in Theorem 6.4.

Let $\{X(t), t \geq 0\}$ be a CTMC on state space $S = \{1, 2, \ldots, N\}$ and let $R = [r_{i.j}]$ be its rate matrix. We have seen before that such a CTMC spends an $\mathrm{Exp}(r_i)$ amount of time in state i $(r_i = \sum_{j=1}^{N} r_{i.j})$ and, if $r_i > 0$, jumps to state j with probability $p_{i.j} = r_{i.j}/r_i$. Now let r be any finite number satisfying

$$r \geq \max_{1 \leq i \leq N} \{r_i\}. \qquad (6.29)$$

Define a matrix $\hat{P} = [\hat{p}_{i.j}]$ as follows:

$$\hat{p}_{i.j} = \begin{cases} 1 - \dfrac{r_i}{r} & \text{if } i = j, \\[2mm] \dfrac{r_{i.j}}{r} & \text{if } i \neq j. \end{cases} \qquad (6.30)$$

Example 6.16. Suppose R is given to be

$$R = \begin{bmatrix} 0 & 2 & 3 & 0 \\ 4 & 0 & 2 & 0 \\ 0 & 2 & 0 & 2 \\ 1 & 0 & 3 & 0 \end{bmatrix}. \tag{6.31}$$

Then we have $r_1 = 5, r_2 = 6, r_3 = 4$, and $r_4 = 4$. Hence, from (6.29), we can choose $r = \max\{5, 6, 4, 4\} = 6$. Then, from (6.30) we get

$$\hat{P} = \begin{bmatrix} \frac{1}{6} & \frac{2}{6} & \frac{3}{6} & 0 \\ \frac{4}{6} & 0 & \frac{2}{6} & 0 \\ 0 & \frac{2}{6} & \frac{2}{6} & \frac{2}{6} \\ \frac{1}{6} & 0 & \frac{3}{6} & \frac{2}{6} \end{bmatrix}. \tag{6.32}$$

Note that $\hat{P}$ is stochastic. □

Using the above notation, the next theorem gives a method of computing $P(t)$.

Theorem 6.4 (The Matrix $P(t)$). *The transition probability matrix $P(t) = [p_{i,j}(t)]$ is given by*

$$P(t) = \sum_{k=0}^{\infty} e^{-rt} \frac{(rt)^k}{k!} \hat{P}^k. \tag{6.33}$$

We present the proof of the above theorem in Appendix A at the end of this chapter. Equation (6.33) provides a very stable method of computing $P(t)$. It is known as the *uniformization method*, since it involves observing the CTMC at rate r that is a uniform upper bound on the transition rate r_i out of state $i \in S$. In numerical computations, we approximate $P(t)$ by using the first M terms of the infinite series in (6.33). In order to make the method numerically implementable we need to describe how to choose M to guarantee a specified accuracy. A good rule of thumb is to choose

$$M \approx \max\{rt + 5 * \sqrt{rt}, 20\}.$$

A more precise algorithm is developed in Appendix B. Since the uniformization algorithm terminates quicker (i.e., uses less terms of the series) if the value of rt is small, it makes sense to use the smallest value of r allowed by (6.29), namely

$$r = \max_{1 \le i \le N} \{r_i\}.$$

This is what we do in the next three examples. However, in some cases it is more convenient to use an r that is larger than the one above. This is illustrated in Example 6.20.

Example 6.17. Let $\{X(t), t \geq 0\}$ be a CTMC with state space $\{1, 2, 3, 4\}$ and the rate matrix given in Example 6.16. Using the values of r and $\hat{P}$ from that example we get

$$P(t) = \sum_{k=0}^{\infty} e^{-6t} \frac{(6t)^k}{k!} \begin{bmatrix} \frac{1}{6} & \frac{2}{6} & \frac{3}{6} & 0 \\ \frac{4}{6} & 0 & \frac{2}{6} & 0 \\ 0 & \frac{2}{6} & \frac{2}{6} & \frac{2}{6} \\ \frac{1}{6} & 0 & \frac{3}{6} & \frac{2}{6} \end{bmatrix}^k. \tag{6.34}$$

We compute $P(t)$ by using the first M terms of the infinite series in (6.34). We use $\epsilon = .00001$ in the algorithm in Appendix B and display the value of $P(t)$ and the value of M.

$$P(.5) = \begin{bmatrix} .2506 & .2170 & .3867 & .1458 \\ .2531 & .2384 & .3744 & .1341 \\ .1691 & .1936 & .4203 & .2170 \\ .1580 & .1574 & .3983 & .2862 \end{bmatrix}, \quad M = 13,$$

$$P(1) = \begin{bmatrix} .2062 & .2039 & .3987 & .1912 \\ .2083 & .2053 & .3979 & .1885 \\ .1968 & .1984 & .4010 & .2039 \\ .1920 & .1940 & .4015 & .2125 \end{bmatrix}, \quad M = 19,$$

$$P(5) = \begin{bmatrix} .2000 & .2000 & .4000 & .2000 \\ .2000 & .2000 & .4000 & .2000 \\ .2000 & .2000 & .4000 & .2000 \\ .2000 & .2000 & .4000 & .2000 \end{bmatrix}, \quad M = 56.$$

It can be numerically verified that $P(1) = P(.5)P(.5)$, thus verifying the Chapman–Kolmogorov equations (6.8). □

Example 6.18 (General Machine Shop). Consider the machine shop of Example 6.7. Suppose the shop has four machines and two repair persons. Suppose the lifetimes of the machines are exponential random variables with mean 3 days, while the repair times are exponential random variables with mean 2 hours. The shop operates 24 hours a day. Suppose all the machines are operating at 8:00 a.m. Monday. What is the expected number of working machines at 5:00 p.m. the same Monday?

Using the notation of Example 6.7, we see that the repair rate is $\lambda = .5$ per hour and the failure rate is $\mu = 1/72$ per hour. (Note that we must use consistent units.) Let $X(t)$ be the number of working machines at time t. Then, from Example 6.7, we see that $\{X(t), t \geq 0\}$

is a CTMC on {0, 1, 2, 3, 4} with the following rate matrix:

$$R = \begin{bmatrix} 0 & 1 & 0 & 0 & 0 \\ \frac{1}{72} & 0 & 1 & 0 & 0 \\ 0 & \frac{2}{72} & 0 & 1 & 0 \\ 0 & 0 & \frac{3}{72} & 0 & \frac{1}{2} \\ 0 & 0 & 0 & \frac{4}{72} & 0 \end{bmatrix}. \tag{6.35}$$

We can take 8:00 a.m. as time 0 (hours). Thus we are given $X(0) = 4$. We are asked to compute the expected number of working machines at 5:00 p.m., i.e., after 9 hours. Thus we need $E(X(9))$. From (6.29) we get $\dot{r} = \frac{74}{72}$, and from (6.30) we get

$$\hat{P} = \begin{bmatrix} \frac{2}{74} & \frac{72}{74} & 0 & 0 & 0 \\ \frac{1}{74} & \frac{1}{74} & \frac{72}{74} & 0 & 0 \\ 0 & \frac{2}{74} & 0 & \frac{72}{74} & 0 \\ 0 & 0 & \frac{3}{74} & \frac{35}{74} & \frac{36}{74} \\ 0 & 0 & 0 & \frac{4}{74} & \frac{70}{74} \end{bmatrix}. \tag{6.36}$$

Using an error tolerance of $\epsilon = .00001$ and $t = 9$ hours in the uniformization algorithm of Appendix B we get

$$P(9) = \begin{bmatrix} .0002 & .0020 & .0136 & .1547 & .8295 \\ .0000 & .0006 & .0080 & .1306 & .8607 \\ .0000 & .0002 & .0057 & .1154 & .8787 \\ .0000 & .0002 & .0048 & .1070 & .8880 \\ .0000 & .0001 & .0041 & .0987 & .8971 \end{bmatrix}. \tag{6.37}$$

The algorithm used the first 26 terms of the infinite series to compute the above matrix. Now we are given that $X(0) = 4$. Hence the pmf of $X(9)$ is given by the last row of the above matrix. Hence the expected number of working machines at 5:00 p.m. is given by

$$E(X(9)) = 0 * .0000 + 1 * .0001 + 2 * .0041 + 3 * .0987 + 4 * .8971 = 3.8928.$$

Note that the probability that all machines are working at 5:00 p.m. is .8971, however this is not the probability that there were no failures from 8:00 a.m. to 5:00 p.m. There may have been one or more failures and the corresponding repairs. □

Example 6.19 (Airplane Reliability). Consider the four-engine jet airplane of Example 6.8. Suppose each engine can give trouble-free service for 200 hours on the average before it needs to be shut down due to any problem. Assume that all four engines are operational at the beginning of a 6-hour flight. What is the probability that this fight lands safely?

We assume that the only trouble on the flight is from the engine failures. Let $\{X(t), t \geq 0\}$ be the CTMC as in Example 6.8. The rate matrix is as given in (6.22) with $\lambda = 1/200 = .005$ per hour. The flight lands safely if the state of the CTMC stays in the set $\{5, 6, 8, 9\}$ for the duration $[0, 6]$. Since this CTMC cannot return to this set of states once it leaves it, the required probability is given by the probability that $X(6)$ is in $\{5, 6, 8, 9\}$ given that $X(0) = 9$, the initial state. Using the uniformization algorithm of Appendix B we get the probability of a safe landing as

$$p_{9.5}(6) + p_{9.6}(6) + p_{9.8}(6) + p_{9.9}(6) = .0033 + .0540 + .0540 + .8869 = .9983.$$

Thus the probability of a crash is .0017. This is far too large for any commercial airline! □

Example 6.20 (Two-State CTMC). Consider the two-state CTMC of Example 6.5 with state space $\{0, 1\}$ and rate matrix as given below:

$$R = \begin{bmatrix} 0 & \lambda \\ \mu & 0 \end{bmatrix}. \tag{6.38}$$

Here $r_0 = \lambda$ and $r_1 = \mu$. Hence (6.29) would suggest $r = \max\{\lambda, \mu\}$. However, it is more convenient to choose

$$r = \lambda + \mu.$$

Then the $\hat{P}$ matrix of (6.30) is given by

$$\hat{P} = \begin{bmatrix} \dfrac{\mu}{\lambda + \mu} & \dfrac{\lambda}{\lambda + \mu} \\ \dfrac{\mu}{\lambda + \mu} & \dfrac{\lambda}{\lambda + \mu} \end{bmatrix}. \tag{6.39}$$

It can be readily verified that

$$\hat{P}^k = \begin{bmatrix} \dfrac{\mu}{\lambda + \mu} & \dfrac{\lambda}{\lambda + \mu} \\ \dfrac{\mu}{\lambda + \mu} & \dfrac{\lambda}{\lambda + \mu} \end{bmatrix}, \qquad k \geq 1. \tag{6.40}$$

Substituting in (6.33) we get

$$P(t) = \sum_{k=0}^{\infty} e^{-rt} \frac{(rt)^k}{k!} \hat{P}^k$$

$$= e^{-rt} \cdot \begin{bmatrix} 1 & 0 \\ 0 & 1 \end{bmatrix} + \sum_{k=1}^{\infty} e^{-rt} \frac{(rt)^k}{k!} \begin{bmatrix} \dfrac{\mu}{\lambda + \mu} & \dfrac{\lambda}{\lambda + \mu} \\ \dfrac{\mu}{\lambda + \mu} & \dfrac{\lambda}{\lambda + \mu} \end{bmatrix}$$

$$
= e^{-rt} \cdot \left(\begin{bmatrix} 1 & 0 \\ 0 & 1 \end{bmatrix} + \begin{bmatrix} \dfrac{\mu}{\lambda+\mu} & \dfrac{\lambda}{\lambda+\mu} \\[2ex] \dfrac{\mu}{\lambda+\mu} & \dfrac{\lambda}{\lambda+\mu} \end{bmatrix} \sum_{k=1}^{\infty} e^{-rt} \dfrac{(rt)^k}{k!} \right)
$$

$$
= e^{-rt} \cdot \left(\begin{bmatrix} 1 & 0 \\ 0 & 1 \end{bmatrix} + \begin{bmatrix} \dfrac{\mu}{\lambda+\mu} & \dfrac{\lambda}{\lambda+\mu} \\[2ex] \dfrac{\mu}{\lambda+\mu} & \dfrac{\lambda}{\lambda+\mu} \end{bmatrix} (1 - e^{-rt}) \right)
$$

$$
= \begin{bmatrix} \dfrac{\mu}{\lambda+\mu} & \dfrac{\lambda}{\lambda+\mu} \\[2ex] \dfrac{\mu}{\lambda+\mu} & \dfrac{\lambda}{\lambda+\mu} \end{bmatrix} + \begin{bmatrix} \dfrac{\lambda}{\lambda+\mu} & -\dfrac{\lambda}{\lambda+\mu} \\[2ex] -\dfrac{\mu}{\lambda+\mu} & \dfrac{\mu}{\lambda+\mu} \end{bmatrix} e^{-rt}. \tag{6.41}
$$

From this we can read

$$
p_{1.1}(t) = \frac{\lambda}{\lambda+\mu} + \frac{\mu}{\lambda+\mu} e^{-(\lambda+\mu)t}.
$$

Thus the probability that a machine is up at time t given that it was up at time 0 is given by the above expression. Note that the machine could have gone through many failures and repairs up to time t. The above probability includes the sum total of all such possibilities.

□

6.8. Occupancy Times

In this section we shall do the short-term analysis of a CTMC $\{X(t), t \geq 0\}$, i.e., we study the behavior of the CTMC over a finite interval $[0, T]$. We concentrate on the *occupancy time* of a given state, i.e., the expected length of time the system spends in that state during a given interval of time. Here we develop a method of computing such quantities.

Let $X(t)$ be the state of a system at time t, and assume that $\{X(t), t \geq 0\}$ is a CTMC with state space $S = \{1, 2, \ldots, N\}$ and rate matrix R. Let $m_{i,j}(T)$ be the expected amount of time the CTMC spends in state j during the interval $[0, T]$, starting in state i. $m_{i,j}(T)$ is called the occupancy time of state j until time T starting from state i. The following theorem shows a way of computing $m_{i,j}(T)$:

Theorem 6.5 (The Occupancy Times). *Let $P(t) = [p_{i,j}(t)]$ be the transition probability matrix of $\{X(t), t \geq 0\}$. Then*

$$
m_{i,j}(T) = \int_0^T p_{i,j}(t)\, dt, \qquad 1 \leq i, j \leq N. \tag{6.42}
$$

Proof. Let i and j be fixed, and define

$$Y_j(t) = \begin{cases} 1 & \text{if } X(t) = j, \\ 0 & \text{otherwise.} \end{cases}$$

The total amount of time spent in state j by the CTMC during $[0, T]$ is then given by

$$\int_0^T Y_j(t)\, dt.$$

Hence we get

$$m_{i.j}(T) = \mathsf{E}\left(\int_0^T Y_j(t)\, dt \mid X(0) = i\right)$$

$$= \int_0^T \mathsf{E}(Y_j(t) \mid X(0) = i)\, dt$$

(integral and E can be interchanged here)

$$= \int_0^T \mathsf{P}(Y_j(t) = 1 \mid X(0) = i)\, dt$$

$$= \int_0^T \mathsf{P}(X(t) = j \mid X(0) = i)\, dt$$

$$= \int_0^T p_{i.j}(t)\, dt$$

as desired.   □

Example 6.21 (Two-State Machine). Consider the two-state machine of Example 6.5. Suppose the expected time until failure of the machine is 10 days, while the expected repair time is 1 day. Suppose the machine is working at the beginning of January. Compute the expected total up time of the machine in the month of January.

Let $\{X(t), t \geq 0\}$ be as in Example 6.5. Using the data given here in time units of days, we get: $\mu = .1$ (days)$^{-1}$ and $\lambda = 1$ (days)$^{-1}$. Since there are 31 days in January, the expected uptime in January is given by $m_{1.1}(31)$. From Example 6.20, we have

$$p_{1.1}(t) = \frac{\lambda}{\lambda + \mu} + \frac{\mu}{\lambda + \mu} e^{-(\lambda + \mu)t}$$

$$= \frac{10}{11} + \frac{1}{11} e^{-1.1t}. \tag{6.43}$$

Substituting in (6.42) we get

$$m_{1.1}(31) = \int_0^{31} p_{1.1}(t)\, dt$$

$$= \int_0^{31} \left(\frac{10}{11} + \frac{1}{11} e^{-1.1t}\right) dt$$

$$= \frac{10}{11} \cdot 31 + \frac{1}{11} \cdot \frac{1}{1.1}(1 - e^{-34.1})$$
$$= 28.26 \text{ days.}$$

Thus the expected uptime during January is 28.26 days. Hence the expected downtime must be $31 - 28.26 = 2.74$ days. The reader should compute $m_{1,0}(31)$ and verify this independently. □

Note that Theorem 6.5 can be used directly if $p_{i,j}(t)$ is known as a function of t in a closed form, as was the case in the above example. However, $p_{i,j}(t)$ is rarely known in closed form, and mostly has to be computed using the infinite series representation of (6.33). How do we compute $m_{i,j}(T)$ in such cases?

First we compute the uniformization constant r from (6.29) and the $\hat{P}$ matrix from (6.30). Next we construct an $N \times N$ *occupancy matrix*

$$M(t) = [m_{i,j}(t)]. \tag{6.44}$$

The next theorem gives an infinite series expression for $M(T)$, analogous to Theorem 6.4.

Theorem 6.6 (The Matrix $M(T)$). *Let Y be a $P(rT)$ random variable. Then*

$$M(T) = \frac{1}{r} \sum_{k=0}^{\infty} P(Y > k)\hat{P}^k, \qquad T \geq 0. \tag{6.45}$$

Proof. From (6.33) we get

$$p_{i,j}(t) = \sum_{k=0}^{\infty} e^{-rt} \frac{(rt)^k}{k!} [\hat{P}^k]_{i,j}.$$

We also have

$$\int_0^T e^{-rt} \frac{(rt)^k}{k!} dt = \frac{1}{r}\left(1 - \sum_{l=0}^{k} e^{-rT} \frac{(rT)^l}{l!}\right)$$
$$= \frac{1}{r}P(Y > k).$$

Substituting in (6.42) we get

$$m_{i,j}(T) = \int_0^T p_{i,j}(t)\,dt$$
$$= \int_0^T \sum_{k=0}^{\infty} e^{-rt} \frac{(rt)^k}{k!} [\hat{P}^k]_{i,j}\,dt$$
$$= \sum_{k=0}^{\infty} \left(\int_0^T e^{-rt} \frac{(rt)^k}{k!}\,dt\right)[\hat{P}^k]_{i,j}$$
$$= \sum_{k=0}^{\infty} \frac{1}{r}P(Y > k)[\hat{P}^k]_{i,j}.$$

Putting the above equation in matrix form yields (6.45). □

The above theorem provides a very stable method of computing the matrix $M(T)$. An algorithm to compute $M(T)$ based on the above theorem is developed in Appendix C. We illustrate the above theorem with several examples below.

Example 6.22 (Single Server Queue). Consider the ATM queue of Example 6.9. Suppose the customers arrive at a rate of 10 per hour, and take on the average 4 minutes at the machine to complete their transactions. Suppose that there is space for at most five customers in front of the ATM and that it is idle at 8:00 a.m. What is the expected amount of time the machine is idle during the next hour?

The arrival process is Poisson with rate $\lambda = 10$ per hour. The processing times are iid exponential with mean 4/60 hours, i.e., with parameter $\mu = 15$ per hour. The capacity is $K = 5$. Let $X(t)$ be the number of customers in the queue at time t. With these parameters, $\{X(t), t \geq 0\}$ is a CTMC with state space $\{0, 1, 2, 3, 4, 5\}$ and rate matrix given by

$$R = \begin{bmatrix} 0 & 10 & 0 & 0 & 0 & 0 \\ 15 & 0 & 10 & 0 & 0 & 0 \\ 0 & 15 & 0 & 10 & 0 & 0 \\ 0 & 0 & 15 & 0 & 10 & 0 \\ 0 & 0 & 0 & 15 & 0 & 10 \\ 0 & 0 & 0 & 0 & 15 & 0 \end{bmatrix}. \tag{6.46}$$

We are given $X(0) = 0$. The expected idle time is thus given by $m_{0,0}(1)$. Using the Uniformization Algorithm of Appendix C with $\epsilon = .00001$, we get

$$M(1) = \begin{bmatrix} .4451 & .2548 & .1442 & .0814 & .0465 & .0280 \\ .3821 & .2793 & .1604 & .0920 & .0535 & .0326 \\ .3246 & .2407 & .2010 & .1187 & .0710 & .0441 \\ .2746 & .2070 & .1780 & .1695 & .1046 & .0663 \\ .2356 & .1806 & .1598 & .1569 & .1624 & .1047 \\ .2124 & .1648 & .1489 & .1492 & .1571 & .1677 \end{bmatrix}.$$

The algorithm uses the first 44 terms of the infinite series to compute the above result. Thus $m_{0,0}(1) = .4451$ hours, i.e., if the ATM is idle at the beginning, it is idle on the average $.4451 * 60 = 26.71$ minutes of the first hour. Similarly the queue is filled to capacity for $m_{0.5} = .0280$ hours $= 1.68$ minutes of the first hour. □

Example 6.23 (Manufacturing). Consider the manufacturing operation described in Example 6.15. Suppose the system operates 24 hours a day, and the demands occur at the rate of five per hour, and the average time to manufacture one item is 10 minutes. The machine is turned on whenever the stock of manufactured items reduces to two, and it stays on until the stock rises to four, at which point it is turned off. It stays off until the stock reduces to two, and so on. Suppose the stock is at four (and the machine is off) at the beginning. Compute the expected amount of time during which the machine is on during the next 24 hours.

Let $X(t)$ be the number of items in stock at time t, and let $Y(t)$ be the state of the machine at time t, 1 if it is on and 0 if it is off. Then $\{(X(t), Y(t)), t \geq 0\}$ is a CTMC with state space

$$S = \{1 = (0, 1), 2 = (1, 1), 3 = (2, 1), 4 = (3, 1), 5 = (4, 0), 6 = (3, 0)\}.$$

The demand rate is $\mu = 5$ per hour and the production rate is $\lambda = 6$ per hour. Following the analysis of Example 6.15 we get the rate matrix of the CTMC as

$$R = \begin{bmatrix} 0 & 6 & 0 & 0 & 0 & 0 \\ 5 & 0 & 6 & 0 & 0 & 0 \\ 0 & 5 & 0 & 6 & 0 & 0 \\ 0 & 0 & 5 & 0 & 6 & 0 \\ 0 & 0 & 0 & 0 & 0 & 5 \\ 0 & 0 & 5 & 0 & 0 & 0 \end{bmatrix}. \tag{6.47}$$

We are given that the initial state of the system is $(X(0), Y(0)) = (4, 0)$, i.e., the initial state is 5. The machine is on whenever the system is in states 1, 2, 3, or 4. Hence the total expected on time during 24 hours is given by $m_{5,1}(24) + m_{5,2}(24) + m_{5,3}(24) + m_{5,4}(24)$. We use the Uniformization Algorithm of Appendix C with $T = 24$ and $\epsilon = .00001$. The algorithm uses the first 331 terms of the infinite series to yield

$$M(24) = \begin{bmatrix} 4.0004 & 4.6322 & 5.4284 & 2.9496 & 3.5097 & 3.4798 \\ 3.8602 & 4.6639 & 5.4664 & 2.9703 & 3.5345 & 3.5047 \\ 3.7697 & 4.5553 & 5.5361 & 3.0084 & 3.5802 & 3.5503 \\ 3.7207 & 4.4965 & 5.4656 & 3.0608 & 3.6431 & 3.6132 \\ 3.7063 & 4.4793 & 5.4448 & 2.9586 & 3.7204 & 3.6906 \\ 3.7380 & 4.5173 & 5.4905 & 2.9835 & 3.5503 & 3.7204 \end{bmatrix}.$$

Hence the total expected on time is given by

$$3.7063 + 4.4793 + 5.4448 + 2.9586 = 16.5890 \quad \text{hours.}$$

If the system had no stock in the beginning, the total expected uptime would be

$$4.0004 + 4.6322 + 5.4284 + 2.9496 = 17.0106 \quad \text{hours.} \qquad \square$$

6.9. Limiting Behavior

In the study of the limiting behavior of DTMCs in Section 5.6 we studied three quantities: the limiting distribution, the stationary distribution, and the occupancy distribution. In this section we study similar quantities for the CTMCs. Let $\{X(t), t \geq 0\}$ be a CTMC with state space $S = \{1, 2, \ldots, N\}$ and rate matrix R. In Section 6.7 we saw how to compute $\mathsf{P}(X(t) = j)$ for $1 \leq j \leq N$. In this section we study

$$p_j = \lim_{t \to \infty} \mathsf{P}(X(t) = j). \tag{6.48}$$

If the above limit exists we call it the *limiting probability* of being in state j. If it exists for all $1 \leq j \leq N$, we call

$$p = [p_1, p_2, \ldots, p_N]$$

the *limiting distribution* of the CTMC. If the limiting distribution exists, we ask if it is unique, and how to compute it if it is unique.

Here we recapitulate some important facts from Appendix A. The reader does not need to follow the appendix if he/she is willing to accept these facts as true. Recall the definition of the uniformization constant from (6.29). We opt to choose strict inequality here as follows:

$$r > \max_{1 \leq j \leq N} \{r_j\}.$$

Construct $\hat{P}$ from R by using (6.30). We have shown in Theorem 6.14 that $\hat{P}$ is a stochastic matrix. Let $\{\hat{X}_n, n \geq 0\}$ be a DTMC on state space $\{1, 2, \ldots, N\}$ with transition probability matrix $\hat{P}$. Let $N(t)$ be a PP(r), and assume that it is independent of $\{\hat{X}_n, n \geq 0\}$. Then we saw in Appendix A that

$$P(X(t) = j) = P(\hat{X}_{N(t)} = j).$$

Hence the limiting distribution of $X(t)$, if it exists, is identical to the limiting distribution of $\hat{X}_n$. We use this fact below.

Suppose $\{\hat{X}_n, n \geq 0\}$ is irreducible, i.e., it can go from any state i to any state j in a finite number of steps. However, $\hat{X}_n = i$ for some n if and only if $X(t) = i$ for some t. This implies that the CTMC $\{X(t), t \geq 0\}$ can go from any state i to any state j in a finite amount of time. This justifies the following definition:

Definition 6.3 (Irreducible CTMC). The CTMC $\{X(t), t \geq 0\}$ is said to be irreducible if the corresponding DTMC $\{\hat{X}_n, n \geq 0\}$ is irreducible.

Using this definition we get the next theorem:

Theorem 6.7 (Limiting Distribution). *An irreducible CTMC $\{X(t), t \geq 0\}$ with rate matrix $R = [r_{i,j}]$ has a unique limiting distribution $p = [p_1, p_2, \ldots, p_N]$. It is given by the solution to*

$$p_j r_j = \sum_{i=1}^{N} p_i r_{i,j}, \qquad 1 \leq j \leq N, \tag{6.49}$$

$$\sum_{i=1}^{N} p_i = 1. \tag{6.50}$$

Proof. The CTMC $\{X(t), t \geq 0\}$ is irreducible, hence the DTMC $\{\hat{X}_n, n \geq 0\}$ is irreducible. The choice of the uniformization constant r implies that $\hat{p}_{j,j} = 1 - r_j/r > 0$ for all j. Thus the DTMC $\{\hat{X}_n, n \geq 0\}$ is aperiodic and irreducible. Hence, from Theorem 5.10, it has a limiting distribution

$$\lim_{n\to\infty} P(\hat{X}_n = j) = \pi_j, \qquad 1 \le j \le N.$$

The limiting distribution satisfies (5.36) and (5.37), as follows:

$$\pi_j = \sum_{i=1}^{N} \pi_i \hat{p}_{i,j}$$

and

$$\sum_{j=1}^{N} \pi_j = 1.$$

Now, $N(t)$ goes to ∞ as t goes to ∞. Hence

$$\begin{aligned}
p_j &= \lim_{t\to\infty} P(X(t) = j) \\
&= \lim_{t\to\infty} P(\hat{X}_{N(t)} = j) \\
&= \lim_{n\to\infty} P(\hat{X}_n = j) \\
&= \pi_j.
\end{aligned} \tag{6.51}$$

Hence, the limiting distribution p exists and is unique, and is equal to the limiting distribution of the DTMC! Substituting $p_j = \pi_j$ in the balance equations satisfied by π, we get

$$p_j = \sum_{i=1}^{N} p_i \hat{p}_{i,j} \tag{6.52}$$

and

$$\sum_{j=1}^{N} p_j = 1.$$

The last equation produces (6.50). Substituting in (6.52) for $\hat{p}_{i,j}$ from (6.30) we get

$$\begin{aligned}
p_j &= \sum_{i=1}^{N} p_i \hat{p}_{i,j} \\
&= p_j \hat{p}_{j,j} + \sum_{i\ne j} p_i \hat{p}_{i,j} \\
&= p_j\left(1 - \frac{r_j}{r}\right) + \sum_{i\ne j} p_i \frac{r_{i,j}}{r}.
\end{aligned}$$

Cancelling p_j from both sides and rearranging we get

$$p_j \frac{r_j}{r} = \sum_{i\ne j} p_i \frac{r_{i,j}}{r}.$$

Cancelling the common factor r and recalling that $r_{i,i} = 0$ we get (6.49). $\square$

We can think of $p_j r_j$ as the rate at which the CTMC leaves state j, and $p_i r_{i,j}$ as the rate at which the CTMC enters state j from state i. Thus, (6.49) says that the limiting probabilities are such that the rate of entering state j from all other states is the same as the rate of leaving state j. Hence it is called the *balance equation* or *rate-in–rate-out* equation.

The balance equations can easily be written down from the rate diagram as follows: imagine that there is a flow from node i to node j at rate $p_i r_{i,j}$ on every arc (i, j) in the rate diagram. Then the balance equations simply say that the total flow into a node from all other nodes in the rate diagram equals the total flow out of that node to all other nodes.

In Chapter 5 we saw that the limiting distribution of a DTMC is also its stationary distribution. Is the same true for CTMCs as well? The answer is given by the next theorem:

Theorem 6.8 (Stationary Distribution). *Let $\{X(t), t \geq 0\}$ be an irreducible CTMC with limiting distribution p. Then it is also the stationary distribution of the CTMC, i.e.,*

$$P(X(0) = j) = p_j \quad \text{for} \quad 1 \leq j \leq N \Rightarrow$$

$$P(X(t) = j) = p_j \quad \text{for} \quad 1 \leq j \leq N, \quad t \geq 0.$$

Proof. From the proof of Theorem 6.7, $p_j = \pi_j$ for all $1 \leq j \leq N$. Suppose $P(X(0) = j) = P(\hat{X}_0 = j) = p_j$ for $1 \leq j \leq N$. Then, from Corollary 5.3

$$P(\hat{X}_k = j) = \pi_j = p_j, \quad 1 \leq j \leq N, \quad k \geq 0,$$

and

$$P(X(t) = j) = \sum_{i=1}^{N} P(X(t) = j | X(0) = i)P(X(0) = i)$$

$$= \sum_{i=1}^{N} P(X(0) = i) \sum_{k=0}^{\infty} e^{-rt} \frac{(rt)^k}{k!} [\hat{P}^k]_{i,j}$$

$$= \sum_{k=0}^{\infty} e^{-rt} \frac{(rt)^k}{k!} \sum_{i=1}^{N} P(X(0) = i) P(\hat{X}_k = j | \hat{X}_0 = i)$$

$$= \sum_{k=0}^{\infty} e^{-rt} \frac{(rt)^k}{k!} \sum_{i=1}^{N} P(\hat{X}_k = j | \hat{X}_0 = i) P(\hat{X}_0 = i)$$

$$= \sum_{k=0}^{\infty} e^{-rt} \frac{(rt)^k}{k!} P(\hat{X}_k = j)$$

$$= \sum_{k=0}^{\infty} e^{-rt} \frac{(rt)^k}{k!} p_j$$

$$= p_j$$

as desired.   □

The next question is about the occupancy distribution. In DTMCs, the limiting distribution is also the occupancy distribution. Does this hold true for the CTMCs? The answer is yes if we define the occupancy of state i to be the expected long-run fraction of the time the CTMC spends in state i. The main result is given in the next theorem:

Theorem 6.9 (Occupancy Distribution). *Let $m_{i,j}(T)$ be the expected total time spent in state j up to time T by an irreducible CTMC starting in state i. Then*

$$\lim_{T \to \infty} \frac{m_{i,j}(T)}{T} = p_j,$$

where p_j is the limiting probability that the CTMC is in state j.

Proof. From Theorem 6.6 we see that

$$M(T) = \frac{1}{r} \sum_{k=0}^{\infty} \mathsf{P}(Y > k) \hat{P}^k, \qquad T \geq 0,$$

where $M(T) = [m_{i,j}(T)]$ and Y is a $\mathsf{P}(rT)$ random variable. We have

$$\mathsf{P}(Y > k) = \sum_{m=k+1}^{\infty} a_m,$$

where

$$a_m = e^{-rT} \frac{(rT)^m}{m!}.$$

Substituting, we get

$$M(T) = \frac{1}{r} \sum_{k=0}^{\infty} \sum_{m=k+1}^{\infty} a_m \hat{P}^k$$

$$= \frac{1}{r} \sum_{m=1}^{\infty} a_m \sum_{k=0}^{m-1} \hat{P}^k$$

$$= \frac{1}{r} \sum_{m=1}^{\infty} e^{-rT} \frac{(rT)^{m-1}}{(m-1)!} \frac{rT}{m} \sum_{k=0}^{m-1} \hat{P}^k.$$

Hence

$$\frac{M(T)}{T} = \sum_{m=1}^{\infty} e^{-rT} \frac{(rT)^{m-1}}{(m-1)!} \frac{1}{m} \sum_{k=0}^{m-1} \hat{P}^k$$

$$= \sum_{m=0}^{\infty} e^{-rT} \frac{(rT)^m}{m!} \frac{1}{m+1} \sum_{k=0}^{m} \hat{P}^k$$

$$= \mathsf{E}\left(\frac{1}{N(T)+1} \sum_{k=0}^{N(T)} \hat{P}^k \right),$$

where $N(T)$ is the number of events in a PP(r) up to time T. From Theorems 5.7 and 5.9 we see that

$$\lim_{n \to \infty} \frac{1}{n+1} \sum_{k=0}^{n} [\hat{P}^k]_{i,j} = \pi_j = p_j.$$

Also, as $T \to \infty$, $N(T) \to \infty$. Using these two observations we get

$$\lim_{T \to \infty} \frac{m_{i,j}(T)}{T} = \lim_{T \to \infty} \mathsf{E}\left(\frac{1}{N(T)+1} \sum_{k=0}^{N(T)} [\hat{P}^k]_{i,j} \right)$$

$$= \lim_{n \to \infty} \frac{1}{n+1} \sum_{k=0}^{n} [\hat{P}^k]_{i,j}$$

$$= p_j$$

as desired.   □

Thus, an irreducible CTMC has a unique limiting distribution, which is also its stationary distribution, and also its occupancy distribution. It can be computed by solving the balance equation (6.49) along with the normalizing equation (6.50). We illustrate the above concepts by means of several examples below.

Example 6.24 (Two-State Machine). Consider the two-state machine as described in Example 6.5. Compute the limiting distribution of the state of the machine.

The two-state CTMC of Example 6.5 is irreducible. Hence it has a unique limiting distribution $[p_0, p_1]$. The two balance equations are

$$\lambda p_0 = \mu p_1,$$

$$\mu p_1 = \lambda p_0.$$

Note that they are the same! The normalizing equation is

$$p_0 + p_1 = 1.$$

The solution is given by

$$p_0 = \frac{\mu}{\lambda + \mu}, \qquad p_1 = \frac{\lambda}{\lambda + \mu}.$$

Note that this is also the stationary distribution and the occupancy distribution of the machine (check this using the $P(t)$ matrix from Example 6.20).

Suppose the expected time until failure of the machine is 10 days, while the expected repair time is 1 day. Thus $\lambda = 1$ and $\mu = .1$. Hence

$$p_0 = \frac{1}{11}, \qquad p_1 = \frac{10}{11}.$$

Thus, in the long run the machine spends 10 out of 11 days in working condition. This makes intuitive sense in view of the fact that the machine stays down for 1 day, followed by being up for 10 days on average!   □

Example 6.25 (Finite Birth and Death Process). Let $\{X(t), t \geq 0\}$ be a birth and death process as described in Example 6.10. Assume that all the birth rates $\{\lambda_i, 0 \leq i < K\}$ and death rates $\{\mu_i, 1 \leq i \leq K\}$ are positive. Then the CTMC is irreducible, and hence has a unique limiting distribution. We compute it below.

The balance equations are as follows:

$$\lambda_0 p_0 = \mu_1 p_1,$$
$$(\lambda_1 + \mu_1) p_1 = \lambda_0 p_0 + \mu_2 p_2,$$
$$(\lambda_2 + \mu_2) p_2 = \lambda_1 p_1 + \mu_3 p_3,$$
$$\vdots \quad \vdots$$
$$(\lambda_{K-1} + \mu_{K-1}) p_{K-1} = \lambda_{K-2} p_{K-2} + \mu_K p_K,$$
$$\mu_K p_K = \lambda_{K-1} p_{K-1}.$$

From the first equation above we get

$$p_1 = \frac{\lambda_0}{\mu_1} p_0.$$

Adding the first two balance equations we get

$$\lambda_1 p_1 = \mu_2 p_2,$$

which yields

$$p_2 = \frac{\lambda_1}{\mu_2} p_1 = \frac{\lambda_0 \lambda_1}{\mu_1 \mu_2} p_0.$$

Proceeding this way (adding the first three balance equations, etc.) we get, in general,

$$p_i = \rho_i p_0, \qquad 0 \leq i \leq K, \tag{6.53}$$

where $\rho_0 = 1$ and

$$\rho_i = \frac{\lambda_0 \lambda_1 \cdots \lambda_{i-1}}{\mu_1 \mu_2 \cdots \mu_i}, \qquad 1 \leq i \leq K. \tag{6.54}$$

The only unknown left is p_0, which can be computed by using the normalization equation as follows:

$$p_0 + p_1 + \cdots + p_{K-1} + p_K = (\rho_0 + \rho_1 + \cdots + \rho_{K-1} + \rho_K) p_0 = 1.$$

This yields

$$p_0 = \frac{1}{\sum_{j=0}^{K} \rho_j}. \tag{6.55}$$

Combining this with (6.53) we get

$$p_i = \frac{\rho_i}{\sum_{j=0}^{K} \rho_j}, \qquad 0 \leq i \leq K. \tag{6.56}$$

This also gives the stationary distribution and the occupancy distribution of the birth and death processes. □

Example 6.26 (The General Machine Shop). Consider the machine shop of Example 6.18, with four machines and two repair persons. Compute the long-run probability that all the machines are up. Also compute the long-run fraction of the time that both repair persons are busy.

Let $X(t)$ be the number of working machines at time t. Then $\{X(t), t \geq 0\}$ is a birth and death process on $\{0, 1, 2, 3, 4\}$ with birth rates $\lambda_0 = 1; \lambda_1 = 1; \lambda_2 = 1; \lambda_3 = \frac{1}{2}$ and death rates $\mu_1 = \frac{1}{72}; \mu_2 = \frac{2}{72}; \mu_3 = \frac{3}{72}; \mu_4 = \frac{4}{72}$. Hence we get

$$\rho_0 = 1, \qquad \rho_1 = 72, \qquad \rho_2 = 2592, \qquad \rho_3 = 62208, \qquad \rho_4 = 559872.$$

Thus

$$\sum_{i=0}^{4} \rho_i = 624745.$$

Hence using (6.56) we get

$$p_0 = \frac{1}{624745} = 1.6001 \times 10^{-6},$$

$$p_1 = \frac{72}{624745} = 1.1525 \times 10^{-4},$$

$$p_2 = \frac{2592}{624745} = 4.149 \times 10^{-3},$$

$$p_3 = \frac{62208}{624745} = 9.957 \times 10^{-2},$$

$$p_4 = \frac{559872}{624745} = 8.962 \times 10^{-1}.$$

Thus, in the long run all the machines are functioning 89.6% of the time. Both repair persons are busy whenever the system is in state 0 or 1 or 2. Hence the long-run fraction of the time that both repair persons are busy is given by $p_0 + p_1 + p_2 = .00427$. □

Example 6.27 (Finite Capacity Single Server Queue). Consider the finite capacity single server queue of Example 6.9. Let $X(t)$ be the number of customers in the queue at time t. Then we have seen that $\{X(t), t \geq 0\}$ is a birth and death process on state space $\{0, 1, \ldots, K\}$ with birth rates

$$\lambda_i = \lambda, \qquad 0 \leq i < K,$$

and death rates

$$\mu_i = \mu, \qquad 1 \leq i \leq K.$$

Now define

$$\rho = \frac{\lambda}{\mu}.$$

Using the results of Example 6.25 we get

$$\rho_i = \rho^i, \qquad 0 \le i \le K,$$

and hence, assuming $\rho \ne 1$,

$$\sum_{i=0}^{K} \rho_i = \sum_{i=0}^{K} \rho^i = \frac{1 - \rho^{K+1}}{1 - \rho}.$$

Substituting in (6.56) we get

$$p_i = \frac{1 - \rho}{1 - \rho^{K+1}} \rho^i, \qquad 0 \le i \le K. \tag{6.57}$$

This gives the limiting distribution of the number of customers in the single server queue. We leave it to the reader to verify that, if $\rho = 1$, the limiting distribution is given by

$$p_i = \frac{1}{K + 1}, \qquad 0 \le i \le K. \tag{6.58}$$

□

Example 6.28 (Leaky Bucket). Consider the leaky bucket of Example 6.13. Suppose the token buffer size is $M = 10$ and the data buffer size is $D = 14$. The tokens are generated at the rate $\mu = 1$ per millisecond, and data packets arrive at the rate $\lambda = 1$ per millisecond. What fraction of the time is the data buffer empty? What fraction of the time is it full?

Let $Y(t)$ be the number of tokens in the token buffer at time t, and let $Z(t)$ be the number of data packets in the data buffer at time t. In Example 6.13 we saw that the process $\{X(t), t \ge 0\}$ defined by

$$X(t) = M - Y(t) + Z(t)$$

behaves like a single server queue with capacity $K = M + D = 24$, arrival rate $\lambda = 1$, and service rate $\mu = 1$. Hence $\rho = \lambda/\mu = 1$. Thus, from (6.58), we get

$$p_i = \frac{1}{25}, \qquad 0 \le i \le 24.$$

Now, from the analysis in Example 6.13, the data buffer is empty whenever the $\{X(t), t \ge 0\}$ process is in states 0 through $M = 10$. Hence the long-run fraction of the time the data buffer is empty is given by

$$\sum_{i=0}^{10} p_i = \frac{11}{25} = .44.$$

We can interpret this to mean that 44% of the packets enter the network with no delay at the leaky bucket. Similarly, the data buffer is full whenever the $\{X(t), t \ge 0\}$ process is in state $M + D = 24$. Hence the long-run fraction of the time the data buffer is full is given by

$$p_{24} = \frac{1}{25} = .04.$$

This means that 4% of the packets will be dropped due to buffer overflow. ☐

Example 6.29 (Manufacturing). Consider the manufacturing setup of Example 6.23. Compute the long-run fraction of the time the machine is off.

Let $X(t)$ and $Y(t)$ be as in Example 6.23. We see that $\{(X(t), Y(t)), t \geq 0\}$ is an irreducible CTMC with state space $\{1, 2, \ldots, 6\}$ and rate matrix as given in (6.47). Hence it has a unique limiting distribution, which is also its occupancy distribution. The machine is off in state $5 = (4, 0)$ and $6 = (3, 0)$. Thus the long-run fraction of the time that the machine is off is given by $p_5 + p_6$. The balance equations and the normalizing equation can be solved numerically to obtain

$$p_1 = .1585, \quad p_2 = .1902, \quad p_3 = .2282,$$
$$p_4 = .1245, \quad p_5 = .1493, \quad p_6 = .1493.$$

Hence the required answer is $.1493 + .1493 = .2986$. Thus the machine is turned off about 30% of the time! ☐

6.10. Cost Models

Following the development in the previous chapter, we now study cost models associated with CTMCs. Recall Example 6.2(a), where we were interested in the expected total cost of operating a machine for T units of time, and also in the long-run cost per unit time of operating the machine. In this section we shall develop methods of computing such quantities.

We assume the following cost model: Let $\{X(t), t \geq 0\}$ be a CTMC with state space $\{1, 2, \ldots, N\}$ and rate matrix R. Whenever the CTMC is in state i it incurs cost continuously at rate $c(i)$, $1 \leq i \leq N$.

6.10.1. Expected Total Cost

In this subsection we study the ETC: the *expected total cost* up to a finite time T, called the horizon. Note that the cost rate at time t is $c(X(t))$. Hence the total cost up to time T is given by

$$\int_0^T c(X(t)) \, dt.$$

The expected total cost up to T, starting from state i, is given by

$$g(i, T) = \mathsf{E}\left(\int_0^T c(X(t))\,dt \,|\, X(0) = i \right), \qquad 1 \le i \le N. \tag{6.59}$$

The next theorem gives a method of computing these costs. First, it is convenient to introduce the following vector notation:

$$c = \begin{bmatrix} c(1) \\ c(2) \\ \vdots \\ c(N-1) \\ c(N) \end{bmatrix},$$

and

$$g(T) = \begin{bmatrix} g(1, T) \\ g(2, T) \\ \vdots \\ g(N-1, T) \\ g(N, T) \end{bmatrix}.$$

Theorem 6.10 (ETC). *Let* $M(T) = [m_{i,j}(T)]$ *be the occupancy matrix as given in Theorem 6.6. Then*

$$g(T) = M(T)c. \tag{6.60}$$

Proof. We have

$$g(i, T) = \mathsf{E}\left(\int_0^T c(X(t))\,dt \,|\, X(0) = i \right)$$

$$= \int_0^T \mathsf{E}(c(X(t)) | X(0) = i)\,dt$$

$$= \int_0^T \sum_{j=1}^N c(j)\mathsf{P}(X(t) = j | X(0) = i)\,dt$$

$$= \int_0^T \sum_{j=1}^N c(j)p_{i,j}(t)\,dt$$

$$= \sum_{j=1}^N c(j) \int_0^T p_{i,j}(t)\,dt$$

$$= \sum_{j=1}^N c(j)m_{i,j}(T),$$

where the last equation follows from Theorem 6.5. In matrix form this yields (6.60).   □

We illustrate with several examples.

Example 6.30 (Two-State Machine). Consider the two-state machine as described in Example 6.5. Suppose the machine produces revenue at rate A per day when it is up, and the repair costs B per day of downtime. What is the expected net income from the machine up to time T (days) if the machine started in up state?

Let $\{X(t), t \geq 0\}$ be a CTMC on $\{0, 1\}$ as described in Example 6.5. (Assume that the parameters λ and μ have units of per day.) Since we are interested in revenues, we set $c(0)$, the revenue rate in state 0, to be $-B$ per day of downtime, and $c(1) = A$ per day of up time. From Example 6.20 we get

$$p_{1,1}(t) = \frac{\lambda}{\lambda + \mu} + \frac{\mu}{\lambda + \mu} e^{-(\lambda + \mu)t},$$

and

$$p_{1,0}(t) = \frac{\mu}{\lambda + \mu} - \frac{\mu}{\lambda + \mu} e^{-(\lambda + \mu)t}.$$

Using Theorem 6.5 we get

$$m_{1,1}(T) = \int_0^T p_{1,1}(t) \, dt$$

$$= \int_0^T \left(\frac{\lambda}{\lambda + \mu} + \frac{\mu}{\lambda + \mu} e^{-(\lambda + \mu)t} \right) dt$$

$$= \frac{\lambda T}{\lambda + \mu} + \frac{\mu}{(\lambda + \mu)^2} (1 - e^{-(\lambda + \mu)T}).$$

Similarly,

$$m_{1,0}(T) = \frac{\mu T}{\lambda + \mu} - \frac{\mu}{(\lambda + \mu)^2} (1 - e^{-(\lambda + \mu)T}).$$

The net income up to T can then be obtained by Theorem 6.10 as

$$g(1, T) = m_{1,0}(T)c(0) + m_{1,1}(T)c(1)$$

$$= \frac{(\lambda A - \mu B)T}{\lambda + \mu} + \frac{\mu(B + A)}{(\lambda + \mu)^2} (1 - e^{-(\lambda + \mu)T}).$$

As a numerical example consider the data from Example 6.21. Assume $A = \$2400$ per day and $B = \$480$ per day. The net revenue in the month of January ($T = 31$) is given by

$$g(1, 31) = -m_{1,0}(31)B + m_{1,1}(31)A.$$

In Example 6.21 we have computed $m_{1,0}(31) = 2.7355$ days, and $m_{1,1}(31) = 28.2645$ days. Substituting in the above equation we get

$$g(1, 31) = -2.7355 \cdot 480 + 28.2645 \cdot 2400 = \$66521.60. \qquad \square$$

Example 6.31 (General Machine Shop). Consider the general machine shop of Example 6.18. Each machine produces a revenue of $50 per hour of up time, and the downtime costs $15 per hour per machine. The repair time costs an additional $10 per hour per machine. Compute the net expected revenue during the first day, assuming all the machines are up in the beginning.

Let $X(t)$ be the number of up machines at time t. Then $\{X(t), t \geq 0\}$ is a CTMC on $\{0, 1, 2, 3, 4\}$ with the rate matrix given in (6.35). The revenue rates in various states are as follows:

$$c(0) = 0 * 50 - 4 * 15 - 2 * 10 = -80,$$
$$c(1) = 1 * 50 - 3 * 15 - 2 * 10 = -15,$$
$$c(2) = 2 * 50 - 2 * 15 - 2 * 10 = 50,$$
$$c(3) = 3 * 50 - 1 * 15 - 1 * 10 = 125,$$
$$c(4) = 4 * 50 - 0 * 15 - 0 * 10 = 200.$$

The net expected revenue vector is given by

$$g(24) = M(24) * c$$

$$= \begin{bmatrix} 1.0143 & 1.0318 & 1.1496 & 3.6934 & 17.1108 \\ 0.0143 & 1.0319 & 1.1538 & 3.7931 & 18.0069 \\ 0.0004 & 0.0320 & 1.1580 & 3.8940 & 18.9155 \\ 0.0001 & 0.0044 & 0.1623 & 3.9964 & 19.8369 \\ 0.0000 & 0.0023 & 0.0876 & 2.2041 & 21.7060 \end{bmatrix} \begin{bmatrix} -80 \\ -15 \\ 50 \\ 125 \\ 200 \end{bmatrix}$$

$$= \begin{bmatrix} 3844.69 \\ 4116.57 \\ 4327.23 \\ 4474.96 \\ 4621.05 \end{bmatrix}.$$

The required answer is given by $g(4, 24) = \$4621.05$.   □

6.10.2.   Long-Run Cost Rates

In Example 6.30, suppose the downtime cost B is given. We want to know how much the revenue rate during uptime should be so that it is economically profitable to operate the machine. If we go by total cost, the answer will depend on the planning horizon T, and also on the initial state of the machine. An alternative is to compute the long-run net revenue per unit time for this machine and insist that this be positive for profitability. This answer will not depend on T, and as we shall see, not even on the initial state of the machine. Hence, computing such long-run cost rates or revenue rates is very useful. We shall show how to compute these quantities in this subsection.

First, define the long-run cost rate as

$$g(i) = \lim_{T \to \infty} \frac{g(i, T)}{T}.$$ (6.61)

The next theorem shows an easy way to compute $g(i)$.

Theorem 6.11 (Long-Run Cost Rate). *Suppose* $\{X(t), t \geq 0\}$ *is an irreducible CTMC with limiting distribution* $p = [p_1, p_2, \ldots, p_N]$. *Then*

$$g = g(i) = \sum_{j=1}^{N} p_j c(j), \qquad 1 \leq i \leq N.$$

Proof. We have

$$g(i) = \lim_{T \to \infty} \frac{g(i, T)}{T}$$

$$= \lim_{T \to \infty} \frac{\sum_{j=1}^{N} m_{i,j}(T) c(j)}{T}$$

(from Theorem 6.10)

$$= \sum_{j=1}^{N} c(j) \lim_{T \to \infty} \frac{m_{i,j}(T)}{T}$$

$$= \sum_{j=1}^{N} c(j) p_j$$

where the last equality follows from Theorem 6.9. This proves the theorem.   □

Note that the theorem is intuitively obvious: the CTMC incurs a cost at rate $c(j)$ per unit time it spends in state j. It spends p_j fraction of the time in state j, regardless of the starting state. Hence it must incur costs at rate $\sum_{j=1}^{N} p_j c(j)$ in the long run. We illustrate this result with several examples.

Example 6.32 (Two-State Machine). Consider the two-state machine of Example 6.30. Suppose the downtime cost of the machine is B per unit time. What is the minimum revenue rate A during the uptime needed to break even in the long run?

From Example 6.24 the machine is up for $p_1 = \lambda/(\lambda + \mu)$ fraction of the time in the long run, and down for $p_0 = \mu/(\lambda + \mu)$ fraction of the time in the long run. Using $c(0) = -B$ and $c(1) = A$ in Theorem 6.11 we obtain the long-run net revenue per unit time as

$$g = \frac{\lambda A - \mu B}{\lambda + \mu}.$$

For breakeven we must have

$$g \geq 0,$$

i.e.,

$$\frac{A}{\mu} \geq \frac{B}{\lambda}.$$

Note that this condition makes sense: A/μ is the expected revenue during one uptime, and B/λ is the expected cost of one downtime. If the former is greater than the latter, the machine is profitable to operate, otherwise it is not! □

Example 6.33 (Telephone Switch). Consider the CTMC model of the telephone switch described in Example 6.11. The switch has a capacity of six calls. The call arrival rate is four calls per minute and the arrival process is Poisson. The average duration of a call is 2 minutes, and the call durations are iid exponential random variables. Compute the expected revenue per minute if each caller is charged 10 cents per minute of the call duration. What is the rate at which revenue is lost due to the switch being full?

Let $X(t)$ be the number of calls carried by the switch at time t. Then $\{X(t), t \geq 0\}$ is a birth and death process with birth rates

$$\lambda_i = 4, \quad 0 \leq i \leq 5,$$

and death rates

$$\mu_i = i/2, \quad 1 \leq i \leq 6.$$

The revenue rate in state i is $c(i) = 10i$ cents per minute. Hence the long-run rate of revenue is

$$g = \sum_{j=0}^{6} p_j c(j) = 10 \sum_{j=0}^{6} j p_j.$$

Note that the last sum is just the expected number of calls in the switch in steady state. Using the results of Example 6.25 we get

$$\rho_i = \frac{8^i}{i!}, \quad 0 \leq i \leq 6,$$

and

$$p_i = \frac{\rho_i}{\sum_{j=0}^{6} \rho_j}, \quad 0 \leq i \leq 6.$$

Hence we get

$$g = 10 \frac{\sum_{i=0}^{6} i \rho_i}{\sum_{j=0}^{6} \rho_j} = 48.8198.$$

Hence the switch earns revenue at a rate of 48.82 cents per minute.

Note that calls arrive at the rate of four per minute, and the average revenue from each call is 20 cents (duration is 2 minutes). Hence if the switch had ample capacity so that no call was rejected, the revenue will be $4 * 20 = 80$ cents per minute. But the actual revenue is only 48.82 cents per minute. Hence the revenue is lost at the rate $80 - 48.82 = 31.18$ cents per minute.   □

6.11. First Passage Times

In this section we study the *first passage times* in CTMCs, following the development in the DTMCs in Section 5.8. Let $\{X(t), t \geq 0\}$ be a CTMC with state space $\{1, \ldots, N\}$ and rate matrix R. The first passage time into state N is defined to be

$$T = \min\{t \geq 0 : X(t) = N\}. \tag{6.62}$$

As in Section 5.8, we study the expected value $\mathsf{E}(T)$.

Let

$$m_i = \mathsf{E}(T | X_0 = i). \tag{6.63}$$

Note that $m_N = 0$. The next theorem gives a method of computing m_i, $1 \leq i \leq N - 1$.

Theorem 6.12 (First Passage Times). $\{m_i, 1 \leq i \leq N - 1\}$ *satisfy the following:*

$$r_i m_i = 1 + \sum_{j=1}^{N-1} r_{i,j} m_j, \qquad 1 \leq i \leq N - 1. \tag{6.64}$$

Proof. We shall prove the theorem by conditioning on the time of the first transition and the state of the CTMC after the first transition. Let $X(0) = i$, $1 \leq i \leq N - 1$, and let $X(Y) = j$, where Y is the sojourn time in the initial state. Then, $T = Y$ if $X(Y) = N$, and $T = Y +$ the first passage time into state N from state j, if $j \neq N$. Using this observation we get

$$\mathsf{E}(T | X(0) = i, X(Y) = j) = \mathsf{E}(Y | X(0) = i) + m_j, \quad 1 \leq i \leq N - 1, \quad 1 \leq j \leq N - 1,$$

and

$$\mathsf{E}(T | X(0) = i, X(Y) = N) = \mathsf{E}(Y | X(0) = i), \qquad 1 \leq i \leq N - 1.$$

Now, Y is an $\mathrm{Exp}(r_i)$ random variable if $X(0) = i$. Hence

$$\mathsf{E}(Y | X(0) = i) = \frac{1}{r_i}.$$

From the properties of CTMCs, we also have

$$\mathsf{P}(X(Y) = j | X(0) = i) = \frac{r_{i,j}}{r_i},$$

where $r_{i,i} = 0$ and $r_i = \sum_{j=1}^{N} r_{i,j}$. Using these facts we get, for $1 \le i \le N-1$,

$$m_i = \mathsf{E}(T|X(0) = i)$$

$$= \sum_{j=1}^{N} \mathsf{E}(T|X(0) = i, X(Y) = j)\mathsf{P}(X(Y) = j|X(0) = i)$$

$$= \sum_{j=1}^{N-1} \left(\frac{1}{r_i} + m_j\right)\frac{r_{i,j}}{r_i} + \frac{1}{r_i}\frac{r_{i,N}}{r_i}$$

$$= \frac{1}{r_i} + \sum_{j=1}^{N-1} \frac{r_{i,j}}{r_i}m_j.$$

Multiplying the last equation by r_i yields (6.64). □

We illustrate the above theorem with several examples.

Example 6.34. Let $\{X(t), t \ge 0\}$ be the CTMC in Example 6.16. Compute the expected time the CTMC takes to reach state 4 starting from state 1.

Equations (6.64) are

$$5m_1 = 1 + 2m_2 + 3m_3,$$
$$6m_2 = 1 + 4m_1 + 2m_3,$$
$$4m_3 = 1 + 2m_2.$$

Solving simultaneously we get

$$m_1 = 1.2727, \qquad m_2 = 1.3182, \qquad m_3 = .9091.$$

Thus it takes on the average 1.2727 units of time to go from state 1 to state 4. □

Example 6.35 (Single Server Queue). Consider the single server queue of Example 6.22. Compute the expected time until the queue becomes empty, given that it has one customer at time 0.

Let $X(t)$ be the number of customers in the queue at time t. With the parameters given in Example 6.22, $\{X(t), t \ge 0\}$ is a CTMC with state space $\{0, 1, 2, 3, 4, 5\}$ and rate matrix given by (6.46). Let T be the first passage time to state 0, and let $m_i = \mathsf{E}(T|X(0) = i)$. Equations (6.64) for this case are

$$m_0 = 0,$$
$$25m_i = 1 + 10m_{i+1} + 15m_{i-1}, \qquad 1 \le i \le 4,$$
$$15m_5 = 1 + 15m_4.$$

Solving numerically, we get

$$m_1 = .1737, \qquad m_2 = .3342, \qquad m_3 = .4749, \qquad m_4 = .5860, \qquad m_5 = .6527.$$

Thus the expected time until the queue becomes empty (starting with one person) is .1737 hours, or 10.42 minutes. □

As in Section 5.8 we now study the expected time to reach a set of states, A. Analogous to (5.64), define:

$$T = \min\{n \geq 0 : X(t) \in A\}. \tag{6.65}$$

Theorem 6.12 can be easily extended to the case of the first passage time defined above. Let $m_i(A)$ be the expected time to reach the set A starting from state $i \notin A$. Note that $m_i(A) = 0$ for $i \in A$. Following the same argument as in the proof of Theorem 6.12 we can show that

$$r_i m_i(A) = 1 + \sum_{j \notin A} r_{i,j} m_j(A), \qquad i \notin A. \tag{6.66}$$

We illustrate with an example below.

Example 6.36 (Airplane Reliability). Consider the four-engine airplane of Example 6.19. Suppose in a testing experiment the airplane takes off with four properly functioning engines, and keeps on flying until it crashes. Compute the expected time of the crash.

Let $\{X(t), t \geq 0\}$ be the CTMC of Example 6.8 with $\lambda = .005$ per hour. The airplane crashes as soon as the CTMC visits the set $\{1, 2, 3, 4, 7\}$. Using this as A in (6.66), we get $m_i = 0$ for $i \in A$ and

$$.02m_9 = 1 + .01m_8 + .01m_6,$$
$$.015m_8 = 1 + .01m_5,$$
$$.015m_6 = 1 + .01m_5,$$
$$.01m_5 = 1.$$

Solving backward recursively, we get

$$m_5 = 100, \qquad m_6 = \frac{400}{3}, \qquad m_8 = \frac{400}{3}, \qquad m_9 = \frac{550}{3}.$$

Thus the expected time to failure is $m_9 = 550/3 = 183$ hours and 20 minutes.   □

Appendix A: Proof Of Theorem 6.4

In this appendix we give the proof of Theorem 6.4, which is restated below.

Theorem 6.13 (Theorem 6.4). *The transition probability matrix $P(t) = [p_{i,j}(t)]$ is given by*

$$P(t) = \sum_{k=0}^{\infty} e^{-rt} \frac{(rt)^k}{k!} \hat{P}^k. \tag{6.67}$$

The proof is developed in stages below. We start with the important properties of the $\hat{P}$ matrix.

Theorem 6.14 (Properties of $\hat{P}$). *Let R be a rate matrix of a CTMC. The $\hat{P}$ matrix as defined in (6.30) is a stochastic matrix, i.e., it is a one-step transition matrix of a DTMC.*

Proof. (1) If $i \neq j$, then $\hat{p}_{i,j} = r_{i,j}/r \geq 0$. For $i = j$ we have

$$\hat{p}_{i,i} = 1 - \frac{r_i}{r} \geq 0$$

since r is the largest r_i.

(2) For $1 \leq i \leq N$, we have

$$\sum_{j=1}^{N} \hat{p}_{i,j} = \hat{p}_{i,i} + \sum_{j \neq i} \hat{p}_{i,j}$$

$$= 1 - \frac{r_i}{r} + \sum_{j \neq i} \frac{r_{i,j}}{r}$$

$$= 1 - \frac{r_i}{r} + \sum_{j=1}^{N} \frac{r_{i,j}}{r}$$

$$= 1 - \frac{r_i}{r} + \frac{r_i}{r} = 1.$$

(We have used the fact that $r_{i,i} = 0$.) This proves the theorem.   □

Now suppose we observe the CTMC $\{X(t), t \geq 0\}$ at times $0 = S_0 < S_1 < S_2 < S_3 < \cdots$, etc., and let $\hat{X}_n = X(S_n)$, i.e., $\hat{X}_n$ is the nth observation. The sequence $\{S_n, n \geq 0\}$ is defined recursively as follows. We start with $S_0 = 0$ and $\hat{X}_0 = X(0)$ as the initial state. Let $n \geq 0$ be a fixed integer. Suppose $\hat{X}_n = i$. Let T_{n+1} be the remaining sojourn time of the CTMC in state i. Let Y_{n+1} be an $\text{Exp}(r - r_i)$ random variable that is independent of T_{n+1}, and the history of $\{X(t), t \geq 0\}$ up to time S_n. The next observation time S_{n+1} is defined as

$$S_{n+1} = S_n + \min\{T_{n+1}, Y_{n+1}\},$$

and the next observation is given by

$$\hat{X}_{n+1} = X(S_{n+1}).$$

The next theorem describes the structure of the observation process $\{\hat{X}_n, n \geq 0\}$.

Theorem 6.15 (The Embedded DTMC). *The stochastic process $\{\hat{X}_n, n \geq 0\}$ is a DTMC with one-step transition probability matrix $\hat{P}$.*

Proof. The Markov property of $\{\hat{X}_n, n \geq 0\}$ follows from the Markov property of $\{X(t), t \geq 0\}$ and the construction of the $\{Y_n, n \geq 1\}$ variables. Next we compute the transition probabilities. Suppose $\hat{X}_n = i$. If $T_{n+1} > Y_{n+1}$ the next observation occurs at time $S_n + Y_{n+1}$. However, the state of the CTMC at this time is the same as at time S_n. Hence, $\hat{X}_{n+1} = i$.

(Conversely, $\hat{X}_{n+1} = i$ implies that $T_{n+1} > Y_{n+1}$.) This yields

$$P(\hat{X}_{n+1} = i|\hat{X}_n = i) = P(T_{n+1} > Y_{n+1}|X(S_n) = i)$$
$$= \frac{r - r_i}{r - r_i + r_i} = 1 - \frac{r_i}{r},$$

where the second equality follows from (6.18). Next, if $T_{n+1} \leq Y_{n+1}$ the next observation occurs at time $S_n + T_{n+1}$. Now, the state of the CTMC at this time is j with probability $r_{i.j}/r$. This yields, for $i \neq j$,

$$P(\hat{X}_{n+1} = j|\hat{X}_n = i) = P(X(S_{n+1}) = j|X(S_n) = i)$$
$$= P(X(S_n + T_{n+1}) = j, T_{n+1} < Y_{n+1}|X(S_n) = i)$$
$$= \frac{r_{i.j}}{r_i} \cdot \frac{r_i}{r_i + r - r_i} = \frac{r_{i.j}}{r}.$$

The theorem follows from this. □

Now let $N(t)$ be the number of observations up to time t. The next theorem describes the structure of the counting process $\{N(t), t \geq 0\}$.

Theorem 6.16 (The Observation Process). $\{N(t), t \geq 0\}$ *is a Poisson process with rate* r, *and is independent of* $\{\hat{X}_n, n \geq 0\}$.

Proof. Given $\hat{X}_n = i$, we see that $T_{n+1} \sim \text{Exp}(r_i)$ and $Y_{n+1} \sim \text{Exp}(r - r_i)$ are independent random variables. Hence, using (6.16), we see that $\min\{T_{n+1}, Y_{n+1}\}$ is an exponential random variable with parameter $r_i + r - r_i = r$. But this is independent of $\hat{X}_n$! Hence the intervals between consecutive observations are iid $\text{Exp}(r)$ random variables. Hence, by the results in Section 6.5, we see that $\{N(t), t \geq 0\}$ is $PP(r)$. This also establishes the independence of $\{N(t), t \geq 0\}$ and $\{\hat{X}_n, n \geq 0\}$. □

Using the results of Theorems 6.15 and 6.16, we see that

$$X(t) = \hat{X}_{N(t)}, \qquad t \geq 0. \tag{6.68}$$

Using the above equation we now prove Theorem 6.4.

Proof of Theorem 6.4. We have

$$p_{i.j}(t) = P(X(t) = j|X(0) = i)$$
$$= \sum_{k=0}^{\infty} P(X(t) = j|X(0) = i, N(t) = k)P(N(t) = k|X(0) = i)$$
$$= \sum_{k=0}^{\infty} P(\hat{X}_{N(t)} = j|\hat{X}_0 = i, N(t) = k)P(N(t) = k|\hat{X}_0 = i)$$

(using (6.68))

$$= \sum_{k=0}^{\infty} \mathsf{P}(\hat{X}_k = j | \hat{X}_0 = i) e^{-rt} \frac{(rt)^k}{k!}$$

(due to the independence of $\{\hat{X}_n, \ n \geq 0\}$ and $\{N(t), \ t \geq 0\}$, and Theorem 6.3)

$$= \sum_{k=0}^{\infty} [\hat{P}^k]_{i,j} e^{-rt} \frac{(rt)^k}{k!}.$$

Here we have used Theorem 5.2 to write

$$\mathsf{P}(\hat{X}_k = j | \hat{X}_0 = i) = [\hat{P}^k]_{i,j}.$$

Putting the above equation in matrix form we get (6.33).   □

Appendix B: Uniformization Algorithm to Compute $P(t)$

Here we develop an algorithm to compute $P(t)$ based on the results of Theorem 6.4. We first develop bounds on the error resulting from approximating the infinite series in (6.33) by a finite series of the first M terms.

Theorem 6.17 (Error Bounds for $P(t)$). *For a fixed $t \geq 0$, let*

$$P^M(t) = [p_{i,j}^M(t)] = \sum_{k=0}^{M} e^{-rt} \frac{(rt)^k}{k!} \hat{P}^k. \tag{6.69}$$

Then

$$|p_{i,j}(t) - p_{i,j}^M(t)| \leq \sum_{k=M+1}^{\infty} e^{-rt} \frac{(rt)^k}{k!}, \tag{6.70}$$

for all $1 \leq i, j \leq N$.

Proof. We have

$$|p_{i,j}(t) - p_{i,j}^M(t)| = \left| \sum_{k=0}^{\infty} e^{-rt} \frac{(rt)^k}{k!} [\hat{P}^k]_{i,j} - \sum_{k=0}^{M} e^{-rt} \frac{(rt)^k}{k!} [\hat{P}^k]_{i,j} \right|$$

$$= \left| \sum_{k=M+1}^{\infty} e^{-rt} \frac{(rt)^k}{k!} [\hat{P}^k]_{i,j} \right|$$

$$\leq \sum_{k=M+1}^{\infty} e^{-rt} \frac{(rt)^k}{k!},$$

where the last inequality follows because $[\hat{P}^k]_{i,j} \leq 1$ for all i, j, k.   □

The error bound in (6.70) can be used as follows. Suppose we want to compute $P(t)$ within a tolerance of ϵ. Choose an M such that

$$\sum_{k=M+1}^{\infty} e^{-rt} \frac{(rt)^k}{k!} \leq \epsilon.$$

Such an M can always be chosen. However, it depends upon t, r, and ϵ. For this M, Theorem 6.17 guarantees that the approximation $P^M(t)$ is within ϵ of $P(t)$. This yields the following algorithm:

The Uniformization Algorithm for $P(t)$. This algorithm computes the transition probability matrix $P(t)$ (within a given numerical accuracy ϵ) for a CTMC with a rate matrix R:

(1) Given $R, t, 0 < \epsilon < 1$.

(2) Compute r by using equality in (6.29).

(3) Compute $\hat{P}$ by using (6.30).

(4) $A = \hat{P}$; $B = e^{-rt} I$; $c = e^{-rt}$; $sum = c$; $k = 1$.

(5) While $sum < 1 - \epsilon$ do:
$$c = c * (rt)/k$$
$$B = B + cA$$
$$A = A\hat{P}$$
$$sum = sum + c$$
$$k = k + 1$$
end.

(6) B is within ϵ of $P(t)$.

Appendix C: Uniformization Algorithm to Compute $M(T)$

Here we develop an algorithm to compute $M(T)$ based on the representation in Theorem 6.6. First, the following theorem gives the accuracy of truncating the infinite series in (6.45) at $k = K$ to approximate $M(T)$:

Theorem 6.18 (Error Bounds for $M(T)$).

$$\left| m_{i,j}(T) - \sum_{k=0}^{K} \frac{1}{r} P(Y > k)[\hat{P}^k]_{i,j} \right| \leq \frac{1}{r} \sum_{k=K+1}^{\infty} p(Y > k) \tag{6.71}$$

for all $1 \leq i, j \leq N$.

Proof. Follows along the same lines as that of Theorem 6.17.   □

Now, from (2.35) we have

$$rT = \mathsf{E}(Y) = \sum_{k=0}^{\infty} \mathsf{P}(Y > k).$$

Hence

$$\frac{1}{r} \sum_{k=K+1}^{\infty} \mathsf{P}(Y > k) = \frac{1}{r} \left(\sum_{k=0}^{\infty} \mathsf{P}(Y > k) - \sum_{k=0}^{K} \mathsf{P}(Y > k) \right)$$

$$= \frac{1}{r} \left(rT - \sum_{k=0}^{K} \mathsf{P}(Y > k) \right)$$

$$= T - \frac{1}{r} \left(\sum_{k=0}^{K} \mathsf{P}(Y > k) \right).$$

Thus, for a given error tolerance $\epsilon > 0$, if we choose K such that

$$\frac{1}{r} \sum_{k=0}^{K} \mathsf{P}(Y > k) > T - \epsilon,$$

then the error in truncating the infinite series at $k = K$ is bounded above by ϵ. This observation is used in the following algorithm:

The Uniformization Algorithm for $M(T)$. This algorithm computes the matrix $M(T)$ (within a given numerical accuracy ϵ) for a CTMC with a rate matrix R:

(1) Given $R, T, 0 < \epsilon < 1$.

(2) Compute r by using equality in (6.29).

(3) Compute $\hat{P}$ by using (6.30).

(4) $A = \hat{P}; k = 0$.

(5) $yek(= \mathsf{P}(Y = k)) = e^{-rT}; \quad ygk(= \mathsf{P}(Y > k)) = 1 - yek; \quad sum(= \sum_{r=0}^{k} \mathsf{P}(Y > r)) = ygk$.

(6) $B = ygk * I$.

(7) While $sum/r < T - \epsilon$ do:
$$k = k + 1;$$
$$yek = yek * (rT)/k; ygk = ygk - yek;$$
$$B = B + ygk * A;$$
$$A = A\hat{P};$$
$$sum = sum + ygk;$$
end.

(8) B/r is within ϵ of $M(T)$, and uses the first $k + 1$ terms of the infinite series in (6.45).

6.12. Problems

CONCEPTUAL PROBLEMS

6.1. Suppose $T_i, i = 1, 2$, the lifetime of machine i, are independent $\text{Exp}(\lambda_i)$ random variables. Suppose machine 1 fails before machine 2. Show that the remaining lifetime of machine 2 from the time of failure of machine 1 is an $\text{Exp}(\lambda_2)$ random variable, i.e., show that

$$P(T_2 > T_1 + t | T_2 > T_1) = e^{-\lambda_2 t}.$$

This is sometimes called the *strong memoryless property*.

6.2. Let T_1 and T_2 be two iid $\text{Exp}(\lambda)$ random variables. Using Conceptual Problem 6.1 or otherwise, show that

$$E(\max(T_1, T_2)) = \frac{1}{2\lambda} + \frac{1}{\lambda}.$$

6.3. Generalize Conceptual Problem 6.2 as follows: Let T_i, $1 \le i \le k$, be k iid $\text{Exp}(\lambda)$ random variables. Show that

$$E(\max(T_1, T_2, \ldots, T_k)) = \frac{1}{\lambda} \sum_{i=1}^{k} \frac{1}{i}.$$

6.4. A weight of L tons is held up by two cables that share the load equally. When one of the two cables breaks, the other cable carries the entire load. When the last cable breaks, we have a failure. The failure rate of a cable subject to M tons of load is λM per year. The lifetimes of the two cables are independent of each other. Let $X(t)$ be the number of cables that are still unbroken at time t. Show that $\{X(t), t \ge 0\}$ is a CTMC and find its rate matrix.

6.5. Generalize Conceptual Problem 6.4 to the case of K independent cables sharing the load equally.

6.6. Customers arrive at a bank according to a Poisson process with rate λ per hour. The bank lobby has enough space for 10 customers. When the lobby is full, an incoming customer goes to another branch, and is lost. The bank manager assigns one teller to customer service as long as the number of customers in the lobby is three or less, she assigns two tellers if the number is more that three but less than eight, else she assigns three tellers. The service times of the customers are iid $\text{Exp}(\mu)$ random variables. Let $X(t)$ be the number of customers in the bank at time t. Model $\{X(t), t \ge 0\}$ as a birth and death process, and derive the birth and death rates.

6.7. A computer has five processing units (PUs). The lifetimes of the PUs are iid $\text{Exp}(\mu)$ random variables. When a PU fails, the computer tries to isolate it automatically and reconfigure the system using the remaining PUs. However this process succeeds with probability c, called the *coverage factor*. If the reconfiguring succeeds the system continues with one less PU. If the process fails, the entire system crashes. Assume that the reconfiguring process is instantaneous, and that once the system crashes it stays down forever. Let $X(t)$ be 0 if the system is down at time t, otherwise it equals the number of working PUs at time t. Model $\{X(t), t \geq 0\}$ as a CTMC and show its rate matrix.

6.8. A service station has three servers indexed 1, 2, and 3. When a customer arrives, he is assigned to an idle server with the lowest index. If all servers are busy the customer goes away. The service times at server i are iid $\text{Exp}(\mu_i)$ random variables. (Different μ's represent different speeds of the servers.) The customers arrive according to a PP(λ). Model this system as a CTMC.

6.9. A single server service station serves two types of customers. Customers of type i, $i = 1, 2$, arrive according to a PP(λ_i), independently of each other. The station has space to handle at most K customers. The service times are iid $\text{Exp}(\mu)$ for both types of customer. The admission policy is as follows: If, at the time of an arrival, the total number of customers in the system is M or less (here $M < K$ is a fixed integer), the arriving customer is allowed to join the queue; else he is allowed to join only if he is of type 1. This creates a preferential treatment for type 1 customers. Let $X(t)$ be the number of customers (of both types) in the system at time t. Show that $\{X(t), t \geq 0\}$ is a birth and death process. Derive its birth and death rates.

6.10. A system consisting of two components is subject to a series of shocks that arrive according to a PP(λ). A shock can cause the failure of component 1 alone with probability p, component 2 alone with probability q, both the components with probability r, or have no effect with probability $1 - p - q - r$. No repairs are possible. The system fails when both the components fail. Model the state of the system as a CTMC.

6.11. A machine shop consists of two borers and two lathes. The lifetimes of the borers are $\text{Exp}(\mu_b)$ random variables, and those of the lathes are $\text{Exp}(\mu_l)$ random variables. The machine shop has two repair persons: Al and Bob. Al can repair both lathes and borers, while Bob can only repair lathes. Repair times for the borers are $\text{Exp}(\lambda_b)$, and for the lathes $\text{Exp}(\lambda_l)$, regardless of who repairs the machines. Borers have priority in repairs. Repairs can be prempted. Making appropriate independence assumptions, model this machine shop as a CTMC.

6.12. An automobile part needs three machining operations performed in a given sequence. These operations are performed by three machines. The part is fed to the first machine, where the machining operation takes an $\text{Exp}(\mu_1)$ amount of time. After the operation is complete the part moves to machine 2, where the machining operation takes $\text{Exp}(\mu_2)$ amount of time; after which it moves to machine 3 where the operation takes $\text{Exp}(\mu_3)$ amount of time. There is no storage room between the two machines, and hence if machine 2 is working, the part from machine 1 cannot be removed even if the operation at machine 1 is complete. We say that machine 1 is blocked in such a case. There is an ample supply of unprocessed parts available so that machine 1 can always process a new part when a completed part moves to machine 2. Model this system as a CTMC. (*Hint*: Note that machine 1 may be working or blocked, machine 2 may be working, blocked, or idle, and machine 3 may be working or idle.)

6.13. A single server queuing system has a finite capacity K. The customers arrive according to a PP(λ) and demand iid Exp(μ) service times. The server is subject to failures and repairs as follows: the server serves for an Exp(θ) amount of time and then fails. Once it fails all the customers in the system go away instantaneously. The server takes an Exp(α) time to get repaired. While the server is down, no new customers are allowed to enter the system. If the system is full, the arriving customers depart immediately without getting served. Model this system as a CTMC.

6.14. Consider the five processor system as described in Conceptual Problem 6.7. Suppose that the system maintenance starts as soon as the system crashes. The amount of time needed to repair the system is an Exp(λ) random variable, at the end of which all five processors are functioning. Model this system as a CTMC.

6.15. Consider the two-component system of Conceptual Problem 6.10. Furthermore, suppose that a single repair person starts the system repair as soon as the system fails. The repair time of each component is an Exp(α) random variable. The system is turned on when both components are repaired. Assume that shocks have no effect on the components unless the system is on. Model this as a CTMC.

6.16. Do Conceptual Problem 6.15, but now assume that repair can start on a component as soon as it fails, even though the system is still up; and that the component is put back in service as soon as it is repaired. The shocks have no effect on a component under repair. The repair person gives priority to component 1 over component 2 for repairs.

6.17. Consider the model of the telephone switch in Example 6.11. Show that the limiting distribution of the number of calls in the switch is given by

$$p_j = \frac{\rho^j / j!}{\sum_{i=0}^{K} \rho^i / i!}, \qquad 0 \le j \le K,$$

where $\rho = \lambda / \mu$. (*Hint*: Use the results of Example 6.25.)

COMPUTATIONAL PROBLEMS

6.1. A communications satellite is controlled by two computers. It needs at least one functioning computer to operate properly. Suppose the lifetimes of the computers are iid exponential random variables with mean 5 years. Once a computer fails, it cannot be repaired. What is the cdf of the lifetime of the satellite? What is the expected lifetime of the satellite?

6.2. A spacecraft is sent to observe Jupiter. It is supposed to take pictures of Jupiter's moons, and send them back to Earth. There are three critical systems involved: the camera, the batteries, and the transmission antenna. These three systems fail independently of each other. The mean lifetime of the battery system is 6 years, that of the camera is 12 years, and that of the antenna is 10 years. Assume all lifetimes are exponential random variables. The spacecraft is to reach Jupiter after 3 years to carry out this mission. What is the probability that the mission is successful?

6.3. In Computational Problem 6.2, suppose the monitoring station on Earth conducts a test run after 2 years, and finds that no data are received, indicating that one of the three systems has failed. What is the probability that the camera has failed?

6.4. A car comes equipped with one spare tire. The lifetimes of the four tires at the beginning of a long-distance trip are iid exponential random variables with a mean of 5000 miles. The spare tire has an exponential lifetime with a mean of 1000 miles. Compute the expected number of miles that can be covered without having to go to a tire shop. (*Hint*: Compute the expected time until the first tire failure, and then until the second tire failure.)

6.5. A customer enters a bank and finds that all three tellers are occupied and he is next in line. Suppose mean transaction times are 5 minutes, and transaction times are independent exponential random variables. What is the expected amount of time that he has to wait before he enters service?

6.6. In Computational Problem 6.5, suppose the bank also has a teller machine outside with one person currently using it and no one waiting behind him. The processing time at the machine is exponentially distributed with mean 2 minutes. Should the customer in Computational Problem 6.5 wait at the machine, or at the inside tellers, if the aim is to minimize the expected waiting time until the start of service? Assume that once he decides to wait in one place, he does not change his mind.

6.7. Compute the transition probability matrix $P(t)$ at $t = 0.20$, for a CTMC on $S = \{1, 2, 3, 4, 5\}$ with the rate matrix given below:

$$R = \begin{bmatrix} 0 & 4 & 4 & 0 & 0 \\ 5 & 0 & 5 & 5 & 0 \\ 5 & 5 & 0 & 4 & 4 \\ 0 & 5 & 5 & 0 & 4 \\ 0 & 0 & 5 & 5 & 0 \end{bmatrix}.$$

6.8. Compute the transition probability matrix $P(t)$ at $t = 0.10$, for a CTMC on $S = \{1, 2, 3, 4, 5, 6\}$ with the rate matrix given below:

$$R = \begin{bmatrix} 0 & 6 & 0 & 0 & 0 & 0 \\ 0 & 0 & 6 & 0 & 0 & 0 \\ 0 & 0 & 0 & 6 & 0 & 0 \\ 0 & 0 & 0 & 0 & 6 & 0 \\ 0 & 0 & 0 & 0 & 0 & 6 \\ 6 & 0 & 0 & 0 & 0 & 0 \end{bmatrix}.$$

6.9. Compute the transition probability matrix $P(t)$ at $t = 0.10$, for a CTMC on $S = \{1, 2, 3, 4, 5, 6\}$ with the rate matrix given below:

$$
R = \begin{bmatrix}
0 & 6 & 6 & 8 & 0 & 0 \\
0 & 0 & 6 & 8 & 0 & 0 \\
0 & 0 & 0 & 6 & 0 & 0 \\
0 & 6 & 8 & 0 & 0 & 0 \\
8 & 0 & 0 & 0 & 0 & 6 \\
6 & 0 & 0 & 0 & 6 & 0
\end{bmatrix}.
$$

6.10. Consider the model of Conceptual Problems 6.4 and 6.5. Suppose the individual cables have a failure rate of .2 per year per ton of load. We need to build a system to support 18 tons of load, to be equally shared by the cables. If we use three cables, what is the probability that the system will last for more than 2 years?

6.11. In Computational Problem 6.5, how many cables are needed to ensure with probability more than .999 that the system will last for more than 2 years?

6.12. Consider Conceptual Problem 6.7. Suppose the mean lifetime of a processor is 2 years, and the coverage factor is .94. What is the probability that the five processor system functions for 5 years without fail? Assume all processors are operating at time 0.

6.13. Consider Conceptual Problem 6.8. Suppose the mean service time at the third server is 8 minutes. Server 2 is twice as fast as server 3, and server 1 is twice as fast as server 2. The customers arrive at a rate of 20 per hour. Compute the probability that all three servers are busy at time $t = 20$ minutes, assuming that the system is empty at time 0.

6.14. Consider the system of Conceptual Problem 6.10. The shocks arrive on average once a day. About half of the shocks cause no damage, one out of ten damages component 1 alone, about one out of five damages component 2 alone, and the rest damage both components. Compute the probability that the system survives for at least 4 weeks, if both components are working initially.

6.15. Consider the Call Center of Example 6.12. Suppose the arrival rate is 60 per hour and each call takes an average of 6 minutes to handle. If there are eight reservation agents, and the system can put four callers on hold at one time, what is the probability that the system is full after 1 hour, assuming it is idle at time 0?

6.16. Consider the single server model of Conceptual Problem 6.13. Suppose the server is working at time 0. Compute the probability that the server is working at time t.

6.17. Compute the occupancy matrix $M(t)$ for the CTMC in Computational Problem 6.7, using the t values given there.

6.18. Compute the occupancy matrix $M(t)$ for the CTMC in Computational Problem 6.8, using the t values given there.

6.19. Compute the occupancy matrix $M(t)$ for the CTMC in Computational Problem 6.9, using the t values given there.

6.20. Consider the model of Conceptual Problem 6.6. Suppose the arrival rate is 20 customers per hour, and the average service time is 4 minutes. Compute the expected amount of time three tellers are working during an 8-hour day, assuming there are no customers in the bank at the beginning of the day.

6.21. Consider Computational Problem 6.13. Compute the expected amount of time the three servers are busy during the first hour.

6.22. Consider the three machine production system described in Conceptual Problem 6.12. Suppose the mean processing times at the three machines are 1 minute, 1.2 minutes, and 1 minute, respectively. Compute the expected amount of time machine 1 is blocked during the first hour, assuming all machines are working at time 0.

6.23. Consider the telephone switch of Example 6.33. Compute the expected amount of time the switch is full during 1 day, assuming it is idle at the beginning of the day.

6.24. Consider the Call Center of Example 6.12. Suppose the arrival rate is 60 per hour and each call takes an average of 6 minutes to handle. If there are eight reservation agents, what is the expected time that all of them are busy during an 8-hour shift, assuming none is busy at the beginning? The system can put four callers on hold at one time.

6.25. In Computational Problem 6.22, compute the expected amount of time machine 3 is working during the first hour, assuming all machines are working at time 0.

6.26. Classify the CTMCs in Computational Problems 6.7, 6.8, and 6.9 as irreducible or reducible.

6.27. Compute the limiting distribution of the CTMC from Computational Problem 6.7.

6.28. Compute the limiting distribution of the CTMC from Computational Problem 6.8.

6.29. Compute the limiting distribution of the state of the system of Computational Problem 6.13.

6.30. Compute the limiting distribution of the state of the system of Conceptual Problem 6.6 assuming the arrival rate is 20 customers per hour, and the average service time is 6 minutes.

6.31. Compute the long-run fraction of the time that the last machine is working in the three machine production system of Computational Problem 6.22.

6.32. Consider the Call Center as described in Example 6.12 with the data given in Computational Problem 6.15. Compute:

(1)   The long-run fraction of the time that all the agents are busy.

(2)   The long-run fraction of the time that the call center has to turn away calls.

(3)   The expected number of busy agents in steady state.

6.33. Consider the telephone switch of Example 6.33. Compute the expected number of calls handled by the switch in steady state.

6.34. Consider the five processor system as described in Conceptual Problem 6.7 with parameters given in Computational Problem 6.12. Suppose that the system maintenance starts as soon as the system crashes. The amount of time needed to repair the system is exponentially distributed with mean 5 days, at the end of which all five processors are functioning. Compute the limiting probability that the system is under repair.

6.35. Consider the two-component system of Conceptual Problem 6.15 with the data given in Computational Problem 6.14. The repair time of each component is an exponential random variable with mean 2 days. Compute the long-run fraction of the time the system is down.

6.36. Do Computational Problem 6.35, but now assume the system behaves as described in Conceptual Problem 6.16.

6.37. Suppose the CTMC in Computational Problem 6.7 incurs costs at the following rates: $c(i) = 2 * i + 1$, $1 \leq i \leq 5$. Compute the ETC incurred over the time interval $[0, 10]$ if the initial state is 2.

6.38. Compute the long-run cost rate for the CTMC in Computational Problem 6.37.

6.39. Suppose the CTMC in Computational Problem 6.8 incurs costs at the following rates: $c(i) = 4 * i^2 + 1$, $1 \leq i \leq 6$. Compute the ETC incurred over the time interval $[0, 20]$ if the initial state is 6.

6.40. Compute the long-run cost rate for the CTMC in Computational Problem 6.39.

6.41. Suppose the CTMC in Computational Problem 6.9 incurs costs at the following rates: $c(i) = 2 * i^2 - 3 * i + 1$, $1 \leq i \leq 6$. Compute the ETC incurred over the time interval $[0, 15]$ if the initial state is 4.

6.42. Consider the Call Center of Example 6.12 with data from Computational Problem 6.15. Suppose each busy agent costs $20 per hour, and each customer in the system produces a revenue of $25 per hour spent in the system. Compute the long-run net revenue per unit time of operating the system.

6.43. Consider Conceptual Problem 6.8 with data from Computational Problem 6.13. Suppose the costs of the three servers are $40, $20, and $10 per hour, respectively. (The server costs money only when it is busy.) Each customer can be charged $c per hour that they spend in the system. What is the smallest value of c that can make the system profitable in the long run?

6.44. Consider the three machine production system described in Conceptual Problem 6.12 with data from Computational Problem 6.22. Each machine costs $40 per hour while it is working on a component. Each machine adds value at a rate of $75 per hour to the part that it works on. The value added, or the cost of operating, is zero when the machine is idle or blocked. Compute the net contribution of the three machines per unit time in the long run.

6.45. Consider the telephone switch of Example 6.33. Suppose the switch is idle at time 0. Compute the expected total revenue in the first hour, if each admitted call is charged 10 cents per minute.

6.46. Consider the five processor system described in Computational Problem 6.34. Suppose each working processor produces revenue at a rate of $100 per hour, while repairs cost $200 per hour. What is the expected revenue produced during the first year if all five processors are working in the beginning?

6.47. The system in Computational Problem 6.46 is available for lease for $2 million per year. Is it worth leasing?

6.48. Consider the two-component system in Computational Problem 6.35. The system produces revenue at a rate of $400 per day per working component. The repair costs $200 per day. Compute the expected total revenue in the first week of a newly installed system.

6.49. Compute the long-run revenue rate of the system in Computational Problem 6.48.

6.50. Consider the single server queue of Example 6.9 with capacity 10. Suppose the customers arrive at a rate of five per hour. We have a choice of using one of two servers: server 1 serves a customer in 8 minutes, while server 2 takes 10 minutes. Each admitted customer pays $10 for service. Server 1 can be hired for $20 per hour, while server 2 is available for $15 per hour. Which server should be hired to maximize the net revenue per hour in the long run? (*Hint*: Argue that server 1 brings in revenue at a rate of $75 per hour spent serving a customer, while server 2 brings in revenue at a rate of $60 per hour when it is busy. Idle servers bring in no revenue.)

6.51. Compute the expected time to go from state 1 to state 5 in the CTMC of Computational Problem 6.7.

6.52. Compute the expected time to go from state 2 to state 4 in the CTMC of Computational Problem 6.8.

6.53. Compute the expected time to go from state 6 to state 4 in the CTMC of Computational Problem 6.9.

6.54. Consider the system described in Computational Problem 6.10. Compute the expected time of failure.

6.55. In the system described in Computational Problem 6.10, how many cables should be used in order to get the expected lifetime of the system to be at least 2 years?

6.56. Compute the expected lifetime of a system described in Computational Problem 6.14.

6.57. Compute the expected lifetime of the five processor system described in Computational Problem 6.12.

7

Generalized Markov Models

7.1. Introduction

The main focus of this book is to study systems that evolve randomly in time. We encountered several applications in Chapter 5 where the system is observed at time $n = 0, 1, 2, 3, \ldots$. In such a case we define X_n as the state of the system at time n, and study the discrete-time stochastic process $\{X_n, n \geq 0\}$. In Chapter 5 we studied the systems that have the Markov property at each time $n = 0, 1, 2, 3, \ldots$, i.e., the future of the system from any time n onward depends on its history up to time n only through the state of the system at time n. We found this property to be immensely helpful in studying the behavior of these systems.

We also encountered several applications in Chapter 6 where the system is observed continuously at time $t \geq 0$, and hence we need to study the continuous-time stochastic process $\{X(t), t \geq 0\}$. In Chapter 6, we studied the systems that have the Markov property at all times $t \geq 0$, i.e., the future of the system from any time t onward depends on its history up to time t only through the state of the system at time t. Again, we found that the Markov property made it possible to study many aspects of such systems.

This creates a natural question: What can be done if the system does not have the Markov property at all the integer points n, or all the real points t? We address this question in this chapter. Essentially, we shall study *generalized Markov models*, i.e., continuous-time stochastic processes $\{X(t), t \geq 0\}$ that have the Markov property at a set of (possibly random) time points $0 = S_0 \leq S_1 \leq S_2 \leq S_3, \ldots$, etc., i.e., the future of the system from time S_n onward depends on the history of the system up to time S_n only via the state of the system at time S_n. This relaxation comes with a price. The transient and short-term analysis of these systems becomes much harder, and hence we shall not attempt it in this book. (We refer the reader to more advanced texts for more information on this.) Here, we shall concentrate on the long-term analysis and on the long-run cost models.

We begin this development with a simple process, called the renewal process, in the next section.

7.2. Renewal Processes

Suppose we are observing a series of events occurring randomly over time. For example, the events may be accidents taking place at a particular traffic intersection, or the births in a maternity ward, or the failures of a repairable system, or the earthquakes in a given state, or the arrivals of customers at a service station, or transactions at a bank, or the transmissions of packets in a data communication network, or the completion of jobs by a computer system, etc.

Let S_n be the time when the nth event occurs. We assume that the events are indexed in increasing time of their occurrence, i.e., $S_n \leq S_{n+1}$. If the nth and $(n+1)$st events occur at the same time, $S_n = S_{n+1}$. We shall also assume that $S_0 = 0$. Define

$$T_n = S_n - S_{n-1}, \qquad n \geq 1.$$

Thus T_n, called the *inter-event time*, is the (random) time interval between the occurrence of the $(n-1)$st and the nth event.

Now let $N(t)$ be the total number of events observed during $(0, t]$. Note that the event at time 0 is not counted in $N(t)$, but any event at time t is counted. The process $\{N(t), t \geq 0\}$ is called a *counting process* generated by $\{T_n, n \geq 1\}$. A typical sample path of a counting process is shown in Figure 7.1. The figure also illustrates the relationship between the random variables S_n, T_n, and $N(t)$.

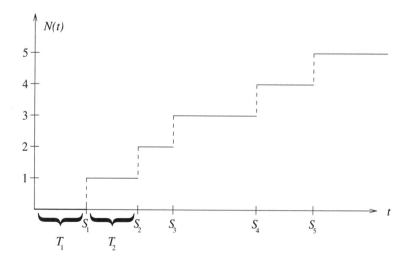

FIGURE 7.1. A typical sample path of a counting process.

Next we define *renewal process* as a special case of a counting process.

Definition 7.1 (Renewal Process). A counting process $\{N(t), t \geq 0\}$ generated by $\{T_n, n \geq 1\}$ is called a renewal process if $\{T_n, n \geq 1\}$ is a sequence of nonnegative iid random variables.

We present several examples of renewal processes below.

Example 7.1 (Poisson Process). A Poisson process with rate λ as defined in Section 6.5 is a renewal process because, by its definition, the inter-event times $\{T_n, n \geq 1\}$ are iid $\text{Exp}(\lambda)$ random variables. $\qquad\square$

Example 7.2 (Battery Replacement). Mr Smith replaces the battery in his car as soon as it dies. We ignore the time required to replace the battery since it is small compared to the lifetime of the battery. Let $N(t)$ be the number of batteries replaced during the first t years of the life of the car, not counting the one that the car came with. Then $\{N(t), t \geq 0\}$ is a renewal process if the lifetimes of the successive batteries are iid.

Now, it is highly inconvenient for Mr Smith to have the car battery die on him at random. To avoid this, he replaces the battery once it becomes 3 years old, even if it hasn't failed yet. Of course if the battery fails before it is 3 years old, he has to replace it anyway. By following this policy, he tries to reduce the unscheduled replacements, which are more unpleasant than the planned replacements. Now let $N(t)$ be the number of batteries replaced up to time t, planned or unplanned. Is $\{N(t), t \geq 0\}$ a renewal process?

By definition, $\{N(t), t \geq 0\}$ is a renewal process if the inter-replacement times $\{T_n, n \geq 1\}$ are iid nonnegative random variables. Let L_n be the lifetime of the nth battery (in years), counting the original battery as battery number 1. Suppose $\{L_n, n \geq 1\}$ are iid nonnegative random variables. Mr Smith will do the first replacement at time $T_1 = 3$ if $L_1 > 3$, or else he has to do it at time $T_1 = L_1$. Thus $T_1 = \min\{L_1, 3\}$. Similarly the time between the first and second replacement is $T_2 = \min\{L_2, 3\}$ and, in general, $T_n = \min\{L_n, 3\}$. Since $\{L_n, n \geq 1\}$ are iid nonnegative random variables, so are the inter-replacement times $\{T_n, n \geq 1\}$. Hence $\{N(t), t \geq 0\}$ is a renewal process. $\qquad\square$

Example 7.3 (Renewal Processes in DTMCs). Let X_n be the state of a system at time n, and suppose $\{X_n, n \geq 0\}$ is a DTMC on state space $\{1, 2, \ldots, N\}$. Suppose the DTMC starts in a fixed state i, $1 \leq i \leq N$. Define S_n as the nth time the process visits state i. Mathematically, S_n's can be recursively defined as follows: $S_0 = 0$ and

$$S_n = \min\{k > S_{n-1} : X_k = i\}, \qquad n \geq 1.$$

In this case $N(t)$ counts the number of visits to state i during $(0, t]$. Note that although a DTMC is a discrete-time process, the $\{N(t), t \geq 0\}$ process is a continuous-time process. Is $\{N(t), t \geq 0\}$ a renewal process?

Let $T_n = S_n - S_{n-1}$ be the time between the $(n-1)$st and nth visits to state i. The DTMC is in state i at time 0, and returns to state i at time T_1. The next visit to state i occurs after T_2 transitions. Due to the Markov property and time homogeneity, the random variable T_2

is independent of T_1 and has the same distribution. By continuing this way, we see that $\{T_n, n \geq 1\}$ is a sequence of iid nonnegative random variables. Hence $\{N(t), t \geq 0\}$ is a renewal process. □

Example 7.4 (Renewal Processes in CTMCs). Let $\{X(t), t \geq 0\}$ be a CTMC on state space $\{1, 2, \ldots, N\}$. Suppose the CTMC starts in state i, $1 \leq i \leq N$. Define S_n as the nth time the process enters state i. In this case $N(t)$ counts the number of visits to state i during $(0, t]$. Is $\{N(t), t \geq 0\}$ a renewal process?

The answer is "yes" by the same argument as in the previous example, since the CTMC has the Markov property at all times $t \geq 0$. □

Next we study the *long-run renewal rate*, i.e., we study

$$\lim_{t \to \infty} \frac{N(t)}{t}.$$

We first need to define what we mean by the above limit, since $N(t)/t$ is a random variable for each t. We do this in the following general definition:

Definition 7.2 (Convergence with Probability One). Let Ω be a sample space and let $Z_n : \Omega \to (-\infty, \infty)$ $(n \geq 1)$ be a sequence of random variables defined on it. The random variables Z_n are said to converge with probability one to a constant c as n tends to ∞, if

$$P(\omega \in \Omega : \lim_{n \to \infty} Z_n(\omega) = c) = 1.$$

In words, convergence with probability one implies that the set of ω's for which the convergence holds, has probability one. There may be some ω's for which the convergence fails, however, the probability of all such ω's is 0. The "convergence with probability one" is also known as "strong convergence" or "almost sure convergence," although we will not use these terms in this book. One of the most important "convergence with probability one" results is stated without proof in the next theorem.

Theorem 7.1 (Strong Law of Large Numbers (SLLN)). *Let $\{T_n, n \geq 1\}$ be a sequence of iid random variables with common mean τ, and let*

$$S_n = T_1 + T_2 + \cdots + T_n.$$

Then

$$\lim_{n \to \infty} \frac{S_n}{n} = \tau \tag{7.1}$$

with probability one.

With these preliminaries we are ready to study the long-run renewal rate. First consider a special case where the inter-event times are all exactly equal to τ, i.e., the nth event takes place at time $n\tau$. Then we see that the renewal process $\{N(t), t \geq 0\}$ is a deterministic process in this case, and the long-run rate at which the events take place is $1/\tau$, i.e.,

$$\lim_{t \to \infty} \frac{N(t)}{t} = \frac{1}{\tau}.$$

Intuitively, we would expect that the above limit should continue to hold when the inter-event times are iid random variables with mean τ. This intuition is made precise in the next theorem.

Theorem 7.2 (Long-Run Renewal Rate). *Let $\{N(t), t \geq 0\}$ be a renewal process generated by a sequence of nonnegative iid random variables $\{T_n, n \geq 1\}$ with common mean $0 < \tau < \infty$. Then*

$$\lim_{t \to \infty} \frac{N(t)}{t} = \frac{1}{\tau} \qquad (7.2)$$

with probability one.

Proof. Let $S_0 = 0$ and

$$S_n = T_1 + T_2 + \cdots + T_n, \qquad n \geq 1.$$

Thus the nth event takes place at time S_n. Since $N(t)$ is the number events that occur up to time t, it can be seen that the last event at or before t occurs at time $S_{N(t)}$, and the first event after t occurs at time $S_{N(t)+1}$. Thus

$$S_{N(t)} \leq t < S_{N(t)+1}.$$

Dividing by $N(t)$ we get

$$\frac{S_{N(t)}}{N(t)} \leq \frac{t}{N(t)} < \frac{S_{N(t)+1}}{N(t)}. \qquad (7.3)$$

Now, since $\tau < \infty$,

$$\lim_{t \to \infty} N(t) = \infty$$

with probability one. Hence,

$$\lim_{t \to \infty} \frac{S_{N(t)}}{N(t)} = \lim_{n \to \infty} \frac{S_n}{n}$$
$$= \tau \qquad (7.4)$$

with probability one, from the SLLN. Similarly,

$$\lim_{t \to \infty} \frac{S_{N(t)+1}}{N(t)} = \lim_{t \to \infty} \frac{S_{N(t)+1}}{N(t)+1} \frac{N(t)+1}{N(t)}$$
$$= \lim_{n \to \infty} \frac{S_{n+1}}{n+1} \lim_{n \to \infty} \frac{n+1}{n}$$
$$= \tau. \qquad (7.5)$$

Taking limits in (7.3) we get

$$\lim_{t \to \infty} \frac{S_{N(t)}}{N(t)} \leq \lim_{t \to \infty} \frac{t}{N(t)} \leq \lim_{t \to \infty} \frac{S_{N(t)+1}}{N(t)}.$$

Substituting from (7.4) and (7.5) in the above we get

$$\tau \leq \lim_{t \to \infty} \frac{t}{N(t)} \leq \tau.$$

Hence, taking reciprocals, and assuming $0 < \tau < \infty$, we get

$$\lim_{t \to \infty} \frac{N(t)}{t} = \frac{1}{\tau}.$$ □

Remark. If $\tau = 0$, or $\tau = \infty$, we can show that (7.2) remains valid if we interpret $1/0 = \infty$ and $1/\infty = 0$.   □

Thus the long-run renewal rate equals the reciprocal of the mean inter-event time, as intuitively expected. This seemingly simple result proves to be tremendously powerful in the long-term analysis of non-Markovian processes. We first illustrate it with several examples.

Example 7.5 (Poisson Process). Let $\{N(t), t \geq 0\}$ be a PP(λ). Then the inter-event times are iid Exp(λ), with mean $\tau = 1/\lambda$. Hence, from Theorem 7.2,

$$\lim_{t \to \infty} \frac{N(t)}{t} = \frac{1}{\tau} = \lambda.$$

Thus the long-run rate of events in a Poisson process is λ, as we have seen before. We have also seen in Section 6.5 that for a fixed t, $N(t)$ is a Poisson random variable with parameter λt. Hence

$$\lim_{t \to \infty} \frac{\mathsf{E}(N(t))}{t} = \frac{\lambda t}{t} = \lambda.$$

Thus the expected number of events per unit time in a PP(λ) is also λ! Note that this does not follow from Theorem 7.2, but happens to be true.   □

Example 7.6 (Battery Replacement). Consider the battery replacement problem of Example 7.2. Suppose the lifetimes of successive batteries are iid random variables that are uniformly distributed over $(1, 5)$ years. Suppose Mr Smith replaces batteries only upon failure. Compute the long-run rate of replacements.

The mean of a U($1, 5$) random variable is 3. (See Example 2.39.) Hence the mean inter-replacement time is 3 years. Using Theorem 7.2, we see that the rate of replacement is $1/3 = .333$ per year.   □

Example 7.7 (Battery Replacement (Continued)). Suppose Mr Smith follows the policy of replacing the batteries upon failure or upon becoming 3 years old. Compute the long-run rate of replacement under this policy.

Let L_i be the lifetime of the ith battery. The ith inter-replacement time is then $T_i = \min\{L_i, 3\}$ as seen in Example 7.2. Hence,

$$\begin{aligned}
\tau &= \mathsf{E}(T_i) \\
&= \mathsf{E}(\min\{L_i, 3\}) \\
&= \int_1^3 x f_{L_i}(x)\,dx + \int_3^5 3 f_{L_i}(x)\,dx \\
&= \frac{1}{4}\left(\int_1^3 x\,dx + \int_3^5 3\,dx \right)
\end{aligned}$$

$$= \tfrac{1}{4}\left(\tfrac{8}{2} + 3*2\right)$$
$$= 2.5 \text{ years.}$$

Hence, from Theorem 7.2, the long-run rate of replacement is $1/2.5 = .4$ under this policy. This is clearly more than $1/3$, the rate of replacement under the "replace upon failure" policy of Example 7.6. Why should Mr Smith accept a higher rate of replacement? Presumably to reduce the rate of unplanned replacements. So, let us compute the rate of unplanned replacements under the new policy.

Let $N_1(t)$ be the number of unplanned replacements up to time t. $\{N_1(t), t \geq 0\}$ is a renewal process (see Conceptual Problem 7.2.) Hence, we can use Theorem 7.2, if we can compute τ, the expected value of T_1, the time of the first unplanned replacement. Now, if $L_1 < 3$, the first replacement is unplanned and hence $T_1 = L_1$. Otherwise the first replacement is a planned one and takes place at time 3. The expected time until the unplanned replacement from then on is still τ. Using the fact that L_1 is U(1, 5), we get

$$\begin{aligned}
\tau &= E(T_1) \\
&= E(T_1|L_1 < 3)P(L_1 < 3) + E(T_1|L_1 \geq 3)P(L_1 \geq 3) \\
&= E(L_1|L_1 < 3)P(L_1 < 3) + (3 + \tau)P(L_1 \geq 3) \\
&= E(\min(L_1, 3)) + \tau P(L_1 \geq 3) \\
&= 2.5 + .5\tau.
\end{aligned}$$

Solving for τ we get

$$\tau = 5.$$

Hence the unplanned replacements occur at a rate of $1/5 = .2$ per year in the long-run. This is less than $1/3$, the rate of unplanned replacements under the "replace upon failure" policy. Thus, if unplanned replacements are really inconvenient, it may be preferable to use this modified policy. We need to wait until the next section to quantify how "inconvenient" the unplanned replacements have to be in order to justify the higher total rate of replacements. □

Example 7.8 (Two-State Machine). Consider the CTMC model of a machine subject to failures and repairs as described in Example 6.5. Suppose the machine is working at time 0. Compute the long-run rate of repair completions for this machine.

Let $X(t)$ be the state of the machine at time t, 1 if it is up and 0 if it is down. $\{X(t), t \geq 0\}$ is a CTMC as explained in Example 6.5. Then, $N(t)$, the number of repair completions up to time t, is the number of times the CTMC enters state 1 up to time t. Since $X(0)$ is given to be 1, we see from Example 7.4 that $\{N(t), t \geq 0\}$ is a renewal process. The expected time until the first repair completion, τ, is the expected time in state 1 plus the expected time in state 0, which is $1/\mu + 1/\lambda$. Hence, using Theorem 7.2, the long-run rate of repair completions is given by

$$\frac{1}{\tau} = \frac{\lambda\mu}{\lambda + \mu}.$$ □

Example 7.9 (Manpower Model). Consider the manpower planning model of Paper-Pushers Inc. as described in Example 5.8. Compute the long-run rate of turnover (the rate at which an existing employee is replaced by a new one) for the company.

Note that the company has 100 employees, each of whom moves from grade to grade according to a DTMC. When the ith employee quits he is replaced by a new employee in grade 1, and is assigned employee id i. Suppose we keep track of the turnover of the employees with id i, $1 \leq i \leq 100$. Let $N_i(t)$ be the number of new employees that have joined with employee id i up to time t, not counting the initial employee. Since each new employee starts in grade 1, and moves from grade to grade according to a DTMC, the lengths of their tenure with the company are iid random variables. Hence $\{N_i(t), t \geq 0\}$ is a renewal process. From Example 5.33, the expected length of stay of an employee is 1.4 years. Hence, using Theorem 7.2, we get

$$\lim_{t \to \infty} \frac{N_i(t)}{t} = \frac{1}{1.4} \text{ per year.}$$

Since there are 100 similar slots, the turnover for the company is given by

$$\lim_{t \to \infty} \sum_{i=1}^{100} \frac{N_i(t)}{t} = \frac{100}{1.4} = 71.43 \text{ per year.}$$

This is very high, and suggests that some steps are needed to bring it down.   □

7.3. Cumulative Processes

Consider a system that incurs costs and earns rewards as time goes on. Thinking of rewards as negative costs, let $C(t)$ be the total net cost (i.e., cost − reward) incurred by a system over the interval $(0, t]$. Since $C(t)$ may increase and decrease with time, and take both positive and negative values, we think of $\{C(t), t \geq 0\}$ as a continuous-time stochastic process with state space $(-\infty, \infty)$.

Now suppose we divide the time into consecutive intervals, called *cycles*, with T_n being the (possibly random, but nonnegative) length of the nth cycle. Define $S_0 = 0$, and

$$S_n = T_1 + T_2 + \cdots + T_n, \qquad n \geq 1.$$

Thus the nth interval is $(S_{n-1}, S_n]$. Define

$$C_n = C(S_n) - C(S_{n-1}), \qquad n \geq 1.$$

Then C_n is the net cost (positive or negative) incurred over the nth cycle.

Definition 7.3 (Cumulative Processes). The stochastic process $\{C(t), t \geq 0\}$ is called a cumulative process if $\{(T_n, C_n), n \geq 1\}$ is a sequence of iid bivariate random variables.

Thus the total cost process is a cumulative process if the successive interval lengths are iid, and the costs incurred over these intervals are iid. However, the cost over the nth interval may depend on its length. Indeed, it is this possibility of dependence that gives the cumulative processes their power! The *cumulative processes* are also known as *renewal reward processes*. These processes are useful in the long-term analysis of systems when we want to compute the long-term cost rates, or the long-run fraction of time spent in different states, etc. We illustrate with a few examples.

Example 7.10 (Battery Replacement). Consider the battery replacement problem described in Example 7.7. Now suppose the cost of replacing a battery is $75, if it is a planned replacement; and $125 if it is an unplanned replacement. Thus the cost of unexpected delay, the towing involved, and disruption to the normal life of Mr Smith, from the sudden battery failure, is figured to be $50! Let $C(t)$ be the total battery replacement cost up to time t. Is $\{C(t), t \geq 0\}$ a cumulative process?

Let L_n be the lifetime of the nth battery, and let $T_n = \min\{L_n, 3\}$ be the nth inter-replacement time. The nth replacement takes place at time S_n and costs $\$C_n$. Think of $(S_{n-1}, S_n]$ as the nth interval. Then C_n is the cost incurred over the nth interval. Now, we have

$$C_n = \begin{cases} 125 & \text{if } L_n < 3, \\ 75 & \text{if } L_n \geq 3. \end{cases}$$

Note that C_n explicitly depends on L_n, and hence on T_n. However, $\{(T_n, C_n), n \geq 1\}$ is a sequence of iid bivariate random variables. Hence, $\{C(t), t \geq 0\}$ is a cumulative process. $\qquad \square$

We have always talked about $C(t)$ as the total cost up to time t. However, t need not represent time, as shown by the next example.

Example 7.11 (Prorated Warranty). A tire company issues the following 50,000 mile prorated warranty on its tires that sell for $95 per tire: Suppose the new tire fails when it has L miles on it. If $L > 50,000$, the customer has to buy a new tire for the full price. If $L \leq 50,000$, the customer gets a new tire for $95 * L/50,000$. Assume that the customer continues to buy the same brand of tire after each failure. Let $C(t)$ be the total cost to the customer over the first t miles. Is $\{C(t), t \geq 0\}$ a cumulative process?

Let T_n be the life (in miles) of the nth tire purchased by the customer. Let C_n be the cost of the nth tire. The pro rata warranty implies that

$$C_n = \$95(\min\{50,000, T_n\})/50,000.$$

Thus the customer incurs a cost C_n after $S_n = T_1 + T_2 + \cdots + T_n$ miles. Think of T_n as the nth interval and C_n as the cost incurred over it. Now suppose that successive tires have iid lifetimes. Then $\{(T_n, C_n), n \geq 1\}$ is a sequence of iid bivariate random variables. Hence, $\{C(t), t \geq 0\}$ is a cumulative process. $\qquad \square$

In the examples so far, the cost was incurred in a lump sum fashion at the end of an interval. In the next example costs and rewards are incurred continuously over time.

Example 7.12 (Two-State Machine). Consider the CTMC model of a machine subject to failures and repairs as described in Example 6.5. Suppose the machine is working at time 0. Suppose the machine produces revenue at rate $\$A$ per unit time when it is up, and costs $\$B$ per unit time for a repair person to repair it. Let $C(t)$ be the net total cost up to time t. Show that $\{C(t), t \geq 0\}$ is a cumulative process.

Let $X(t)$ be the state of the machine at time t, 1 if it is up and 0 if it is down. Let U_n be the nth uptime and let D_n be the nth downtime. Then $\{U_n, n \geq 1\}$ is a sequence of iid $\text{Exp}(\lambda)$ random variables, and it is independent of $\{D_n, n \geq 1\}$, a sequence of iid $\text{Exp}(\mu)$ random variables. Since the machine goes through an up–down cycle repeatedly, it seems natural to take the nth interval to be $T_n = U_n + D_n$. The cost over the nth interval is

$$C_n = BD_n - AU_n.$$

Now, the independence of the successive up and down times implies that $\{(T_n, C_n), n \geq 1\}$ is a sequence of iid bivariate random variables, although C_n does depend upon T_n. Hence $\{C(t), t \geq 0\}$ is a cumulative process. $\qquad\square$

Example 7.13 (Compound Poisson Process). In Section 6.5 we studied the Poisson process as a stochastic process that counts events occurring one at a time, with iid exponential inter-event times. Here we shall introduce a generalization called a Compound Poisson Process, or CPP for short.

Definition 7.4. Let $\{N(t), t \geq 0\}$ be a Poisson process with rate λ. Let $\{C_n, n \geq 0\}$ be a sequence of iid random variables, that is also independent of the Poisson process. Let

$$C(t) = \sum_{n=1}^{N(t)} C_n, \qquad t \geq 0. \tag{7.6}$$

The stochastic process $\{C(t), t \geq 0\}$ is called a *Compound Poisson Process*.

A CPP arises as a natural model of a batch-arrival process, where T_n represents the inter-batch-arrival time and C_n represents the size of the nth batch. Let S_n be the time of the nth event in the Poisson process (with $S_0 = 0$). Let $T_n = S_n - S_{n-1}$, $n \geq 1$. We know that $\{T_n, n \geq 1\}$ are iid $\text{Exp}(\lambda)$ random variables. Furthermore, from (7.6) we see that

$$C_n = C(S_n) - C(S_{n-1}), \qquad n \geq 1.$$

From the definition of the CPP, we see that $\{(T_n, C_n), n \geq 1\}$ is a sequence of iid bivariate random variables. (In fact, C_n is also independent of T_n!) Hence it follows that the CPP $\{C(t), t \geq 0\}$ is a cumulative process.

Next we compute $\mathsf{E}(C(t))$. Let

$$\mathsf{E}(C_n) = \tau, \qquad n \geq 1.$$

From Section 6.5 we see that $N(t)$ is a Poisson random variable with parameter λt, and hence

$$\mathsf{E}(N(t)) = \lambda t.$$

Note that $C(t)$ is a random sum of random variables. Using Example 4.15, we get

$$\mathsf{E}(C(t)) = \mathsf{E}\left(\sum_{n=1}^{N(t)} C_n\right)$$
$$= \mathsf{E}(C_1)\mathsf{E}(N(t))$$
$$= \tau \lambda t. \tag{7.7}$$

As a numerical example, consider the process of customers arriving at a restaurant. Suppose the customers arrive in parties (or batches) of variable sizes. The successive party sizes are iid random variables and are binomially distributed with parameters $n = 10$ and $p = .4$. (Thus the mean size of a party is $\tau = np = 4$. Note that a binomial random variable can take a value 0 with probability $(1 - p)^n$. What does it mean to get a batch arrival of size 0?) The parties themselves arrive according to a Poisson process with a rate of 10 per hour. Let $C(t)$ be the total number of arrivals up to time t, assuming $C(0) = 0$. Then $\{C(t), t \geq 0\}$ is a compound Poisson process and

$$\mathsf{E}(C(t)) = \tau \lambda t = (4)(10)t = 40t.$$

Suppose the above process is a valid model of the arrival process over the time interval from 7 p.m. to 10 p.m. Then the expected total number of arrivals over those 3 hours will be given by $40 * 3 = 120$. □

In general, it is very difficult to compute $\mathsf{E}(C(t))$ for a general cumulative process. One case where this calculation is easy is the compound Poisson process as worked out in Example 7.13 above. Hence we shall restrict our attention to the long-term analysis. In our long-term analysis we are interested in the long-run cost rate, i.e., in

$$\lim_{t \to \infty} \frac{C(t)}{t}.$$

The next theorem gives the main result.

Theorem 7.3 (Long-Run Cost Rate). *Let $\{C(t), t \geq 0\}$ be a cumulative process with the corresponding sequence of iid bivariate random variables $\{(T_n, C_n), n \geq 1\}$. Then*

$$\lim_{t \to \infty} \frac{C(t)}{t} = \frac{\mathsf{E}(C_1)}{\mathsf{E}(T_1)} \tag{7.8}$$

with probability one.

Proof. We shall prove the theorem under the assumption that the cost C_n over the nth cycle is incurred as a lump sum cost at the end of the cycle. We refer the reader to more advanced texts for the proof of the general case.

Let $N(t)$ be the number of cycles that are completed by time t. Then $\{N(t), t \geq\}$ is a renewal process generated by $\{T_n, n \geq 1\}$. The total cost up to time t can be written as

$$C(t) = \begin{cases} 0 & \text{if } N(t) = 0, \\ \sum_{n=1}^{N(t)} C_n & \text{if } N(t) > 0. \end{cases}$$

Assuming $N(t) > 0$, we have

$$\frac{C(t)}{t} = \frac{\sum_{n=1}^{N(t)} C_n}{t}$$

$$= \frac{\sum_{n=1}^{N(t)} C_n}{N(t)} \cdot \frac{N(t)}{t}.$$

Taking limits as $t \to \infty$, and using Theorem 7.2, we get

$$\lim_{t \to \infty} \frac{C(t)}{t} = \lim_{t \to \infty} \frac{\sum_{n=1}^{N(t)} C_n}{N(t)} \cdot \lim_{t \to \infty} \frac{N(t)}{t}$$

$$= \lim_{k \to \infty} \frac{\sum_{n=1}^{k} C_n}{k} \cdot \frac{1}{\mathsf{E}(T_1)}$$

$$\text{(follows from Theorem 7.2)}$$

$$= \mathsf{E}(C_1) \cdot \frac{1}{\mathsf{E}(T_1)},$$

where the last step follows from Theorem 7.1. This yields the theorem.   □

We show the application of the above theorem with a few examples.

Example 7.14 (Battery Replacement). Consider the battery replacement problem of Example 7.10. Compute the long-run cost per year of following the preventive replacement policy.

Let $C(t)$ be the total cost up to time t. We have seen in Example 7.10 that $\{C(t), t \geq 0\}$ is a cumulative process with the iid bivariate sequence $\{(T_n, C_n), n \geq 1\}$ given there. Hence the long-run cost rate can be computed from Theorem 7.3. We have already computed $\mathsf{E}(T_1) = 2.5$ in Example 7.7. Next we compute

$$\mathsf{E}(C_1) = 75 * \mathsf{P}(L_1 > 3) + 125 * \mathsf{P}(L_1 \leq 3)$$

$$= 75 * .5 + 125 * .5$$

$$= 100.$$

Hence the long-run cost rate is

$$\lim_{t \to \infty} \frac{C(t)}{t} = \frac{\mathsf{E}(C_1)}{\mathsf{E}(T_1)} = \frac{100}{2.5} = 40$$

dollars per year.

What is the cost rate of the "replace upon failure" policy? In this case the total cost is a cumulative process with $\{(T_n, C_n), n \geq 1\}$ with $T_n = L_n$ and $C_n = 125$ (since all replacements are unplanned) for all $n \geq 1$. Hence, from Theorem 7.3,

$$\lim_{t \to \infty} \frac{C(t)}{t} = \frac{125}{3} = 41.67$$

dollars per year. Thus the preventive maintenance policy is actually more economical than the "replace upon failure!"   □

Example 7.15 (Prorated Warranty). Consider the prorated warranty of Example 7.11. The lifetimes of the tires are iid random variables with common pdf

$$f(x) = 2 * 10^{-10}x \qquad \text{for} \quad 0 \le x \le 100{,}000 \text{ miles.}$$

Suppose the customer has an option of buying the tire without warranty for $90. Based upon the long-run cost per mile, should the customer get the warranty? (We assume that the customer either always gets the warranty, or never gets the warranty.)

First consider the case of no warranty. The nth cycle length is T_n, the lifetime of the nth tire. The cost over the nth cycle is $90. Clearly, $\{(T_n, C_n), n \ge 1\}$ is a sequence of iid bivariate random variables. Hence the total cost is a cumulative process, and the long-run cost rate is given by Theorem 7.3. We have

$$E(T_1) = \int_0^{100.000} 2 * 10^{-10}x^2 \, dx$$

$$= 2 * 10^{-10}\frac{x^3}{3} \Big|_0^{100.000}$$

$$= \tfrac{2}{3} \cdot 10^5.$$

Hence the long-run cost rate of no warranty is

$$\frac{E(C_1)}{E(T_1)} = 135 * 10^{-5}$$

dollars per mile, or $1.35 per 1000 miles.

Next consider the policy of always buying the tires under warranty. The nth cycle here is as in the no warranty case. However, the cost C_n is given by

$$C_n = 95\frac{\min\{50{,}000, T_n\}}{50{,}000}.$$

Hence

$$E(C_1) = \int_0^{100.000} 2 * 10^{-10} * 95\frac{\min\{50{,}000, x\}}{50{,}000}x \, dx$$

$$= \int_0^{50.000} 2 * 10^{-10} * \frac{95}{50{,}000}x^2 \, dx + \int_{50.000}^{100.000} 95 * 2 * 10^{-10}x \, dx$$

$$= \frac{95}{6} + \frac{95 * 3}{4}$$

$$= 87.08333.$$

Hence the long-run cost under warranty is

$$\frac{87.08333}{.666 * 10^5} = 130.625 * 10^{-5}$$

dollars per mile, or $1.31 per 1000 miles. This is less than the cost of not buying the warranty. Hence the customer should buy the warranty. □

Example 7.16 (Two-State Machine). Consider the two-state machine described in Example 7.12. Compute the long-run net cost per unit time.

The machine produces revenue at rate $\$A$ per unit time when it is up, and costs $\$B$ per unit time to repair it. $C(t)$ is the net total cost up to time t. We have already shown that $\{C(t), t \geq 0\}$ is a cumulative process. Hence, from Theorem 7.3 we get

$$\lim_{t \to \infty} \frac{C(t)}{t} = \frac{\mathsf{E}(C_1)}{\mathsf{E}(T_1)}$$

$$= \frac{\mathsf{E}(BD_1 - AU_1)}{\mathsf{E}(U_1 + D_1)},$$

where U_1 is the first uptime and D_1 is the first downtime of the machine. We have

$$\mathsf{E}(D_1) = \frac{1}{\lambda}, \qquad \mathsf{E}(U_1) = \frac{1}{\mu}.$$

Substituting we get the long-run cost rate as

$$\lim_{t \to \infty} \frac{C(t)}{t} = \frac{B/\lambda - A/\mu}{1/\lambda + 1/\mu}$$

$$= B\frac{\mu}{\lambda + \mu} - A\frac{\lambda}{\lambda + \mu}. \tag{7.9}$$

This is consistent with the result in Example 6.32, where we have computed the long-run revenue rate using the CTMC methods. □

Example 7.17 (A General Two-State Machine). In Examples 7.8, 7.12, and 7.16 we studied a two-state machine whose up and down times are independent exponential random variables. Here we consider a machine whose up and down times are generally distributed. Let U_n be the nth uptime and let D_n be the nth downtime. We assume that $\{(U_n, D_n), n \geq 1\}$ are iid bivariate random variables. Thus the nth downtime may be dependent on the nth uptime. Compute the long-run fraction of the time that the machine is up, assuming that the machine is up initially.

Let $C(t)$ be the total time the machine is up during $(0, t]$. Then $\{C(t), t \geq 0\}$ is a cumulative process with nth cycle length $T_n = U_n + D_n$ and the nth "cost" $C_n = U_n$. Hence, using Theorem 7.3, we get

$$\lim_{t \to \infty} \frac{C(t)}{t} = \frac{\mathsf{E}(C_1)}{\mathsf{E}(T_1)} = \frac{\mathsf{E}(U_1)}{\mathsf{E}(U_1) + \mathsf{E}(D_1)}.$$

The above result is intuitive, however it is surprising that it is not influenced by the dependence between U_1 and D_1.

As a numerical example, suppose the uptimes are uniformly distributed over $[4, 6]$ weeks, and that the nth downtime is exactly 20% of the nth uptime, i.e., $D_n = .2U_n$. In this case, the long-run fraction of the time the machine is up is given by

$$\frac{\mathsf{E}(U_1)}{\mathsf{E}(U_1) + \mathsf{E}(D_1)} = \frac{5}{5 + .2 * 5} = \frac{5}{6}.$$

□

Example 7.18 (Compound Poisson Process). Let $\{C(t), t \geq 0\}$ be a compound Poisson process as defined in Example 7.13. We have seen that this is a cumulative process with

$$\mathsf{E}(C_1) = \tau, \qquad \mathsf{E}(T_1) = 1/\lambda.$$

We have seen in Example 7.13 that

$$\mathsf{E}(C(t)) = \tau \lambda t, \qquad t \geq 0.$$

Hence

$$\frac{\mathsf{E}(C(t))}{t} = \frac{\tau \lambda t}{t} = \frac{\tau}{1/\lambda} = \frac{\mathsf{E}(C_1)}{\mathsf{E}(T_1)}.$$

Thus Theorem 7.3 holds. Indeed the cost rate over any time interval equals the long-run cost rate for the CPP!   □

Armed with Theorem 7.3, we are ready to attack the non-Markovian world!

7.4. Semi-Markov Processes: Examples

Consider a system with state space $\{1, 2, \ldots, N\}$. Suppose the system enters the initial state X_0 at time $S_0 = 0$. It stays there for a nonnegative random amount of time and jumps to another state X_1 (which could be the same as X_0) at time S_1. It stays in the new state for another nonnegative random amount of time and then jumps to the next state X_2 (which could be the same as X_1) at time S_2, and continues this way forever. Thus S_n is the time of the nth transition and X_n is the nth state visited by the system. Let $X(t)$ be the state of the system at time t. Then $X(S_n) = X_n$ for $n \geq 0$.

Definition 7.5 (Semi-Markov Process (SMP)). A stochastic process $\{X(t), t \geq 0\}$ described above is called a semi-Markov process if it has the Markov property at every transition epoch S_n, i.e., the evolution of the process from time $t = S_n$ onward depends on the history of the process up to time S_n only via X_n.

In other words, if $\{X(t), t \geq 0\}$ is an SMP, the process $\{X(t + S_n), t \geq 0\}$, given the entire history $\{X(t), 0 \leq t \leq S_n\}$ and $X(S_n) = i$, is independent of $\{X(t), 0 \leq t < S_n\}$, and is probabilistically identical to $\{X(t), t \geq 0\}$ given $X(0) = i$. The definition of an SMP also explains why such processes are called *semi*-Markov: they have the Markov property only at transition epochs, and not at all times. Notice that the future from time S_n cannot depend upon n either. In that sense the above definition forces the SMP to be "time homogeneous"!

Now, note that the Markov property of the SMP $\{X(t), t \geq 0\}$ at each transition epoch $S_n, n \geq 0$, implies the Markov property of $\{X_n, n \geq 0\}$ at every $n \geq 0$. Thus $\{X_n, n \geq 0\}$ is a time-homogeneous DTMC with state space $\{1, 2, \ldots, N\}$, and is called the *embedded DTMC* of the SMP. Let $P = [p_{i.j}]$ be its transition probability matrix, i.e.,

$$p_{i.j} = \mathsf{P}(X_{n+1} = j | X_n = i), \qquad 1 \leq i, j \leq N.$$

Note that $p_{i,i}$ may be positive, implying that an SMP can jump from state i to itself in one step. Next define

$$w_i = \mathsf{E}(S_1 | X_0 = i), \qquad 1 \le i \le N, \tag{7.10}$$

to be the mean sojourn time in state i. Since the SMP has the Markov property at each transition epoch, we see that the expected sojourn time in state i is w_i on any visit to state i. Let

$$w = [w_1, w_2, \ldots, w_N]$$

be the vector of expected sojourn times.

With the transition matrix P and sojourn time vector w, we can visualize the evolution of the SMP as follows. The SMP starts in state i. It stays there for a w_i amount of time on average and then jumps to state j with probability $p_{i,j}$. It then stays in state j for a w_j amount of time on average and jumps to state k with probability $p_{j,k}$, and so on. A typical sample path of an SMP is shown in Figure 7.2.

Recall that a DTMC is described by its one-step transition probability matrix, and a CTMC by its rate matrix. The natural question is: Are P and w sufficient to describe the SMP? The answer depends upon what questions we plan to ask about an SMP. As stated earlier, the transient and occupancy-time analysis of SMPs is beyond the scope of this book and we shall concentrate mainly on the long-term analysis. We shall show in the next section that P and w are sufficient for this purpose. They are not sufficient if we are interested in transient and occupancy-time analysis.

Next we discuss several examples of SMPs. In each case we describe the SMP by giving the transition probability matrix P of the embedded DTMC, and the vector w of the expected sojourn times.

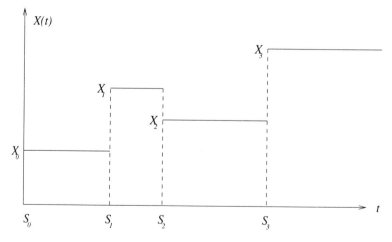

FIGURE 7.2. A typical sample path of a semi-Markov process.

Example 7.19 (Two-State Machine). Consider the two-state machine described in Example 7.17. The machine alternates between "up (1)" and "down (0)" states, with U_n being the nth uptime, and D_n being the nth downtime. In Example 7.17 we had assumed that $\{(U_n, D_n), n \geq 1\}$ is a sequence of iid bivariate random variables. Here we make a further assumption that U_n is also independent of D_n. Let the mean uptime be $\mathsf{E}(U)$ and the mean downtime be $\mathsf{E}(D)$. Suppose $X(t)$ is the state of the machine at time t. Is $\{X(t), t \geq 0\}$ an SMP?

Suppose the machine starts in state $X_0 = 1$. Then it stays there for U_1 amount of time and then moves to state $X_1 = 0$. It stays there for D_1 amount of time, independent of history, and then moves back to state $X_2 = 1$, and proceeds this way forever. This shows that $\{X(t), t \geq 0\}$ is an SMP with state space $\{0, 1\}$. The embedded DTMC has a transition probability matrix given by

$$P = \begin{bmatrix} 0 & 1 \\ 1 & 0 \end{bmatrix}. \tag{7.11}$$

Also

$$w_0 = \mathsf{E}(S_1 | X_0 = 0) = \mathsf{E}(D),$$
$$w_1 = \mathsf{E}(S_1 | X_0 = 1) = \mathsf{E}(U). \qquad \square$$

Example 7.20 (A CTMC Is an SMP). Let $\{X(t), t \geq 0\}$ be a CTMC with state space $\{1, 2, \ldots, N\}$ and rate matrix R. Show that $\{X(t), t \geq 0\}$ is an SMP and compute the matrix P and vector w.

Since a CTMC has the Markov property at all times, it certainly has it at the transition epochs. Hence a CTMC is an SMP. Suppose the CTMC starts in state i. From the properties of the CTMC, we know that S_1, the time the CTMC spends in state i before it jumps, is an $\text{Exp}(r_i)$ random variable, where

$$r_i = \sum_{j=1}^{N} r_{i,j}.$$

Hence

$$w_i = \frac{1}{r_i}.$$

The next state X_1 is j with probability $r_{i,j}/r_i$, independently of S_1. (Recall that $r_{i,i} = 0$.) Thus we have

$$p_{i,j} = \mathsf{P}(X_1 = j | X_0 = i) = \frac{r_{i,j}}{r_i}. \tag{7.12}$$

Note that $p_{i,i} = 0$ in this case, signifying that a CTMC does not jump from state i to itself in one step. $\qquad \square$

Example 7.21 (Series System). Consider a series system of N components. The system fails as soon as any one of the components fails. Suppose the lifetime of the ith ($1 \leq i \leq N$)

component is an $\text{Exp}(v_i)$ random variable, and that the lifetimes of the components are independent. When a component fails, the entire system is shut down, and the failed component is repaired. The mean repair time of the ith component is τ_i. The repair times are mutually independent as well. When a component is under repair no other component can fail since the system is down. Model this system as an SMP.

We first decide the state space of the system. We say that the system is in state 0 if it is functioning, i.e., all components are up; and it is in state i if the system is down and the ith component is under repair. Let $X(t)$ be the state of the system at time t. The state space of $\{X(t), t \geq 0\}$ is $\{0, 1, 2, \ldots, N\}$. The memoryless property of the exponential random variables and the independence assumptions about the lifetimes and the repair times imply that the system has the Markov property every time it changes state. Hence it is an SMP. Next we derive its P matrix and w vector.

Suppose the system starts in state 0. Then it stays there until one of the N components fails, and if the ith component is first to fail it moves to state i. Hence, using the properties of independent exponential random variables (see Section 6.3), we get

$$p_{0,i} = \mathsf{P}(X_1 = i | X_0 = 0) = \frac{v_i}{v}, \qquad 1 \leq i \leq N,$$

where

$$v = \sum_{i=1}^{N} v_i.$$

If the system starts in state i, then it stays there until the repair of the ith component is complete and then it moves to state 0. Hence we get

$$p_{i,0} = \mathsf{P}(X_1 = 0 | X_0 = i) = 1, \qquad 1 \leq i \leq N.$$

Thus the transition probability matrix of the embedded DTMC $\{X_n, n \geq 0\}$ is given by

$$P = \begin{bmatrix} 0 & v_1/v & v_2/v & \cdots & v_N/v \\ 1 & 0 & 0 & \cdots & 0 \\ 1 & 0 & 0 & \cdots & 0 \\ \vdots & \vdots & \vdots & \ddots & \vdots \\ 1 & 0 & 0 & \cdots & 0 \end{bmatrix}. \tag{7.13}$$

Also, the mean sojourn time in state 0 is given by

$$w_0 = \frac{1}{v},$$

and the mean sojourn time in state i is given by

$$w_i = \tau_i, \qquad 1 \leq i \leq N. \qquad \square$$

Example 7.22 (Machine Maintenance). Consider the following maintenance policy for a machine: If the machine fails before it reaches age v (a fixed positive number) it is sent for repair, otherwise it is replaced upon failure with no attempt to repair it. If the machine is sent for repair, and the repair takes less than u (another fixed positive number) amount of time, the machine becomes as good as new, and is put back into use. If the repair takes longer than u, the repair attempt is abandoned and the machine is replaced by a new one. The successive lifetimes of the machine are iid with common cdf $A(\cdot)$ and mean a, the repair times are iid with common cdf $B(\cdot)$, and the replacement times are iid with common cdf $C(\cdot)$ and mean c. Let $X(t)$ be the state of the machine at time t (1 if it is up, 2 if it is under repair, and 3 if it is under replacement). Is $\{X(t), t \geq 0\}$ an SMP, and if it is, what is its P matrix and w vector?

The $\{X(t), t \geq 0\}$ process is an SMP since it has the Markov property at every transition epoch due to the independence assumptions about the lifetimes, repair times, and replacement times. We compute the P matrix of the embedded DTMC next.

Suppose a new machine is put into use at time 0, i.e., $X(0) = X_0 = 1$. Then S_1, the amount of time it spends in state 1, is the lifetime of the machine. If $S_1 \leq v$, then the system moves to state $X_1 = 2$; otherwise it moves to state $X_1 = 3$. Using this, we get

$$p_{1,2} = \mathsf{P}(X_1 = 2|X_0 = 1) = \mathsf{P}(S_1 \leq v) = A(v),$$

and

$$p_{1,3} = \mathsf{P}(S_1 > v) = 1 - A(v).$$

A similar argument in the remaining states yields the following transition probability matrix for the embedded DTMC $\{X_n, n \geq 0\}$:

$$P = \begin{bmatrix} 0 & A(v) & 1 - A(v) \\ B(u) & 0 & 1 - B(u) \\ 1 & 0 & 0 \end{bmatrix}. \tag{7.14}$$

Also, we get the following expected sojourn times:

$$w_1 = \text{expected lifetime} = a,$$

$$w_2 = \mathsf{E}(\min(u, \text{repair time})) = \int_0^u (1 - B(x))\,dx = b(u) \text{ (say)},$$

$$w_3 = \text{expected replacement time} = c. \qquad \square$$

The above examples should suffice to show the usefulness of SMPs. We study the long-term analysis (this includes the limiting behavior and the average cost models) in the next section.

7.5. Semi-Markov Processes: Long-Term Analysis

We start this section with a study of the first passage times in SMPs leading to the long-term analysis of SMPs.

7.5.1. Mean Inter-Visit Times

Let $\{X(t), t \geq 0\}$ be an SMP on state space $\{1, 2, \ldots, N\}$ with transition matrix P and sojourn time vector w. In this subsection we shall compute the expected time between two consecutive visits, called the *inter-visit time*, to state j. This expected value will be useful in the next two subsections. First we need some further notation.

Let Y_j be the first time the SMP *enters* state j, i.e.,

$$Y_j = \min\{S_n : n > 0, X_n = j\}.$$

(Note the strict inequality $n > 0$.) Also let

$$m_{i,j} = \mathsf{E}(Y_j | X_0 = i), \qquad 1 \leq i, j \leq N.$$

The $m_{i,j}$, $i \neq j$, is the mean first passage time from state i to state j. The mean inter-visit time of state j is given by $m_{j,j}$. The next theorem gives the equations satisfied by the mean first passage times.

Theorem 7.4 (Mean First Passage Times). *The mean first passage times, $m_{i,j}$'s, satisfy*

$$m_{i,j} = w_i + \sum_{k=1, k \neq j}^{N} p_{i,k} m_{k,j}, \qquad 1 \leq i, j \leq N. \tag{7.15}$$

Proof. We derive the result by conditioning on X_1. Suppose $X_0 = i$. If $X_1 = j$, then $Y_j = S_1$, and hence

$$\mathsf{E}(Y_j | X_0 = i, X_1 = j) = \mathsf{E}(S_1 | X_0 = i, X_1 = j).$$

On the other hand, if $X_1 = k \neq j$, then the SMP enters state k at time S_1 and the mean time to enter state j from then on is $m_{k,j}$, due to the Markov property, and time homogeneity. Hence, for $k \neq j$,

$$\mathsf{E}(Y_j | X_0 = i, X_1 = k) = \mathsf{E}(S_1 | X_0 = i, X_1 = k) + m_{k,j}.$$

Combining these two equations we get

$$m_{i,j} = \mathsf{E}(Y_j | X_0 = i)$$
$$= \sum_{k=1}^{N} \mathsf{E}(Y_j | X_0 = i, X_1 = k) \mathsf{P}(X_1 = k | X_0 = i)$$
$$= \mathsf{E}(Y_j | X_0 = i, X_1 = j) \mathsf{P}(X_1 = j | X_0 = i)$$
$$+ \sum_{k=1, k \neq j}^{N} \mathsf{E}(Y_j | X_0 = i, X_1 = k) \mathsf{P}(X_1 = k | X_0 = i)$$
$$= \mathsf{E}(S_1 | X_0 = i, X_1 = j) \mathsf{P}(X_1 = j | X_0 = i)$$
$$+ \sum_{k=1, k \neq j}^{N} [\mathsf{E}(S_1 | X_0 = i, X_1 = k) + m_{k,j}] \mathsf{P}(X_1 = k | X_0 = i)$$

$$= \sum_{k=1}^{N} \mathsf{E}(S_1 | X_0 = i, X_1 = k) \mathsf{P}(X_1 = k | X_0 = i)$$

$$+ \sum_{k=1.k \neq j}^{N} m_{k.j} \mathsf{P}(X_1 = k | X_0 = i)$$

$$= \mathsf{E}(S_1 | X_0 = i) + \sum_{k=1.k \neq j}^{N} m_{k.j} \mathsf{P}(X_1 = k | X_0 = i).$$

We have

$$\mathsf{P}(X_1 = k | X_0 = i) = p_{i.k},$$

and from (7.10) we get

$$\mathsf{E}(S_1 | X_0 = i) = w_i.$$

Using these results in the last equation we get (7.15). □

Our main interest is in $m_{j.j}$, the mean inter-visit time of state j. Using the above theorem to compute these quantities is rather inefficient, since we need to solve different sets of N equations in N unknowns to obtain $m_{j.j}$ for different j's. The next theorem gives a more efficient method of computing all $m_{j.j}$'s by solving a single set of N equations in N unknowns, under some restrictions.

Theorem 7.5 (Mean Inter-Visit Times). *Assume that the embedded DTMC $\{X_n, n \geq 0\}$ is irreducible, and let $\pi = [\pi_1, \pi_2, \ldots, \pi_N]$ be a solution to the balance equations*

$$\pi_j = \sum_{i=1}^{N} \pi_i p_{i.j}, \quad 1 \leq j \leq N.$$

Then

$$m_{j.j} = \frac{\sum_{i=1}^{N} \pi_i w_i}{\pi_j}, \quad 1 \leq j \leq N. \tag{7.16}$$

Proof. It follows from the results of Section 5.6 that there is a (nonzero) solution to the balance equations. Now, multiply both sides of (7.15) by π_i and sum over all i. We get

$$\sum_{i=1}^{N} \pi_i m_{i.j} = \sum_{i=1}^{N} \pi_i w_i + \sum_{i=1}^{N} \pi_i \sum_{k=1.k \neq j}^{N} p_{i.k} m_{k.j}$$

$$= \sum_{i=1}^{N} \pi_i w_i + \sum_{k=1.k \neq j}^{N} m_{k.j} \sum_{i=1}^{N} \pi_i p_{i.k}$$

$$= \sum_{i=1}^{N} \pi_i w_i + \sum_{k=1.k \neq j}^{N} m_{k.j} \pi_k$$

(using the balance equations)

$$= \sum_{i=1}^{N} \pi_i w_i + \sum_{i=1, i \neq j}^{N} \pi_i m_{i,j}$$

(replacing k by i in the last sum)

$$= \sum_{i=1}^{N} \pi_i w_i + \sum_{i=1}^{N} \pi_i m_{i,j} - \pi_j m_{j,j}.$$

Cancelling the second sum on the right-hand side with the sum on the left-hand side, we get (7.16). □

Note that we do not need the embedded DTMC to be aperiodic. We also do not need a normalized solution to the balance equations, any solution will do. The condition of irreducibility in the theorem ensures that none of the π_i's will be zero.

Example 7.23 (Two-State Machine). Compute the mean time between two repair completions in the case of the two-state machine described in Example 7.19.

The embedded DTMC $\{X_n, n \geq 0\}$ has the transition probability matrix given in (7.11). This DTMC is irreducible, and a solution to the balance equation is

$$\pi_0 = 1; \quad \pi_1 = 1.$$

The mean sojourn times $[w_0, w_1]$ are given in Example 7.19. Hence, using Theorem 7.5, the mean time between two consecutive repair completions is given by

$$m_{1,1} = \frac{\pi_0 w_0 + \pi_1 w_1}{\pi_1} = \mathsf{E}(U) + \mathsf{E}(D).$$

This is what we expect. □

Example 7.24 (Machine Maintenance). Consider the machine maintenance policy of Example 7.22. Compute the expected time between two consecutive repair commencements.

Let $\{X(t), t \geq 0\}$ be the SMP model developed in Example 7.22. The embedded DTMC has the probability transition matrix given in (7.14). The DTMC is irreducible. The balance equations are

$$\pi_1 = \pi_2 B(u) + \pi_3,$$

$$\pi_2 = \pi_1 A(v),$$

$$\pi_3 = (1 - A(v))\pi_1 + (1 - B(u))\pi_2.$$

One possible solution is

$$\pi_1 = 1, \quad \pi_2 = A(v), \quad \pi_3 = 1 - A(v)B(u).$$

The mean sojourn times $[w_1, w_2, w_3]$ are given in Example 7.22. The time between two consecutive repair commencements is the same as the inter-visit time of state 2. Substituting in (7.16), we get

$$m_{2.2} = \frac{\pi_1 w_1 + \pi_2 w_2 + \pi_3 w_3}{\pi_2}$$

$$= \frac{a + A(v)b(u) + c(1 - A(v)B(u))}{A(v)}. \tag{7.17}$$

As a numerical example, suppose the lifetimes are uniformly distributed over $[10, 40]$ days, the repair times are exponentially distributed with mean 2 days, and the replacement times are exactly 1 day. Suppose the cutoff numbers are $u = 30$ days, and $v = 3$ days. The relevant quantities are

$$A(30) = \mathsf{P}(\text{lifetime} \leq 30) = .6667,$$

$$B(3) = \mathsf{P}(\text{repair time} \leq 3) = 1 - e^{-3/2} = .7769,$$

$$a = \text{expected lifetime} = 25,$$

$$b(3) = \mathsf{E}(\min(3, \text{repair time})) = \frac{1}{.5}(1 - e^{-.5*3}) = 1.5537,$$

$$c = \text{expected replacement time} = 1.$$

Substituting in (7.17), we get

$$m_{2.2} = \frac{25 + (.6667) * (1.5537) + (1)(1 - (.6667)(.7769))}{.6667}$$

$$= \frac{26.5179}{.6667}$$

$$= 39.7769 \text{ days.} \qquad \square$$

With Theorem 7.4, we are ready to study the limiting behavior of the SMP in the next subsection.

7.5.2. Occupancy Distributions

In this subsection we study the limiting behavior of the SMPs. In our study of the limiting behavior of DTMCs and CTMCs, we encountered three related quantities:

(1) the limiting distribution;

(2) the stationary distribution; and

(3) the occupancy distribution.

In the case of the SMPs we shall concentrate only on the last quantity. Study of the first two quantities requires more advanced techniques, and is outside the scope of this book.

Let $M_j(t)$ be the total (random) amount of time the SMP spends in state j during the interval $[0, t]$. We are interested in computing the long-run fraction of the time the SMP spends in state j, i.e., we want to compute

$$\lim_{t \to \infty} \frac{M_j(t)}{t}.$$

If the above limit exists (with probability one), we denote it by p_j, and we call the vector $[p_1, p_2, \ldots, p_N]$ the *occupancy distribution* of the SMP. The next theorem shows when the occupancy distribution exists and how to compute it.

Theorem 7.6 (Occupancy Distribution). *Suppose $\{X(t), t \geq 0\}$ is an SMP with an irreducible embedded DTMC $\{X_n, n \geq 0\}$ with transition probability matrix P. The occupancy distribution of the SMP exists, is independent of the initial state of the SMP, and is given by*

$$p_j = \frac{\pi_j w_j}{\sum_{i=1}^{N} \pi_i w_i}, \qquad 1 \leq j \leq N, \tag{7.18}$$

where $[\pi_1, \pi_2, \ldots, \pi_N]$ is a solution to the balance equations

$$\pi_j = \sum_{i=1}^{N} \pi_i p_{i.j}, \qquad 1 \leq j \leq N.$$

and w_i, $1 \leq i \leq N$, is the mean sojourn time in state i.

Proof. Suppose the SMP starts in state j. We first show that $\{M_j(t), t \geq 0\}$ is a cumulative process as defined in Section 7.3. Let $\hat{S}_n$ be the time of the nth entry into state j. ($\hat{S}_0 = 0$.) Let $\hat{C}_n$ be the total time the SMP spends in state j during the nth cycle $(\hat{S}_{n-1}, \hat{S}_n]$. Thus

$$\hat{C}_n = M_j(\hat{S}_n) - M_j(\hat{S}_{n-1}).$$

Since the SMP has the Markov property at every transition epoch, its behavior from time $\hat{S}_n$ onward depends on the history only via its state at that time. However the SMP is in state j at time $\hat{S}_n$ for all $n \geq 0$, by definition. Hence its behavior over the intervals $(\hat{S}_{n-1}, \hat{S}_n]$ $(n \geq 1)$ is iid. Thus $\{(\hat{T}_n, \hat{C}_n), n \geq 1\}$ is a sequence of iid bivariate random variables, where $\hat{T}_n = \hat{S}_n - \hat{S}_{n-1}$. Then, from the definition of cumulative processes in Section 7.3, it follows that $\{M_j(t), t \geq 0\}$ is a cumulative process.

Using Theorem 7.3, we get

$$\lim_{t \to \infty} \frac{M_j(t)}{t} = \frac{\mathsf{E}(\hat{C}_1)}{\mathsf{E}(\hat{T}_1)}.$$

Now, $\mathsf{E}(\hat{T}_1)$ is just the mean inter-visit time $m_{j.j}$ to state j, and is given by (7.16). Hence

$$\mathsf{E}(\hat{T}_1) = \mathsf{E}(\hat{S}_1) = \frac{\sum_{i=1}^{N} \pi_i w_i}{\pi_j}.$$

Furthermore, during the interval $(0, \hat{S}_1]$ the SMP spends exactly one sojourn time in state j. Hence

$$\mathsf{E}(\hat{C}_1) = w_j.$$

Using these two equations we get

$$\lim_{t \to \infty} \frac{M_j(t)}{t} = \frac{\pi_j w_j}{\sum_{i=1}^{N} \pi_i w_i}.$$

Thus the theorem is proved if the initial state is j. Now consider the case where the initial state is $i \neq j$. Define $\hat{S}_n$, $\hat{T}_n$, and $\hat{C}_n$ as before. Now, $\{(\hat{T}_n, \hat{C}_n), n \geq 2\}$ is a sequence of iid bivariate random variables, and is independent of $(\hat{S}_1, \hat{C}_1)$. We can follow the proof of Theorem 7.3 to show that

$$\lim_{t \to \infty} \frac{M_j(t)}{t} = \frac{\mathsf{E}(\hat{C}_2)}{\mathsf{E}(\hat{T}_2)}.$$

Since the SMP enters state j at time $\hat{S}_1$, it follows that

$$\mathsf{E}(\hat{T}_2) = \frac{\sum_{i=1}^{N} \pi_i w_i}{\pi_j},$$

and

$$\mathsf{E}(\hat{C}_2) = w_j.$$

Thus the theorem is valid for all initial states $1 \leq i \leq N$, and hence for any initial distribution.

$\square$

Note that we have defined the occupancy distribution as the long-run fraction of the time spent in state j, $1 \leq j \leq N$. But in the case of CTMCs and DTMCs, the occupancy distribution was defined as the long-run *expected* fraction of the time spent in state j, $1 \leq j \leq N$. So, strictly speaking, we should have studied

$$\lim_{t \to \infty} \frac{\mathsf{E}(M_j(t))}{t}$$

to be consistent. It can be shown that, under the hypothesis of Theorem 7.6, the above limit also exists and is given by p_j of (7.18); however, it is beyond the scope of this book to do so. We illustrate the theorem by several examples.

Example 7.25 (Two-State Machine). Compute the long-run fraction of the time the machine of Example 7.19 is up.

Using the results of Example 7.19 we see that the state of the machine is described by a two-state SMP $\{X(t), t \geq 0\}$. The embedded DTMC $\{X_n, n \geq 0\}$ has the P matrix given in (7.11), and is irreducible. Hence Theorem 7.6 can be applied. The relevant quantities are given in Examples 7.19 and 7.23. Hence, the long-run fraction of the time the machine is up is given by

$$\begin{aligned} p_1 &= \frac{\pi_1 w_1}{\pi_0 w_0 + \pi_1 w_1} \\ &= \frac{\mathsf{E}(U)}{\mathsf{E}(D) + \mathsf{E}(U)}. \end{aligned}$$

This agrees with our intuition, and with the result in Example 7.17 for a more general two-state machine. $\qquad\square$

Example 7.26 (Series System). Consider the series system described in Example 7.21. Compute the long-run fraction of the time the system is up.

The state of the system is described by an SMP in Example 7.21. The embedded DTMC has a transition probability matrix given in (7.13). The DTMC is irreducible, and the balance equations are given by

$$\pi_0 = \sum_{i=1}^{N} \pi_i,$$

$$\pi_j = \frac{v_j}{v}\pi_0, \qquad 1 \le j \le N,$$

where $v = \sum_{i=1}^{N} v_i$. A solution to the above equations is given by

$$\pi_0 = 1,$$

$$\pi_j = \frac{v_j}{v}, \qquad 1 \le j \le N.$$

The mean sojourn times are computed in Example 7.21. Using (7.18), the long-run fraction of the time the system is up is given by

$$p_0 = \frac{\pi_0 w_0}{\sum_{i=0}^{N} \pi_i w_i}$$

$$= \frac{1/v}{1/v + \sum_{i=1}^{N}(v_i/v)\tau_i}$$

$$= \frac{1}{1 + \sum_{i=1}^{N} v_i \tau_i}. \qquad\square$$

Example 7.27 (Machine Maintenance). Compute the long-run fraction of the time a machine is working if it is maintained by using the policy of Example 7.22.

The state of the machine under the maintenance policy of Example 7.22 is an SMP with the transition matrix of the embedded DTMC as given in (7.14). This DTMC is irreducible, and hence Theorem 7.6 can be used to compute the required quantity as

$$p_1 = \frac{\pi_1 w_1}{\sum_{i=1}^{3} \pi_i w_i}. \tag{7.19}$$

Using the relevant quantities as computed in Examples 7.22 and 7.24 we get

$$p_1 = \frac{a}{a + A(v)b(u) + c(1 - A(v)B(u))}.$$

For the numerical values given in Example 7.24, we get

$$p_1 = \frac{25}{26.5179} = .9428.$$

Thus the machine is up 94% of the time. $\qquad\square$

7.5.3. Long-Run Cost Rates

In this subsection we shall study cost models for systems modeled by SMPs, as we did for the CTMCs in Section 6.10. In the DTMC and the CTMC cases we studied two types of cost models: the total expected cost over a finite horizon, and the long-run expected cost per unit time. In the case of SMPs we shall only study the latter, since the former requires results for the short-term analysis of SMPs, which we have not developed here. We begin with the following cost model.

Let $\{X(t), t \geq 0\}$ be an SMP on state space $\{1, 2, \ldots, N\}$. Suppose that the system incurs cost at a rate of $c(i)$ per unit time while in state i. Now let $C(T)$ be the total cost incurred by the system up to time T. We are interested in the long-run cost rate defined as

$$g = \lim_{T \to \infty} \frac{C(T)}{T},$$

assuming this limit exists with probability one. The next theorem shows when this limit exists, and how to compute it if it does exist.

Theorem 7.7 (Long-Run Cost Rates). *Suppose the hypothesis of Theorem 7.6 holds, and let $[p_1, p_2, \ldots, p_N]$ be the occupancy distribution of the SMP. Then the long-run cost rate exists, is independent of the initial state of the SMP, and is given by*

$$g = \sum_{i=1}^{N} p_i c(i). \tag{7.20}$$

Proof. Let $M_i(T)$ be the total amount of time the SMP spends in state i during $[0, T]$. Since the SMP incurs cost at a rate of $c(i)$ per unit time while in state i, it follows that

$$C(T) = \sum_{i=1}^{N} c(i) M_i(T). \tag{7.21}$$

Hence

$$\lim_{T \to \infty} \frac{C(T)}{T} = \lim_{T \to \infty} \sum_{i=1}^{N} c(i) \frac{M_i(T)}{T}$$

$$= \sum_{i=1}^{N} c(i) p_i,$$

where the last equation follows from Theorem 7.6. This proves the theorem.   □

Although we have assumed a specific cost structure in the above derivation, Theorem 7.7 can be used for a variety of cost structures. For example, suppose the system incurs a lump sum cost d_i whenever it visits state i. We can convert this into the cost structure of Theorem 7.7 by assuming that the per unit time cost rate $c(i)$ in state i is such that the expected total cost incurred in one sojourn time in state i is d_i. Since the expected sojourn time in state i is w_i, we must have $c(i)w_i = d_i$, which yields

$$c(i) = \frac{d_i}{w_i}.$$

Hence the long-run cost rate under the lump sum cost structure can be derived from Theorem 7.7 to be

$$g = \sum_{i=1}^{N} p_i \frac{d_i}{w_i}.$$

Substituting for p_i from (7.18), we get

$$g = \frac{\sum_{i=1}^{N} \pi_i d_i}{\sum_{i=1}^{N} \pi_i w_i}. \tag{7.22}$$

We illustrate with several examples.

Example 7.28 (Two-State Machine). Consider the two-state machine of Example 7.19. Suppose the machine produces a net revenue of $\$A$ per unit time when the machine is up, while it costs $\$B$ per unit time to repair the machine when it is down. Compute the long-run cost rate of the machine.

Using the results of Example 7.19 we see that the state of the machine is described by a two-state SMP $\{X(t), t \geq 0\}$. We have $c(0) = B$ and $c(1) = -A$. From Example 7.25 we have

$$p_0 = \frac{E(D)}{E(D) + E(U)},$$

and

$$p_1 = \frac{E(U)}{E(D) + E(U)}.$$

Substituting in (7.20) we get the long-run cost rate as

$$g = Bp_0 - Ap_1 = \frac{BE(D) - AE(U)}{E(D) + E(U)}.$$

This is consistent with the results of Example 6.32.   □

Example 7.29 (Series System). Consider the series system of Example 7.21 with three components. Suppose the mean lifetime of the first component is 5 days, that of the second component is 4 days, and the third component is 8 days. The mean repair time is 1 day for component 1, .5 day for component 2, and 2 days for component 3. It costs $200 to repair component 1, $150 to repair component 2, and $500 to repair component 3. When the system is working it produces revenues at a rate of R per unit time. What is the minimum value of R that makes it worthwhile to operate the system?

The state of the system is described by an SMP in Example 7.21 with state space $\{0, 1, 2, 3\}$. Recall that the lifetime of component i is an $\text{Exp}(v_i)$ random variable, with mean $1/v_i$. Using the data given above we get

$$v_1 = .2, \qquad v_2 = .25, \qquad v_3 = .125.$$

The mean repair times are

$$\tau_1 = 1, \qquad \tau_2 = .5, \qquad \tau_3 = 2.$$

The cost structure is a mixture of lump sum costs and continuous costs. We first convert it to the standard cost structure. We have

$$c(0) = -R, \qquad c(1) = 200/1 = 200, \qquad c(2) = 150/.5 = 300, \qquad c(3) = 500/2 = 250.$$

Using the results of Example 7.26 we have

$$p_0 = \frac{1}{1 + \sum_{i=1}^{3} v_i \tau_i} = \frac{1}{1.575},$$

$$p_1 = \frac{v_1 \tau_1}{1 + \sum_{i=1}^{3} v_i \tau_i} = \frac{.2}{1.575},$$

$$p_2 = \frac{v_2 \tau_2}{1 + \sum_{i=1}^{3} v_i \tau_i} = \frac{.125}{1.575},$$

$$p_3 = \frac{v_3 \tau_3}{1 + \sum_{i=1}^{3} v_i \tau_i} = \frac{.25}{1.575}.$$

Substituting in (7.20) we get

$$g = \sum_{i=0}^{3} c(i) p_i = \frac{-R + 40 + 37.5 + 62.5}{1.575}.$$

For break even, we must have $g \leq 0$. Hence we must have

$$R \geq 140$$

dollars per day.   □

7.6. Problems

CONCEPTUAL PROBLEMS

7.1. Consider the battery replacement problem of Example 7.2. Let $N(t)$ be the number of planned replacements up to time t. Show that $\{N(t), t \geq 0\}$ is a renewal process.

7.2. Consider the battery replacement problem of Example 7.2. Let $N(t)$ be the number of unplanned replacements up to time t. Show that $\{N(t), t \geq 0\}$ is a renewal process.

7.3. Let $\{N(t), t \geq 0\}$ be a renewal process, and let $X(t)$ be the integer part of $N(t)/2$. Is $\{X(t), t \geq 0\}$ a renewal process?

Definition 7.6 (Delayed Renewal Process). A counting process generated by a sequence of nonnegative random variables $\{T_n, n \geq 1\}$ is called a delayed renewal process if $\{T_n, n \geq 2\}$ is a sequence of iid random variables, and is independent of T_1.

7.4. Let $\{N(t), t \geq 0\}$ be a renewal process, and let $X(t)$ be the integer part of $(N(t) + 1)/2$. Is $\{X(t), t \geq 0\}$ a renewal process? Is it a delayed renewal process?

7.5. Let $\{X(t), t \geq 0\}$ be a CTMC on state space $\{1, 2, \ldots, N\}$. Let $N(t)$ be the number of times the CTMC enters state i over $(0, t]$. Show that $\{N(t), t \geq 0\}$ is a delayed renewal process if $X(0) \neq i$. (We have shown in Example 7.4 that it is a renewal process if $X(0) = i$.)

7.6. Prove Theorem 7.2 for a delayed renewal process, with $\tau = \mathsf{E}(T_n)$, $n \geq 2$.

7.7. Tasks arrive at a receiving station one at a time, the inter-arrival times being iid with common mean τ. As soon as K tasks are accumulated they are instantly dispatched to a workshop. Let $Y(t)$ be the number of tasks received by the receiving station by time t. Show that $\{Y(t), t \geq 0\}$ is a renewal process. Let $Z(t)$ be the number of tasks received by the workshop by time t. Show that $\{Z(t), t \geq 0\}$ is a cumulative process. Compute the long-run rate at which jobs are received by the receiving station, and by the workshop.

7.8. Suppose the tasks in Conceptual Problem 7.7 arrive according to a $PP(\lambda)$. Does this make the $\{Z(t), t \geq 0\}$ process defined there a Compound Poisson Process? Why or why not?

7.9. A machine produces parts in a deterministic fashion at a rate of one per hour. They are stored in a warehouse. Trucks leave from the warehouse according to a Poisson process (i.e., the inter-departure times are iid exponential random variables) with rate λ. The trucks are sufficiently large and the parts are small so that each truck can carry away all the parts produced since the previous truck's departure. Let $Z(t)$ be the number of parts that have left the warehouse by time t. Is $\{Z(t), t \geq 0\}$ a CPP? Why or why not?

7.10. Consider Conceptual Problem 7.7. Let $X(t)$ be the number of tasks in the receiving station at time t. Show that $\{X(t), t \geq 0\}$ is an SMP, and compute its transition probability matrix, and the vector of expected sojourn times.

7.11. Consider the series system described in Example 7.21. Compute the long-run fraction of the time that the component i is down.

7.12. Consider the series system described in Example 7.21. Let $Y(t)$ be 1 if the system is up and 0 if it is down at time t. Show that $\{Y(t), t \geq 0\}$ is an SMP. Compute the transition probability matrix of the embedded DTMC and the vector of expected sojourn times.

7.13. Consider the series system described in Example 7.21. Compute the expected time between two consecutive failures of component i, $1 \leq i \leq N$.

7.14. (Parallel System). Consider a parallel system of N components. The system fails as soon as all the components fail. Suppose the lifetimes of the components are iid $Exp(v)$ random variables. The system is repaired when it fails. The mean repair time of the system is τ. Let $X(t)$ be the number of functioning components at time t. Show that $\{X(t), t \geq 0\}$ is an SMP. Compute the transition matrix P and the sojourn time vector w.

7.15. Consider the parallel system of Conceptual Problem 7.14. Compute the long-run fraction of the time that the system is down.

7.16. Consider the parallel system described in Conceptual Problem 7.14. Compute the expected time between two consecutive failures of the system.

7.17. Consider the parallel system described in Conceptual Problem 7.14. Let $Y(t)$ be 1 if the system is up and 0 if it is down at time t. Show that $\{Y(t), t \geq 0\}$ is an SMP. Compute the transition probability matrix of the embedded DTMC and the vector of expected sojourn times.

7.18. Customers arrive at a public telephone booth according to a PP(λ). If the telephone is available, an arriving customer starts using it. The call durations are iid random variables with common distribution $A(\cdot)$. If the telephone is busy when a customer arrives, he/she simply goes away in search of another public telephone. Let $X(t)$ be 0 if the phone is idle, and 1 if it is busy at time t. Show that $\{X(t), t \geq 0\}$ is an SMP. Compute the P matrix and the w vector.

7.19. A critical part of a machine is available from two suppliers. The lifetime of the parts from the ith ($i = 1, 2$) supplier are iid random variables with common cdf $A_i(\cdot)$. Suppose the following replacement policy is followed: if the component currently in use lasts longer than T amount of time (T is a fixed positive constant), it is replaced upon failure by another component from the same supplier that provided the current component; or else it is replaced upon failure by a component from the other supplier. Replacement is instantaneous. Let $X(t) = i$ if the component in use at time t is from supplier i ($i = 1, 2$). Show that $\{X(t), t \geq 0\}$ is an SMP, and compute its P matrix and w vector.

7.20. A department in a university has several full-time faculty positions. Whenever a faculty member leaves the department, he or she is replaced by a new member at the assistant professor level. Approximately 20% of the assistant professors leave (or are asked to leave) at the end of the fourth year (without being considered for tenure), 30% of the assistant professors are considered for tenure at the end of the fifth year, 30% at the end of the sixth year, and the remaining at the end of the seventh year. The probability of getting a tenure at the end of 5 years is .4, at the end of 6 years is .5, and at the end of 7 years is .6. If a tenure is granted, the assistant professor becomes an associate professor with tenure, otherwise he/she leaves the department. An associate professor spends a minimum of 3 years and a maximum of 7 years in that position before becoming a full professor. At the end of each of the years 3, 4, 5, 6, and 7, there is a 20% probability that the associate professor is promoted to full professor, independent of everything else. If the promotion does not come through in a given year, the associate professor leaves with probability .2, and continues for another year with probability .8. If no promotion comes through even at the end of the seventh year, the associate professor leaves the department. A full professor stays with the department for 6 years on average and then leaves. Let $X(t)$ be the position of a faculty member at time t (1 if assistant professor, 2 if associate professor, and 3 if full professor). Note that if the faculty member leaves at time t, then $X(t) = 1$, since the new faculty member replacing the departing one starts as an assistant professor. Show that $\{X(t), t \geq 0\}$ is an SMP. Compute its P matrix and w vector.

7.21. Redo Conceptual Problem 7.20 with the following modification: a departing faculty member is replaced by an assistant professor with probability .6, an associate professor with probability .3, and a full professor with probability .1.

COMPUTATIONAL PROBLEMS

7.1. Consider the battery replacement problem of Example 7.2. Suppose the battery lifetimes are iid Erlang with parameters $k = 3$ and $\lambda = 1$ (per year). Compute the long-run replacement rate if Mr Smith replaces the batteries upon failure.

7.2. Consider Computational Problem 7.1. Suppose Mr Smith replaces the batteries upon failure or upon reaching the expected lifetime of the battery. Compute the long-run rate of replacements.

7.3. Consider Computational Problem 7.2. Compute the long-run rate of planned replacements.

7.4. Consider Computational Problem 7.2. Compute the long-run rate of unplanned replacements.

7.5. Consider the machine reliability problem of Example 5.4. Suppose a repair has just been completed at time 0, and the machine is up. Compute the number of repair completions per day in the long run.

7.6. Consider the inventory system of Example 5.6. Compute the number of orders placed per week (regardless of the size of the orders) in the long run.

7.7. Consider the five processor computer system described in Conceptual Problem 6.7 and Computational Problem 6.12. Suppose the computer system is replaced by a new one upon failure. Replacement time is negligible. Successive systems are independent. Compute the number of replacements per unit time in the long run.

7.8. Consider the two-component system described in Conceptual Problem 6.10 and Computational Problem 6.14. Suppose the system is replaced instantaneously upon failure by a new one. Successive systems are independent. Compute the number of replacements per unit time in the long run.

7.9. Consider Computational Problem 7.2. Suppose it costs $75 to replace a battery. It costs an additional $75 if the battery fails during operation. Compute the long-run cost rate of the planned replacement policy followed by Mr Smith. Compare the cost rate of the planned replacement policy to the "replace upon failure" policy.

7.10. The lifetime of a machine is an Erlang random variable with parameters $k = 2$ and $\lambda = .2$ (per day). When the machine fails, it is repaired. The repair times are $\text{Exp}(\mu)$ with mean 1 day. Suppose the repair costs $10 per hour. The machine produces revenues at a rate of $200 per day when it is working. Compute the long-run net revenue per day.

7.11. In Computational Problem 7.10, what minimum revenue per working day of the machine is needed to make the operation profitable?

7.12. The Quick-Toast brand of toaster ovens sell for $80 per unit. The lifetime (in years) of the ovens have the following pdf:

$$f(t) = \frac{t^2}{21}, \qquad 1 \le t \le 4.$$

The Eat-Well restaurant uses this oven, and replaces it by a new Quick-Toast oven upon failure. Compute the yearly cost in the long run.

7.13. The Eat-Well restaurant of Computational Problem 7.12 faces the following decision: The toaster oven manufacturer has introduced a maintenance policy on the Quick-Toast toaster ovens. For $10 a year, the manufacturer will replace, free of cost, any toaster oven that fails in the first 3 years of its life. If a toaster oven fails after 3 years, a new one has to be purchased for the full price. Is it worth signing up for the maintenance contract?

7.14. The Quick-Toast toaster ovens of Computational Problem 7.12 can be purchased with a prorated warranty that works as follows. If the oven fails after T years, the manufacturer will buy it back for $80*(4 − T)/3$. The price of the toaster ovens with the warranty is $90. Suppose the Eat-Well restaurant decides to buy the ovens with this warranty. What is the yearly cost of this policy?

7.15. Consider the general two-state machine of Example 7.17. The mean uptime of the machine is 5 days, while the mean downtime is 1 day. The machine produces 100 items per day when it is working, and each item sells for $5. The repair costs $10 per hour. The owner of this machine is considering replacing the current repair person by another who works twice as fast, but costs $20 per hour of repair time. Is it worth switching to the more efficient repair person?

7.16. The lifetime of an oil pump is an exponential random variable with mean 5 days, and the repair time is an exponential random variable with mean 1 day. When the pump is working, it produces 500 gallons of oil per day. The oil is consumed at a rate of 100 gallons a day. The excess oil is stored in a tank of unlimited capacity. Thus the oil reserves increase by 400 gallons a day when the pump is working. When the pump fails, the oil reserves decrease by 100 gallons a day. When the pump is repaired, it is turned on as soon as the tank is empty. Suppose it costs 5 cents per gallon per day to store the oil. Compute the storage cost per day in the long run.

7.17. Consider Computational Problem 7.16 with the following modification: the oil tank has a finite capacity of 1000 gallons. When the tank becomes full the output of the oil pump is reduced to 100 gallons per day, so that the tank remains full until the pump fails. The lifetime of the oil pump is unaffected by the output rate of the pump. Now compute the storage cost per day in the long run.

7.18. In Computational Problem 7.16, what fraction of the time is the pump idle (repaired but not turned on) in the long-run? (*Hint:* Suppose it costs $1 per day to keep the pump idle. Then the required quantity is the long-run cost per day.)

7.19. In Computational Problem 7.17, what fraction of the time is the pump idle (repaired but not turned on) in the long run? (See the hint for Computational Problem 7.18.)

7.20. In Computational Problem 7.17, what fraction of the time is the tank full in the long run? (See the hint for Computational Problem 7.18.)

7.21. Customers arrive at a bank according to a Poisson process at the rate of 20 per hour. Sixty percent of the customers make a deposit, while 40% of the customers make a withdrawal. The deposits are distributed uniformly over [50, 500] dollars, while the withdrawals are distributed uniformly over [100, 200] dollars. Let $Z(t)$ be the total amount of funds (deposits − withdrawals) received by the bank up to time t. (This can be a negative quantity.) What assumptions are needed to make $\{Z(t), t \geq 0\}$ a CPP? Using those assumptions compute the expected funds received by the bank in an 8-hour day.

7.22. Customers arrive at a department store according to a Poisson process at the rate of 80 per hour. The average purchase is $35. Let $Z(t)$ be the amount of money spent by all the customers who arrived by time t. What assumtions are needed to make $\{Z(t), t \geq 0\}$ a CPP? Using those assumptions compute the expected revenue during a 16-hour day.

7.23. Travellers arrive at an airport in batches. The successive batch sizes are iid random variables with the following pmf: $p(1) = .4$, $p(2) = .2$, $p(3) = .2$, $p(4) = .1$, $p(5) = .05$, $p(6) = .03$, $p(8) = .01$, $p(10) = .01$. The batches arrive according to a Poisson process at the rate of 35 per hour. Compute the expected number of travellers that arrive during 3 hours.

7.24. Consider Computational Problem 7.16. Let $X(t)$ be the state of the oil pump at time t (1 if it is up, 2 if it is under repair, and 3 if it is repaired but idle). Is $\{X(t), t \geq 0\}$ an SMP?

7.25. Consider the series system described in Example 7.21 with three components. The mean lifetimes of the three components are 2, 3, and 4 days, respectively, while the mean repair times are .1, .2, and .3 days. Compute the long-run fraction of the time that the system is up.

7.26. Consider the series system described in Computational Problem 7.25. Compute the expected time between two consecutive failures of the first component.

7.27. Consider the series system of Computational Problem 7.25. Compute the long-run fraction of the time that the third component is down.

7.28. Consider the series system of Computational Problem 7.25. Suppose the three components produce revenues at the rate of $10, $20, and $30 per day (respectively) when the system is up. The repair costs of the three components are $15, $20, and $25 per day (respectively) when the component is down. Compute the long-run net revenue per day for the system.

7.29. Consider the parallel system described in Conceptual Problem 7.14 with three components. The mean lifetimes of each components is 4 days, while the mean repair time is 1 day. Compute the long-run fraction of the time that the system is down.

8

Queueing Models

8.1 Queueing Systems

Queues (or waiting lines) are an unavoidable component of modern life. We are required to stand physically in queues in grocery stores, banks, department stores, amusement parks, movie theaters, etc. Although we don't like standing in a queue, we appreciate the fairness that it imposes. Even when we use phones to conduct business, often we are put on hold, and served in a first-come–first-served fashion. Thus we face a queue even if we are in our own home!

Queues are not just for humans, however. Modern communication systems transmit messages (like emails) from one computer to another by queuing them up inside the network a complicated fashion. Modern manufacturing systems maintain queues (called inventories) of raw materials, partly finished goods, and finished goods throughout the manufacturing process. "Supply chain management" is nothing but the management of these queues!

In this chapter we shall study some simple models of queues. Typically, a queueing system consists of a stream of customers (humans, finished goods, messages) that arrive at a service facility, get served according a given service discipline, and then depart. The service facility may have one or more servers, and finite or infinite capacity. In practice we are interested in studying a queueing system, namely, its capacity, number of servers, service discipline, etc. Queueing theory will help us do this by answering the following questions (and many others):

How many customers are there in the queue on average?

How much time does a typical customer spend in the queue?

How many customers are rejected or lost due to capacity limitations?

How busy are the servers?

We start by introducing a standard nomenclature for single-station queues, i.e., queueing systems where customers form a single queue. Such a queue is described as follows:

(1) Arrival Process

We assume that customers arrive one at a time, and the successive inter-arrival times are iid. (Thus the arrival process is a renewal process studied in Chapter 7.) It is described by the distribution of the inter-arrival times, represented by special symbols as follows:

- M: Exponential;
- G: General;
- D: Deterministic;
- E_k: Erlang with k phases, etc.

Note that the Poisson arrival process is represented by M (for memoryless, i.e., exponential inter-arrival times).

(2) Service Times

We assume that the service times of successive customers are iid. They are represented by the same letters as the inter-arrival times.

(3) Number of Servers

Typically denoted by s. All the servers are assumed to be identical, and that any customer can be served by any server.

(4) Capacity

Typically denoted by K. It includes the customers in service. If an arriving customer finds K customers in the system, he/she is permanently lost. If capacity is not mentioned, it is assumed to be infinite.

Example 8.1 (Nomenclature). In an $M/M/1$ queue customers arrive according to a Poisson process, request iid exponential service times and are served by a single server. The capacity is infinite. In an $M/G/2/10$ queue, the customers arrive according to a Poisson process and demand iid service times with general distribution. The service facility has two servers and a capacity to hold 10 customers.   □

In this chapter we shall always assume a first-come–first-served (FCFS) service discipline.

8.2. Single-Station Queues: General Results

In this section we state several general results for single-station queueing systems. Let A_n be the time of the nth arrival. We allow for the possibility that not all arrivals may actually enter the system. Hence we define E_n to be the time of the nth entry. (If all arriving customers enter the system, we have $A_n = E_n$ for all $n = 0, 1, 2, \ldots$.) Let D_n be the time of the nth departure. If the service discipline is FCFS, then the nth entering customer enters at time E_n and leaves at time D_n. Hence the time spent by the nth customer in the system is given by $W_n = D_n - E_n$. (See Figure 8.1.)

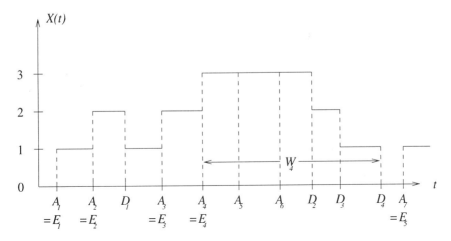

FIGURE 8.1. A typical sample path of the queue-length process.

Let $N(t)$ be the number of customers that arrived by time t, let $D(t)$ be the number of customers that departed by time t, and let $X(t)$ be the number of customers in the system at time t. Furthermore, let

$$X_n = X(D_n+),$$

$$X_n^* = X(E_n-),$$

$$\hat{X}_n = X(A_n-).$$

Thus, X_n is the number of customers left behind by the nth departure, X_n^* is the number of customers in the system as seen by the nth *entry* (not including the entering customer) into the system, and $\hat{X}_n$ is the number of customers in the system as seen by the nth *arrival* (not including the arriving customer) to the system. Now (for $j = 0, 1, 2, \ldots$) define the following limits, assuming they exist:

$$p_j = \lim_{t \to \infty} P(X(t) = j), \tag{8.1}$$

$$\pi_j = \lim_{n \to \infty} P(X_n = j), \tag{8.2}$$

$$\pi_j^* = \lim_{n \to \infty} P(X_n^* = j), \tag{8.3}$$

$$\hat{\pi}_j = \lim_{n \to \infty} P(\hat{X}_n = j). \tag{8.4}$$

We can interpret p_j as the long-run fraction of the time that the system has j customers in it, π_j as the long-run fraction of departures that leave behind j customers in the system, π_j^* as the long-run fraction of entering customers that see j customers ahead of them in the system, and $\hat{\pi}_j$ as the long-run fraction of arrivals that see j customers ahead of them in the system. Since all these quantities relate to a common queueing system, we expect them to be related to each other. The next two theorems state these relations.

Theorem 8.1 (When Is $\pi_j = \pi_j^*$?). *Suppose the arrivals and departures take place one at a time. If the limits in (8.2) or (8.3) exist,*

$$\pi_j = \pi_j^*, \qquad j \geq 0.$$

Idea of Proof. The theorem is a deterministic result with no probabilistic assumptions. Suppose $X(0) = i$. Then using a careful counting argument (using the fact that the arrivals and departures take place one at a time) we can show that

$$\{X_{n+i} \leq j\} \quad \Leftrightarrow \quad \{X_{n+j+1}^* \leq j\}$$

for all $n \geq 0$, and all $j \geq 0$. (See Kulkarni [1995].) Hence

$$P(X_{n+i} \leq j) = P(X_{n+j+1}^* \leq j).$$

Assuming the limit exists on either the left or the right side, we get

$$\lim_{n \to \infty} P(X_{n+i} \leq j) = \lim_{n \to \infty} P(X_{n+j+1}^* \leq j).$$

which yields

$$\lim_{n \to \infty} P(X_n \leq j) = \lim_{n \to \infty} P(X_n^* \leq j).$$

However, this implies

$$\sum_{i=0}^{j} \pi_i = \sum_{i=0}^{j} \pi_i^*$$

and hence

$$\pi_j = \pi_j^*.$$

Hence the theorem follows. $\square$

Theorem 8.2 (Poisson Arrivals See Time Averages: PASTA). *Suppose the arrival process is Poisson and $\{N(t + s), s \geq 0\}$ is independent of $\{X(u), 0 \leq u \leq t\}$. If the limits in (8.1) or (8.4) exist*

$$p_j = \hat{\pi}_j, \qquad j \geq 0.$$

Intuition. The technical condition about independence says that the arrival process from any point t onward is independent of the history of the system up to time t. Now, assuming this condition, and using the properties of the Poisson processes given in Section 6.5, we see that the probability that exactly one arrival occurs in the interval $[t, t + h]$ is independent of the history up to time t. If h is sufficiently small, we can say that the arrival occurs at time t. By assumption, $X(t)$ is independent of what happens to the arrival process over $(t, t + h)$. Hence the conditional distribution of $X(t)$, given that there is an arrival in the interval $(t, t + h)$, is the same as the distribution of $X(t)$. Thus the distribution of the state as seen by an arrival at time t is the same as the distribution of $X(t)$. Thus the long-run fraction of the arrivals that see the system in state j (namely $\hat{\pi}_j$) must be the same as the fraction of the time the system spends in state j (namely p_j). This result is sometimes called PASTA (Poisson Arrivals See Time Averages.) □

We illustrate the above theorems by means of an example.

Example 8.2 ($M/M/1/1$ Queue). Consider an $M/M/1/1$ queue with arrival rate λ and service rate μ. In such a queue the arrival process is Poisson with rate λ and service times are iid exponential with parameter μ. However, since the capacity is 1, an arriving customer can enter the system only if the server is idle; or else the arrival is lost. The $\{X(t), t \geq 0\}$ process for such a queue is modeled as a CTMC on state space $\{0, 1\}$ in Example 6.9 with $K = 1$. The steady-state distribution can be computed from the results of Example 6.27 as follows:

$$p_0 = \frac{\mu}{\lambda + \mu}, \qquad p_1 = \frac{\lambda}{\lambda + \mu}.$$

Since the arrival process is Poisson, it follows from Theorem 8.2 that

$$\hat{\pi}_0 = p_0, \qquad \hat{\pi}_1 = p_1.$$

Thus the long-run probability that a customer is turned away is given by $\hat{\pi}_1 = \lambda/(\lambda + \mu)$. Since a departing customer always leaves an empty system, we have $X_n = 0$ for all $n \geq 0$. Hence

$$\pi_0 = 1, \qquad \pi_1 = 0.$$

Similarly, since a customer can enter only if the system is empty, we have $X_n^* = 0$ for all $n \geq 0$. Hence,

$$\pi_0^* = 1, \qquad \pi_1^* = 0.$$

Thus Theorem 8.1 is verified. Note that the last two observations are valid for any $G/G/1/1$ queue. □

Now define

L = the expected number of customers in the system in steady state;

λ = the arrival rate of customers;

W = the expected time spent in the system by a customer in steady state.

The relation between these quantities is the celebrated Little's Law, and is stated in the following theorem:

Theorem 8.3 (Little's Law). *If the quantities L, λ, and W defined above exist and are finite, they satisfy*

$$L = \lambda W.$$

Intuition. Suppose each arriving customer pays the system $1 per unit time that the customer spends in the system. First we compute the long-run rate, R_{customer}, at which the money is spent by the customers. Customers arrive at the rate of λ per unit time. Each customer spends on average W time in the system and hence pays $$W$ to the system. Hence we must have

$$R_{\text{customer}} = \lambda W.$$

Next we compute the long-run rate, R_{system} at which the system earns money. Since each customer pays the system $1 per unit time, and there are L customers in the system on average, we must have

$$R_{\text{system}} = L.$$

Now we expect that R_{customer} will equal R_{system} since no money is created or destroyed in the transaction. Hence we get

$$L = \lambda W$$

which is Little's Law.   □

Little's Law is also a sample path result, i.e., it does not make any probabilistic assumptions about the system. Care has to be taken in applying it in a consistent manner as described in the following example:

Example 8.3 (Little's Law and $M/M/1/1$ Queue). Consider the $M/M/1/1$ queue of Example 8.2. The expected number of customers in the system in steady state can be computed as

$$L = 0 \cdot p_0 + 1 \cdot p_1 = \frac{\lambda}{\lambda + \mu}. \tag{8.5}$$

Little's Law can be applied to the stream of arriving customers, or entering customers. We discuss these two cases below.

Arriving Customers. With probability p_0, the arriving customer finds the server idle, in which case he/she spends $1/\mu$ amount of time in the system; with probability p_1, the arriving customer finds the server busy, in which case he spends 0 amount of time in the system (since he leaves instantaneously). Hence the expected time spent by a customer in the system is

$$W = \frac{1}{\mu} \cdot p_0 + 0 \cdot p_1 = \frac{1}{\lambda + \mu}. \tag{8.6}$$

The customers arrive according to a PP(λ), hence the arrival rate is λ. Hence, from (8.5) and (8.6) we see that $L = \lambda W$, i.e., Little's Law holds.

Entering Customers. An entering customer always spends $1/\mu$ amount of time in the system on average. Hence

$$W_e = \frac{1}{\mu}. \tag{8.7}$$

In order to verify Little's Law we need to compute λ_e, the rate at which customers enter the system. Since the successive inter-entry times are iid with mean $\tau = 1/\mu + 1/\lambda = (\lambda + \mu)/\lambda\mu$ (mean service time followed by mean idle time), we see that the rate at which the customers enter the system per unit time is given by (see Theorem 7.2) $\lambda_e = 1/\tau = \lambda\mu/(\lambda + \mu)$. Hence, we have

$$L = \frac{\lambda}{\lambda + \mu} = \frac{\lambda\mu}{\lambda + \mu} \cdot \frac{1}{\mu} = \lambda_e W_e.$$

Thus Little's Law is verified.   □

Example 8.4 (Variation of Little's Law). We can use the intuition behind Little's Law to create many variations by changing the payment rules for the customers. For example, suppose each customer pays \$1 per unit time that he spends in the queue (but not in service). Then the expected amount paid by a customer is W_q, the expected time spent in the queue (not including service). Also, the rate at which the system collects revenue is L_q, the expected number of customers in the queue (not including those in service). Hence the same intuitive argument that yielded Little's Law yields

$$L_q = \lambda W_q,$$

where λ is the arrival rate of customers.   □

Example 8.5 (Another Variation of Little's Law). Now consider the following payment scheme for the customers. Each customer pays \$1 per unit time that he spends in service. Then the expected amount paid by a customer is τ, the expected time spent in service. Also, the rate at which the system collects revenue is B, the expected number of customers in service, which is the same as the expected number of busy servers. (In a single-server queue this is also the fraction of the time that the server is busy.) Hence the same intuitive argument that yielded Little's Law yields

$$B = \lambda\tau,$$

where λ is the arrival rate of customers. However, B cannot exceed s, the number of servers at the station. Hence we must have

$$B = \min(s, \lambda\tau).$$ □

The above example creates an interesting question: What happens when $\lambda\tau > s$? We discuss this question in Theorem 8.4 below. When there is an unlimited waiting room space (i.e., when the capacity is infinite) the state space of the $\{X(t), t \geq 0\}$ process is $\{0, 1, 2, \ldots\}$. So far in this book we have always considered finite state spaces. New issues arise in the study of stochastic processes with infinite state space that we did not encounter in the finite state space stochastic processes. In particular, it is possible that $X(t)$ may tend to infinity as t goes to infinity. In such a case, customers may end up waiting an infinite amount of time in the system in the long run. If this occurs, we call the process unstable. For example, if the arrival rate of customers at a queue is larger than the server's ability to serve, we would expect the queue length to build up without bounds, thus making the queue-length process unstable. We formally define the concept below.

Definition 8.1 (Stability). A single-station queue with unlimited capacity is called *stable* if

$$\sum_{j=0}^{\infty} p_j = 1,$$

where p_j is as defined in (8.1). Otherwise it is called *unstable*.

In an unstable queue we have

$$\sum_{j=0}^{\infty} p_j < 1.$$

This can be interpreted to mean that, in the long run, the queue length goes to infinity with a positive probability given by $1 - \sum_{j=0}^{\infty} p_j$. Obviously, having infinite queue lengths is bad for the customers. Hence it makes sense to call such a system unstable. The next theorem gives a condition for stability.

Theorem 8.4 (Condition of Stability). *Consider a single-station queue with s servers and infinite capacity. Suppose the customers enter at rate λ and the mean service time is τ. The queue is stable if*

$$\lambda \tau < s. \tag{8.8}$$

Intuition. From Example 8.5 we get

$$B = \text{expected number of busy servers} = \lambda \tau, \tag{8.9}$$

where λ is the arrival rate of entering customers. However the number of busy servers cannot exceed s, the total number of servers. Hence, for the argument to be valid, we must have

$$\lambda \tau \leq s.$$

What goes wrong when $\lambda \tau > s$? It was implicitly assumed in the derivation of (8.9) that each entering customer eventually enters service. Intuitively, when $\lambda \tau > s$, some of the customers never get to enter service, i.e., they wait in the system indefinitely. This implies that the number in the system must be infinity, i.e., the system must be unstable. It is more difficult to prove that the system can be unstable even if $\lambda \tau = s$. Essentially, this is due to the variance of either the inter-arrival times or the service times. Thus we can safely say that the queue is stable if $\lambda \tau < s$. □

8.3.   Birth and Death Queues with Finite Capacity

Let $X_K(t)$ be the number of customers at time t in a single-station queue with finite capacity K. In this section we consider the case when $\{X_K(t), t \geq 0\}$ is a birth and death process on state space $\{0, 1, 2, \ldots, K\}$, with birth parameters $\{\lambda_i, \ i = 0, 1, 2, \ldots, K - 1\}$ and death parameters $\{\mu_i, \ i = 1, 2, 3, \ldots, K\}$. We have studied such processes in Example 6.10 and their limiting behavior in Example 6.25. We restate the result about the limiting probabilities

$$p_i(K) = \lim_{t \to \infty} \mathsf{P}(X_K(t) = i), \qquad 0 \leq i \leq K.$$

Let $\rho_0 = 1$ and

$$\rho_i = \frac{\lambda_0 \lambda_1 \cdots \lambda_{i-1}}{\mu_1 \mu_2 \cdots \mu_i}, \qquad 1 \leq i \leq K. \tag{8.10}$$

Then, from Example 6.25 we get

$$p_i(K) = \frac{\rho_i}{\sum_{j=0}^{K} \rho_j}, \qquad 0 \leq i \leq K. \tag{8.11}$$

We consider several special cases below.

8.3.1.   *M/M/1/K* Queue

Consider a service station where customers arrive according to a PP(λ). The service times are iid random variables with Exp(μ) distribution. (Equivalently, we say the service rate is μ.) They are served in an FCFS fashion by a single server. The system has a capacity to hold K customers. Thus an incoming customer who sees K customers in the system ahead of him is permanently lost.

We have already seen this system in Example 6.9 as a model of a bank queue. Let $X_K(t)$ be the number of customers in the system at time t. Then we know (see the discussion following Example 6.10) that $\{X_K(t), t \geq 0\}$ is a birth and death process on state space $\{0, 1, \ldots, K\}$ with birth parameters

$$\lambda_i = \lambda, \qquad i = 0, 1, \ldots, K - 1,$$

and death parameters

$$\mu_i = \mu, \qquad i = 1, 2, \ldots, K.$$

Substituting in (8.10) we get

$$\rho_i = \left(\frac{\lambda}{\mu}\right)^i, \qquad 0 \leq i \leq K. \tag{8.12}$$

Using

$$\rho = \frac{\lambda}{\mu}$$

we get

$$\sum_{i=0}^{K} \rho_i = \begin{cases} \dfrac{1 - \rho^{K+1}}{1 - \rho} & \text{if } \rho \neq 1, \\[2ex] K + 1 & \text{if } \rho = 1. \end{cases}$$

Substituting in (8.11) we get

$$p_i(K) = \begin{cases} \dfrac{1 - \rho}{1 - \rho^{K+1}} \rho^i & \text{if } \rho \neq 1, \\[2ex] \dfrac{1}{K + 1} & \text{if } \rho = 1, \end{cases} \qquad 0 \leq i \leq K. \qquad (8.13)$$

We compute several relevant quantities with the help of this limiting distribution.

Server Idleness. The long-run fraction of the time that the server is idle is given by

$$p_0(K) = \frac{1 - \rho}{1 - \rho^{K+1}}.$$

Blocking Probability. The long-run fraction of the time that the system is full is given by

$$p_K(K) = \frac{1 - \rho}{1 - \rho^{K+1}} \rho^K. \qquad (8.14)$$

Since the arrival process is Poisson, we can use Theorem 8.2 to get the system distribution as seen by an arriving (not necessarily entering) customer as

$$\hat{\pi}_i(K) = p_i(K), \qquad i = 0, 1, 2, \ldots, K.$$

Thus an arriving customer sees the system full with probability $p_K(K)$. Hence, $p_K(K)$ also represents the long-run blocking probability and long-run fraction of the arriving customers that are blocked (or lost).

Entering Customers. Now we compute $\pi_i^*(K)$, the probability that an entering customer sees i customers in the system ahead of him. We have $\pi_K^*(K) = 0$ since a customer cannot enter if the system is full. For $0 \leq i \leq K - 1$, we have

$$\begin{aligned} \pi_i^*(K) &= \mathsf{P}(\text{entering customer sees } i \text{ in the system}) \\ &= \mathsf{P}(\text{arriving customer sees } i \text{ in the system} | \text{arriving customer enters}) \\ &= \frac{\mathsf{P}(\text{arriving customer sees } i \text{ in the system, and arriving customer enters})}{\mathsf{P}(\text{arriving customer enters})} \\ &= \frac{\mathsf{P}(\text{arriving customer sees } i \text{ in the system})}{\mathsf{P}(\text{arriving customer enters})} \\ &= \frac{p_i(K)}{1 - p_K(K)}, \qquad 0 \leq i \leq K - 1, \end{aligned}$$

where $p_i(K)$ are as given in (8.13).

Expected Number in the System. The expected number in the system in steady state is computed as follows: If $\rho = 1$, we have

$$L = \sum_{i=0}^{K} i p_i(K) = \frac{1}{K+1} \sum_{i=0}^{K} i = \frac{K}{2}.$$

If $\rho \neq 1$, more tedious calculations yield

$$L = \sum_{i=0}^{K} i p_i(K)$$

$$= \frac{1-\rho}{1-\rho^{K+1}} \sum_{i=0}^{K} i \rho^i$$

$$= \frac{\rho}{1-\rho} \cdot \frac{1 - (K+1)\rho^K + K\rho^{K+1}}{1-\rho^{K+1}}. \tag{8.15}$$

Note that L increases from 0 at $\rho = 0$ to K as $\rho \to \infty$. This stands to reason since as $\rho \to \infty$, the arrival rate is increasing relative to the service rate. Hence, eventually, as the arrival rate becomes very large, the system will always be full, thus leading to $L = K$.

Expected Waiting Time: Arrivals. We can compute the expected time spent by an arrival in the system by using Little's Law as stated in Theorem 8.3. Here the arrival rate is λ and the mean number in the system is L as given in (8.15). Hence W, the expected time spent in the system by an arriving customer is given by

$$W = L/\lambda.$$

Note that an arriving customer spends no time in the system if he finds the system full.

Expected Waiting Time: Entries. Next we compute the expected time spent by an entering customer in the system by using Little's Law as stated in Theorem 8.3. Here the arrival rate of entering customers is

$$\lambda P(\text{arriving customer enters}) = \lambda(1 - p_K(K)).$$

The mean number in the system is L as given in (8.15). Hence W, the expected time spent in the system by an entering customer is given by

$$W = L/(\lambda(1 - p_K(K))). \tag{8.16}$$

Example 8.6 (ATM Queue). Consider the model of the queue in front of an ATM queue as described in Example (6.9). The data given in Example 6.22 implies that this is an $M/M/1/5$ queue with an arrival rate of $\lambda = 10$ per hour, and a service rate of $\mu = 15$ per hour. Thus

$$\rho = \lambda/\mu = 2/3.$$

(1) What is the probability that an incoming customer finds the system full?
Using (8.14) we get the desired answer as

$$p_5(5) = .0481.$$

(2) What is the expected amount of time a customer spends in the system? Using (8.16) we get the desired answer as

$$W = .1494 \text{ hr} = 8.97 \text{ min}.$$

□

8.3.2.   *M/M/s/K* Queue

Let $X_K(t)$ be the number of customers in an $M/M/s/K$ queue at time t. (We shall assume that $K \geq s$, since it does not make sense to have more servers than there is room for customers.) We have seen this queue in Example 6.12 as a model of a call center. We have seen there that $\{X_K(t), t \geq 0\}$ is a birth and death process on $\{0, 1, \ldots, K\}$ with birth parameters

$$\lambda_i = \lambda, \qquad 0 \leq i \leq K - 1,$$

and death parameters

$$\mu_i = \min(i, s)\mu, \qquad 0 \leq i \leq K.$$

The next theorem gives the limiting distribution of an $M/M/s/K$ queue in terms of the dimensionless quantity defined below

$$\rho = \frac{\lambda}{s\mu}. \tag{8.17}$$

Theorem 8.5 (Limiting Distribution of an $M/M/s/K$ Queue). *The limiting distribution of an $M/M/s/K$ queue is given by*

$$p_i(K) = p_0(K)\rho_i, \qquad i = 0, 1, 2, \ldots, K,$$

where

$$\rho_i = \begin{cases} \dfrac{1}{i!}\left(\dfrac{\lambda}{\mu}\right)^i, & \text{if } 0 \leq i \leq s - 1, \\[3mm] \dfrac{s^s}{s!}\rho^i, & \text{if } s \leq i \leq K, \end{cases}$$

and

$$p_0(K) = \left[\sum_{i=0}^{s-1} \frac{1}{i!}\left(\frac{\lambda}{\mu}\right)^i + \frac{1}{s!}\left(\frac{\lambda}{\mu}\right)^s \cdot \frac{1 - \rho^{K-s+1}}{1 - \rho}\right]^{-1}.$$

Proof. The expressions for ρ_i follow from (8.10), and that for $p_i(K)$ follows from (8.11). Furthermore, we have

$$\sum_{i=0}^{K} \rho_i = \sum_{i=0}^{s-1} \frac{1}{i!}\left(\frac{\lambda}{\mu}\right)^i + \sum_{i=s}^{K} \frac{s^s}{s!}\rho^i$$

$$= \sum_{i=0}^{s-1} \frac{1}{i!} \left(\frac{\lambda}{\mu}\right)^i + \frac{s^s}{s!} \rho^s \sum_{i=s}^{K} \rho^{i-s}$$

$$= \sum_{i=0}^{s-1} \frac{1}{i!} \left(\frac{\lambda}{\mu}\right)^i + \frac{s^s}{s!} \rho^s \sum_{i=0}^{K-s} \rho^{i}$$

$$= \sum_{i=0}^{s-1} \frac{1}{i!} \left(\frac{\lambda}{\mu}\right)^i + \frac{s^s}{s!} \rho^s \frac{1 - \rho^{K-s+1}}{1-\rho}.$$

Substituting in (8.11) we get the theorem. □

Using the above limiting distribution we can study the same quantities as in the $M/M/1/K$ case.

Example 8.7 (Call Center). Consider the call center described in Example 6.12. Using the data in Computational Problem 6.15 we get an $M/M/6/10$ queue with an arrival rate of $\lambda = 60$ per hour, and a service rate of $\mu = 10$ per hour per server.

(1) Compute the limiting distribution of the number of calls in the system in steady state. From Theorem 8.5 we get

$$p(K) = [.0020 \quad .0119 \quad .0357 \quad .0715 \quad .1072 \quad .1286 \quad .1286 \quad .1286$$
$$.1286 \quad .1286 \quad .1286].$$

(2) Compute the expected number of calls on hold. This is same as the number of customers in the system not being served. It is given by

$$\sum_{i=7}^{10} (i - 6) p_i(K) = 1.2862.$$

(3) What fraction of the calls are lost? The desired quantity is given by

$$p_{10}(10) = .1286.$$

Thus 12.86% of the incoming calls are lost. □

8.3.3. M/M/K/K Queue

An $M/M/s/K$ queue, with $s = K$, produces the $M/M/K/K$ queue as a special case. Such a system is called a *loss system*, since any customer that cannot enter service immediately upon arrival is lost. We have seen such a queue as a model of a telephone switch in Example 6.11. By setting $s = K$ in Theorem 8.5 we get the limiting distribution of an $M/M/K/K$ queue as given in the following theorem:

Theorem 8.6 (Limiting Distribution of an $M/M/K/K$ Queue). *Let $X(t)$ be the number of customers in an $M/M/K/K$ queue at time t, and*

$$p_i(K) = \lim_{t \to \infty} P(X(t) = i), \qquad 0 \le i \le K.$$

Then

$$p_i(K) = \frac{\rho_i}{\sum_{j=0}^{K} \rho_j}, \qquad 0 \le i \le K,$$

where

$$\rho_i = \frac{(\lambda/\mu)^i}{i!}, \qquad 0 \le i \le K.$$

It is possible to interpret the limiting distribution in the above theorem as the distribution of a truncated (at K) Poisson random variable. To be precise, let Y be a Poisson random variable with parameter λ/μ. Then, for $0 \le i \le K$,

$$
\begin{aligned}
P(Y = i | Y \le K) &= \frac{P(Y = i, Y \le K)}{P(Y \le K)} \\
&= \frac{P(Y = i)}{P(Y \le K)} \\
&= \frac{e^{-\lambda/\mu}(\lambda/\mu)^i/i!}{\sum_{j=0}^{K} e^{-\lambda/\mu}(\lambda/\mu)^j/j!} \\
&= \frac{(\lambda/\mu)^i/i!}{\sum_{j=0}^{K}(\lambda/\mu)^j/j!} = p_i(K).
\end{aligned}
$$

The quantity $p_K(K)$ is an important quantity, called the blocking probability. It is the fraction of the time that the system is full, or the fraction of the customers that are lost, in the long run.

Example 8.8 (Telephone Switch). Recall the model of a telephone switch developed in Example 6.11. Using the data in Example 6.33 we see that the number of calls handled by the switch is an $M/M/6/6$ queue with an arrival rate of $\lambda = 4$ per minute, and a service rate of $\mu = .5$ per minute per server.

(1) What is the expected number of calls in the switch in steady state?
    Using Theorem 8.6 we compute the limiting distribution as

$$p(K) = [.0011 \quad .0086 \quad .0343 \quad .0913 \quad .1827 \quad .2923 \quad .3898].$$

Hence the expected number of calls is given by

$$\sum_{i=0}^{6} i p_i(K) = 4.8820.$$

(2) How large should the switch capacity be if we desire to lose no more than 10% of the incoming calls?

The fraction of calls lost is given by $p_K(K)$. At $K = 6$ this quantity is .3898. By computing $p_K(K)$ for larger values of K, we get

$$p_7(7) = .3082, \qquad p_8(8) = .2356, \qquad p_9(9) = .1731,$$
$$p_{10}(10) = .1217, \qquad p_{11}(11) = .0813.$$

Hence the telephone switch needs at least 11 lines to lose less than 10% of the traffic.

$\square$

8.4. Birth and Death Queues with Infinite Capacity

Let $X(t)$ be the number of customers in a single-station queue with infinite capacity. Here we consider the cases where the $\{X(t), t \geq 0\}$ is a birth and death process on state space $\{0, 1, 2, \ldots\}$ with birth parameters $\{\lambda_i, i \geq 0\}$ and death parameters $\{\mu_i, i \geq 1\}$. We are interested in the condition under which such a queue is stable. If the queue is stable, we are interested in its limiting distribution

$$p_i = \lim_{t \to \infty} P(X(t) = i), \qquad i \geq 0.$$

We have seen a birth and death queue with finite state space $\{0, 1, 2, \ldots, K\}$ in the previous section. We treat the infinite state birth and death queue as the limiting case of it as $K \to \infty$.

Define $\rho_0 = 1$ and

$$\rho_i = \frac{\lambda_0 \lambda_1 \cdots \lambda_{i-1}}{\mu_1 \mu_2 \cdots \mu_i}, \qquad i \geq 1. \tag{8.18}$$

Theorem 8.7 (Infinite Birth and Death Process). *A birth and death queue with infinite state space is stable if and only if*

$$\sum_{j=0}^{\infty} \rho_j < \infty. \tag{8.19}$$

When the queue is stable its limiting distribution is given by

$$p_i = \frac{\rho_i}{\sum_{j=0}^{\infty} \rho_j}, \qquad i \geq 0. \tag{8.20}$$

Proof. Let $p_i(K)$ be as in (8.11). We have

$$p_i = \lim_{K \to \infty} p_i(K)$$

$$= \begin{cases} \dfrac{\rho_i}{\sum_{j=0}^{\infty} \rho_j} & \text{if } \sum_{j=0}^{\infty} \rho_j < \infty, \\[3ex] 0 & \text{if } \sum_{j=0}^{\infty} \rho_j = \infty. \end{cases}$$

Thus we have

$$\sum_{i=0}^{\infty} p_i = \begin{cases} \sum_{i=0}^{\infty} \rho_i / \sum_{j=0}^{\infty} \rho_j & \text{if } \sum_{j=0}^{\infty} \rho_j < \infty, \\ 0 & \text{if } \sum_{j=0}^{\infty} \rho_j = \infty. \end{cases}$$

$$= \begin{cases} 1 & \text{if } \sum_{j=0}^{\infty} \rho_j < \infty, \\ 0 & \text{if } \sum_{j=0}^{\infty} \rho_j = \infty. \end{cases}$$

Hence it follows from Definition 8.1 that the queue is stable if and only if (8.19) holds, and that the limiting distribution of a stable birth and death queue is given by (8.20). □

We study several special cases of the birth and death queues with infinite state spaces below.

8.4.1. *M/M/1* Queue

In an $M/M/1$ queueing system the customers arrive according to a PP(λ) and the service times are iid Exp(μ) random variables. There is a single server and an unlimited waiting room capacity. The next theorem gives the main result for the $M/M/1$ queue. Define

$$\rho = \frac{\lambda}{\mu}. \tag{8.21}$$

The quantity ρ is called the *traffic intensity* of the $M/M/1$ queue.

Theorem 8.8 (Limiting Distribution of an $M/M/1$ Queue). *An $M/M/1$ queue with traffic intensity ρ is stable if and only if*

$$\rho < 1. \tag{8.22}$$

If the queue is stable its limiting distribution is given by

$$p_i = (1 - \rho)\rho^i, \qquad i \geq 0. \tag{8.23}$$

Proof. Let $X(t)$ be the number of customers in the system at time t. $\{X(t), t \geq 0\}$ is a birth and death process on $\{0, 1, 2, \ldots\}$ with birth parameters

$$\lambda_i = \lambda, \qquad i \geq 0,$$

and death parameters

$$\mu_i = \mu, \qquad i \geq 1.$$

Substituting in (8.18) we get

$$\rho_i = \left(\frac{\lambda}{\mu}\right)^i = \rho^i, \qquad i \geq 0.$$

Now

$$\sum_{i=0}^{\infty} \rho_i = \sum_{i=0}^{\infty} \rho^i = \begin{cases} \dfrac{1}{1-\rho} & \text{if } \rho < 1, \\ \\ \infty & \text{if } \rho \geq 1. \end{cases}$$

Thus, from Theorem 8.7 we see that the condition of stability is as given in (8.22). Substituting in (8.20) we get the limiting distribution in (8.23). ☐

Note that the condition of stability in (8.22) is consistent with the general condition in (8.8) (using $\tau = 1/\mu$). Using the definition of traffic intensity ρ, the condition of stability can also be written as

$$\lambda < \mu.$$

In this form it makes intuitive sense: if the arrival rate (λ) is less than the service rate (μ), the system is able to handle the load, and the queue does not build up to infinity. Hence it is stable. If not, the system is unable to handle the load and the queue builds to infinity! Hence it is unstable. It is curious that even in the case $\lambda = \mu$ (or, equivalently, $\rho = 1$) the queue is unstable. This is due to the effect of randomness. Even if the average service capacity of the system is perfectly matched with the average load on the system, it has no leftover capacity to absorb the random buildups that will invariably occur in a stochastic service system.

We shall compute several quantities using the limiting distribution given in Theorem 8.8.

Server Idleness. The probability that the server is idle is given by

$$p_0 = 1 - \rho, \tag{8.24}$$

and the probability that the server is busy is given by

$$1 - p_0 = \rho.$$

Hence the expected number of busy servers is ρ. This is consistent with the general result in Example 8.5.

Embedded Distributions. The queue length in an $M/M/1$ queue changes by ± 1 at a time. The arrival process is Poisson and all arriving customers enter the system. Hence Theorems 8.1 and 8.2 imply that

$$\hat{\pi}_i = \pi_i^* = \pi_i = p_i = (1 - \rho)\rho^i, \qquad i \geq 0.$$

Expected Number in the System. The expected number in the system is given by

$$\begin{aligned} L &= \sum_{i=0}^{\infty} i p_i \\ &= \sum_{i=0}^{\infty} i(1-\rho)\rho^i \\ &= \frac{\rho}{1-\rho}. \end{aligned} \tag{8.25}$$

The last equation is a result of tedious calculations.

Expected Waiting Time. Using Little's Law we get

$$W = \frac{L}{\lambda} = \frac{1}{\mu}\frac{1}{1-\rho} = \frac{1}{\mu - \lambda}. \qquad (8.26)$$

Note that, as expected, both L and W tend to infinity as $\rho \to 1$.

Example 8.9 (Taxi Stand). Customers arrive at a taxi stand according to a Poisson process at a rate of 15 per hour and form a queue for the taxis. Taxis arrive at the stand according a Poisson process with the rate of 18 per hour. Each taxi carries away one passenger. If no customers are waiting when a taxi arrives, it leaves right away without picking up any customers. Let $X(t)$ be the number of customers waiting at the taxi stand at time t. Due to the memoryless property of the exponential distribution, we see that the customer at the head of the line has to wait for an exp(18) amount of time for the next taxi. Thus we can think of $\{X(t), t \geq 0\}$ as an $M/M/1$ queue with an arrival rate of $\lambda = 15$ per hour and a service rate of $\mu = 18$ per hour.

(1) Is this queue stable?
The traffic intensity is

$$\rho = \lambda/\mu = 15/18 = 5/6.$$

Since this is less than 1, the queue is stable, from Theorem 8.8.

(2) On average, how long does a customer wait before getting into a taxi?
This is given by the average waiting time

$$W = 1/(\mu - \lambda) = .3333 \text{ hr.}$$

Thus the average wait for a taxi is 20 minutes.   □

8.4.2. *M/M/s* Queue

Next we study an $M/M/s$ queueing system. Customers arrive according to a PP(λ). The service times are iid Exp(μ) random variables. There are s identical servers and unlimited waiting room capacity. The customers form a single line and get served by the next available server on an FCFS basis. The next theorem gives the main result for the $M/M/s$ queue.

Theorem 8.9 (Limiting Distribution of an $M/M/s$ Queue). *An $M/M/s$ queue with an arrival rate of λ and s identical servers each with service rate μ is stable if and only if*

$$\rho = \frac{\lambda}{s\mu} < 1. \qquad (8.27)$$

If the queue is stable its limiting distribution is given by

$$p_i = p_0\rho_i, \quad i \geq 0,$$

where

$$\rho_i = \begin{cases} \dfrac{1}{i!}\left(\dfrac{\lambda}{\mu}\right)^i & \text{if } 0 \le i \le s-1, \\[3mm] \dfrac{s^s}{s!}\rho^i & \text{if } i \ge s, \end{cases} \tag{8.28}$$

and

$$p_0 = \left[\sum_{i=0}^{s-1}\frac{1}{i!}\left(\frac{\lambda}{\mu}\right)^i + \frac{s^s}{s!}\cdot\frac{\rho^s}{1-\rho}\right]^{-1}.$$

Proof. Let $X(t)$ be the number of customers in the $M/M/s$ queue at time t. $\{X(t), t \ge 0\}$ is a birth and death process on $\{0, 1, 2, \ldots\}$ with birth parameters

$$\lambda_i = \lambda, \quad i \ge 0,$$

and death parameters

$$\mu_i = \min(s, i)\mu, \quad i \ge 1.$$

Using

$$\rho = \frac{\lambda}{s\mu} \tag{8.29}$$

in (8.18) we get (8.28). Now

$$\sum_{i=0}^{\infty}\rho_i = \sum_{i=0}^{s-1}\rho_i + \sum_{i=s}^{\infty}\rho_i$$
$$= \sum_{i=0}^{s-1}\rho_i + \frac{s^s}{s!}\frac{\rho^s}{1-\rho} \quad \text{if } \rho < 1.$$

If $\rho \ge 1$ the above sum diverges. Thus, from Theorem 8.7 we see that the condition of stability is as given in (8.27). The limiting distribution then follows from (8.20). □

Note that the condition of stability in (8.27) can also be written as

$$\lambda < s\mu.$$

In this form it makes intuitive sense: if the arrival rate (λ) is less than the maximum service rate ($s\mu$) (which occurs when all servers are busy), the system is able to handle the load, and the queue does not build up to infinity. Hence it is stable. If not, the system is unable to handle the load and the queue builds to infinity! Hence it is unstable. As in the $M/M/1$ queue, the queue is unstable when $\lambda = s\mu$.

We shall compute several quantities using the limiting distribution given in Theorem 8.9.

Embedded Distributions. The queue length in an $M/M/s$ queue changes by ± 1 at a time. The arrival process is Poisson and all arriving customers enter the system. Hence Theorems 8.1 and 8.2 imply that

$$\hat{\pi}_i = \pi_i^* = \pi_i = p_i.$$

Probability of Waiting. An incoming customer has to wait for service if all the servers are busy. This probability is given by

$$\sum_{i=s}^{\infty} p_i = \sum_{i=s}^{\infty} p_0 \frac{s^s}{s!} \rho^i$$

$$= p_0 \frac{s^s}{s!} \rho^s \sum_{i=s}^{\infty} \rho^{i-s}$$

$$= p_s \sum_{i=0}^{\infty} \rho^i$$

$$= \frac{p_s}{1 - \rho}.$$

Expected Number of Busy Servers. If there are i customers in the system, the number of busy servers is given by $\min(i, s)$. Hence B, the expected number of busy servers, is given by

$$B = \sum_{i=0}^{\infty} \min(i, s) p_i.$$

The above sum can be simplified to

$$B = \frac{\lambda}{\mu}. \tag{8.30}$$

This matches the general result in Example 8.5.

Expected Number in the System. The expected number in the system is given by

$$L = \sum_{i=0}^{\infty} i p_i.$$

This can be simplified to

$$L = \frac{\lambda}{\mu} + p_s \frac{\rho}{(1 - \rho)^2}. \tag{8.31}$$

Expected Waiting Time. Using Little's Law we get

$$W = \frac{L}{\lambda} = \frac{1}{\mu} + \frac{1}{s\mu} \frac{p_s}{(1 - \rho)^2}. \tag{8.32}$$

Note that, as expected, both L and W tend to infinity as $\rho \to 1$.

Example 8.10 (How Many Service Windows?). The US Postal Service has announced that no customer will have to wait more than 5 minutes in line in a post office. It wants to know how many service windows should be kept open to keep this promise. You are hired as a consultant to help decide this. You realize immediately that this is an impossible promise to keep, since the service times and arrival rates of customers are beyond the control of the Post Office. Hence you decide to concentrate on the average wait in the queue. Thus you want to decide how many service windows to keep open so that the mean queueing time (excluding service time) is less than 5 minutes.

Suppose the arrival rate is λ, service times are iid $\exp(\mu)$, and s windows are kept open. Then, using (8.32), the queueing time is given by

$$W_q(s) = W - 1/\mu = \frac{1}{s\mu} \frac{p_s}{(1-\rho)^2}.$$

You estimate the average service time to be 3.8 minutes, and the arrival rate to be 47.2 per hour. Thus

$$\mu = 60/3.8 = 15.7895, \qquad \lambda = 47.2.$$

Using the stability condition of Theorem 8.9, we see that the queue is unstable for $s = 1$ and 2. Using the above formula for $s \geq 3$, we get

$$W_q(3) = 5.8977 \text{ hr} = 353.86 \text{ min}, \qquad W_q(4) = .0316 \text{ hr} = 1.8986 \text{ min}.$$

Hence you recommend that the Post Office should keep four windows open.   □

8.4.3. M/M/∞ Queue

By letting $s \to \infty$ we get the $M/M/\infty$ queue as a limiting case of the $M/M/s$ queue. The limiting distribution of the $M/M/\infty$ queue thus follows from Theorem 8.9 as given below:

Theorem 8.10 (Limiting Distribution of an $M/M/\infty$ Queue). *An $M/M/\infty$ queue with an arrival rate of λ and an unlimited number of identical servers each with service rate μ is always stable, and its limiting distribution is given by*

$$p_i = e^{-\lambda/\mu} \frac{(\lambda/\mu)^i}{i!}, \qquad i \geq 0.$$

Thus, in steady state, the number of customers in an $M/M/\infty$ queue is a Poisson random variable with parameter λ/μ. Hence the expected number of customers in steady state is given by

$$L = \frac{\lambda}{\mu}.$$

Then, using Little's Law, we get

$$W = \frac{1}{\mu},$$

which is equal to the mean service time, since customers enter service immediately upon arrival.

Example 8.11 (Library). Users arrive at a library according to a Poisson process at a rate of 200 per hour. Each user is in the library for an average of 24 minutes. Assume that the time spent by a user in the library is exponentially distributed and independent of the other users. (This implies that the time spent in the checkout line is negligible.) How many users are there in the library on average?

Let $X(t)$ be the number of users in the library at time t. Since each user stays in the library independently of each other, $\{X(t), t \geq 0\}$ can be thought of as an $M/M/\infty$ queue. The parameters are

$$\lambda = 200 \text{ per hour}, \qquad \mu = 60/24 \text{ per hour}.$$

Hence, using Theorem 8.10, the number of users in the library in steady state is a Poisson random variable with parameter $\lambda/\mu = 80$. Hence the expected number of users in the library in steady state is 80. □

8.5. *M/G/1* Queue

Consider a single-station queueing system where customers arrive according to a PP(λ) and require iid service times with mean τ, variance σ^2, and second moment $s^2 = \sigma^2 + \tau^2$. The service times may not be exponentially distributed. The queue is serviced by a single server, and has infinite waiting room. Such a queue is called an $M/G/1$ queue.

Let $X(t)$ be the number of customers in the system at time t. $\{X(t), t \geq 0\}$ is a continuous time stochastic process with state space $\{0, 1, 2, \ldots\}$. Knowing the current state $X(t)$ does not provide enough information about the remaining service time of the customer in service (unless the service times are exponentially distributed), and hence we cannot predict the future based solely on $X(t)$. Hence $\{X(t), t \geq 0\}$ is not a CTMC. Hence we will not be able to study an $M/G/1$ queue in as much detail as the $M/M/1$ queue. Instead we shall satisfy ourselves with results about the expected number and expected waiting time of customers in the $M/G/1$ queue in steady state.

Stability. We begin with the question of stability. Let

$$\rho = \lambda\tau \tag{8.33}$$

be the traffic intensity. Then, from Theorem 8.4, it follows that the $M/G/1$ queue is stable if

$$\rho < 1.$$

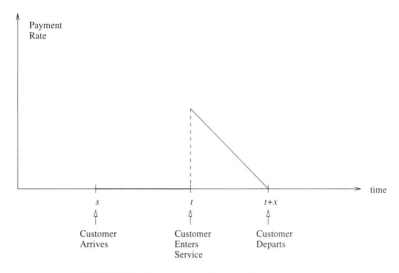

FIGURE 8.2. Payment rate for a typical customer.

Indeed, it is possible to show that the $M/G/1$ queue is unstable if $\rho \geq 1$. Thus $\rho < 1$ is a necessary and sufficient condition of stability. We shall assume that the queue is stable.

Server Idleness. Suppose the queue is stable. It follows from Example 8.5 that, in steady state, the expected number of busy servers is ρ. Since the system is served by a single server, in the long run we must have

$$\mathsf{P}(\text{server is busy}) = \rho,$$

and

$$\mathsf{P}(\text{server is idle}) = 1 - \rho.$$

Remaining Service Time. Consider the following payment scheme: Each customer pays money to the system at rate 0 while waiting for service, and at a rate equal to the remaining service time while in service. For example, consider a customer with service time x. Suppose this customer arrives at time s, and enters service at time t, and departs at time $t + x$. The customer payment rate $c(\cdot)$ is shown in Figure 8.2 as a function of time.

It is clear from the figure that the total amount paid by a customer with service time x is

$$\int_{u=s}^{t+x} c(u)\, du = x^2/2.$$

Hence the expected payment made by a customer with random service time X is

$$E(X^2/2) = \frac{s^2}{2}.$$

Since the arrival rate is λ, the rate at which customers pay the system is

$$R_{\text{customer}} = \lambda\frac{s^2}{2}.$$

Now let us compute the rate at which the system collects revenue. Note that if the server is idle, the system collects no revenue. If the server is busy, the system collects revenue at a rate equal to the remaining service time of the customer in service. Let $S(t)$ be the rate at which the system collects revenue at time t. Then the long-run rate at which the server collects revenue is given by

$$R_{\text{system}} = \lim_{t\to\infty} E(S(t)) = E(S).$$

Since the queue is assumed to be stable, we expect these two rates to be equal, i.e.,

$$E(S) = \lambda\frac{s^2}{2}. \tag{8.34}$$

Note that we can also interpret $S(t)$ as follows: $t + S(t)$ is the first time after t that the server can start serving a new customer. This interpretation is important in the next calculation.

Expected Time in the Queue. Assume that the service discipline is First-Come–First-Served. Let L_q be the expected number of customers in the queue (not including those in service), and let W_q be the expected waiting time in the queue (not including the time spent in service). Then from Example 8.4, we get

$$L_q = \lambda W_q. \tag{8.35}$$

An arriving customer's queueing time has two components: the time until the current customer in service finishes, followed by the service times of all the customers in the queue. Now, by using the interpretation of $S(t)$ given earlier, we see that the expected time until the current customer in service finishes is given by $E(S)$ in steady state. The second component is the sum of the service times of the random number of customers waiting in the queue at the time of arrival of this customer. Using Theorems 8.1 and 8.2 we see that L_q is the number of customers in the queue as seen by an arrival in steady state. Since the service times are iid, and independent, and also independent of the number of customers in the queue, we can use the results of Example 4.15 to see that the expected value of the sum of service times of all the customers in the queue is given by $L_q\tau$. Hence, we have

$$W_q = E(S) + L_q\tau.$$

Substituting from (8.33), (8.34), and (8.35), we get

$$W_q = \frac{\lambda s^2}{2} + \lambda W_q\tau = \frac{\lambda s^2}{2} + \rho W_q,$$

which can be simplified to

$$W_q = \frac{\lambda s^2}{2(1-\rho)}. \tag{8.36}$$

Expected Number in the Queue. Using (8.35) and (8.36), we get the expected number in the queue as

$$L_q = \frac{\lambda^2 s^2}{2(1-\rho)}. \tag{8.37}$$

Expected Time in the System. Let W be the expected time spent in the system by a customer in steady state. Since the time in the system is simply the time in the queue plus the time in service, we get

$$W = \tau + W_q. \tag{8.38}$$

Expected Number in the System. Using Little's Law and (8.38), we get the expected number in the queue as

$$L = \lambda(\tau + W_q) = \rho + L_q = \rho + \frac{\lambda^2 s^2}{2(1-\rho)}. \tag{8.39}$$

Equations (8.36), (8.37), (8.38), and (8.39) are variations of what is known as the *Pollaczec–Khintchine Formula*, named after the two probabilists who first studied the *M/G/1* queue. It is interesting that the first moments of the queue length and the waiting times are affected by the second moments of the service times! This is highly counterintuitive. We can think of the mean service time as a measure of the *efficiency* of the server: the more efficient the server, the smaller the mean service time τ. Similarly, the variance of the service time is an indicator of the *consistency* of the server: the more consistent the server, the smaller the variance σ^2. Thus, among two equally efficient servers, the more consistent server will lead to smaller mean waiting times and smaller mean queue lengths!

Example 8.12 (Telecommunications). Packets arrive at an infinite capacity buffer according to a Poisson process with a rate of 400 per second. All packets are exactly 512 bytes long. The buffer is emptied at a rate of 2 megabits per second. Compute the expected amount of time a packet waits in the buffer before transmission.

The time to transmit a single packet is

$$(512 * 8)/2000000 = .002048 \text{ seconds.}$$

Since the packet lengths are identical, the variance of the transmission time is zero. Thus the buffer can be modeled as an *M/G/1* queue with arrival rate

$$\lambda = 400 \text{ per second,}$$

and iid service times with mean

$$\tau = .002048 \text{ seconds,}$$

and variance 0, i.e., second moment

$$s^2 = (.002048)^2.$$

Hence, the traffic intensity is given by

$$\rho = \lambda\tau = .8192.$$

Since this is less than 1, the queue is stable. Using (8.36), we get the expected time in the queue as

$$W_q = \frac{\lambda s^2}{2(1-\rho)} = \frac{400 * (.002048)^2}{2 * .8192} = .00464 \text{ seconds.}$$

Thus the expected wait before transmission is 4.64 milliseconds. The expected number of packets waiting for transmission can be computed by using (8.37) as

$$L_q = \frac{\lambda^2 s^2}{2(1-\rho)} = 1.8559.$$

The expected number of packets in the buffer (including any in transmission) can be computed as

$$L = \rho + L_q = .8192 + 1.8559 = 2.6751. \qquad \square$$

Example 8.13 ($M/M/1$ as a Special Case of $M/G/1$). Consider an $M/G/1$ queue with iid $\text{Exp}(\mu)$ service times. In this case, the $M/G/1$ queue reduces to an $M/M/1$ queue with

$$\tau = 1/\mu, \qquad \sigma^2 = 1/\mu^2, \qquad s^2 = 2/\mu^2.$$

Using (8.36), (8.37), (8.38), and (8.39) we get

$$W = \frac{1}{\mu}\frac{1}{1-\rho},$$

$$L = \frac{\rho}{1-\rho}.$$

These results match with the corresponding results for the $M/M/1$ queue derived in Section 8.4. $\qquad \square$

8.6. *G/M/1* Queue

Consider a single-station queueing system where customers arrive according to a renewal process with iid inter-arrival times having common cdf $G(\cdot)$, and mean $1/\lambda$. The inter-arrival times may not be exponentially distributed. The queue is serviced by a single server, and has an infinite waiting room. The service times are iid $\text{Exp}(\mu)$ random variables. Such a queue is called a $G/M/1$ queue. If the inter-arrival times are exponentially distributed, it reduces to an $M/M/1$ queue.

Let $X(t)$ be the number of customers in the system at time t. $\{X(t), t \geq 0\}$ is a continuous-time stochastic process with state space $\{0, 1, 2, \ldots\}$. Knowing the current state $X(t)$ does not provide enough information about the time until the next arrival (unless the inter-arrival times are exponentially distributed), and hence we cannot predict the future based solely on $X(t)$. Hence $\{X(t), t \geq 0\}$ is not a CTMC. Hence we will state some of the main results without proof and derive others from it. First define

$$\rho = \frac{\lambda}{\mu} \tag{8.40}$$

as the traffic intensity of the $G/M/1$ queue.

Stability of the *G/M/1* Queue. We begin with the study of stability of the $G/M/1$ queue. From Theorem 8.4, it follows that the $G/M/1$ queue is stable if

$$\rho < 1.$$

Indeed, it is possible to show that (we refer the reader to advanced books for a proof) the $G/M/1$ queue is unstable if $\rho \geq 1$. Thus $\rho < 1$ is a necessary and sufficient condition of stability. We shall assume that the queue is stable in the remaining analysis.

Functional Equation. Here we study a key functional equation that arises in the study of the limiting behavior of a $G/M/1$ queue. Let A represent an inter-arrival time, i.e., A is a random variable with cdf G. Define, for $s \geq 0$,

$$\tilde{G}(s) = \mathsf{E}(e^{-sA}).$$

The key functional equation of a $G/M/1$ queue is as given below:

$$u = \tilde{G}(\mu(1 - u)). \tag{8.41}$$

Before we discuss the solutions of this key functional equation, we describe how to compute $\tilde{G}(s)$. If A is a continuous random variable with pdf $g(x)$,

$$\tilde{G}(s) = \int_0^\infty e^{-sx} g(x)\,dx,$$

and if A is a discrete random variable with pmf $p(x_i) = \mathsf{P}(X = x_i)$, $i = 1, 2, \ldots,$

$$\tilde{G}(s) = \sum_{i=1}^\infty e^{-sx_i} p(x_i).$$

We illustrate with an example below.

Example 8.14 (Computation of $\tilde{G}$).

(1) Exponential Distribution: Suppose the inter-arrival times are Exp(λ). Then the pdf is

$$g(x) = \lambda e^{-\lambda x}, \qquad x \geq 0.$$

Hence

$$\tilde{G}(s) = \int_0^\infty e^{-sx} g(x) \, dx$$

$$= \int_0^\infty e^{-sx} \lambda e^{-\lambda x} \, dx$$

$$= \frac{\lambda}{s + \lambda}.$$

(2) Erlang Distribution: Suppose the inter-arrival times are Erl(k, λ). Then the pdf is

$$g(x) = \lambda e^{-\lambda x} \frac{(\lambda x)^{k-1}}{(k-1)!}, \qquad x \geq 0.$$

Hence

$$\tilde{G}(s) = \int_0^\infty e^{-sx} g(x) \, dx$$

$$= \int_0^\infty e^{-sx} \lambda e^{-\lambda x} \frac{(\lambda x)^{k-1}}{(k-1)!} \, dx$$

$$= \left(\frac{\lambda}{s + \lambda} \right)^k,$$

where the last integral follows from a standard table of integrals.

(3) Hyperexponential Distribution: Suppose the inter-arrival times are Hex(k, λ, p). Then the pdf is

$$g(x) = \sum_{i=1}^k p_i \lambda_i e^{-\lambda_i x}, \qquad x \geq 0.$$

Hence

$$\tilde{G}(s) = \int_0^\infty e^{-sx} g(x) \, dx$$

$$= \int_0^\infty e^{-sx} \sum_{i=1}^k p_i \lambda_i e^{-\lambda_i x} \, dx$$

$$= \sum_{i=1}^k \frac{p_i \lambda_i}{s + \lambda_i}.$$

(4) Degenerate Distribution: Suppose the inter-arrival times are constant, equal to c. Then A takes a value c with probability 1. Hence,

$$\tilde{G}(s) = \sum_{i=1}^\infty e^{-sx_i} p(x_i) = e^{-sc}.$$

(5) Geometric Distribution: Suppose the inter-arrival times are iid G(p) random variables. Then the pmf is given by

$$p(i) = \mathsf{P}(A = i) = (1 - p)^{i-1} p, \qquad i = 1, 2, \dots .$$

Hence

$$\tilde{G}(s) = \sum_{i=1}^{\infty} e^{-si} (1 - p)^{i-1} p$$

$$= pe^{-s} \sum_{i=1}^{\infty} (e^{-s}(1 - p))^{i-1}$$

$$= pe^{-s} \sum_{i=0}^{\infty} (e^{-s}(1 - p))^{i}$$

$$= \frac{pe^{-s}}{1 - e^{-s}(1 - p)}. \qquad \square$$

Now we discuss the solutions to the key functional equation. Note that $u = 1$ is always a solution to the key functional equation. The next theorem discusses other solutions to this equation in the interval $(0, 1)$. We refer the reader to advanced texts for a proof.

Theorem 8.11 (Solution of the Key Functional Equation). *If $\rho \geq 1$ there is no solution to the key functional equation in the interval $(0, 1)$. If $\rho < 1$, there is a unique solution $\alpha \in (0, 1)$ to the key functional equation.*

Example 8.15 (Exponential Distribution). Suppose the inter-arrival times are iid Exp(μ). Then, using the results of Example 8.14, the key functional equation becomes

$$u = \frac{\lambda}{\mu(1 - u) + \lambda}.$$

This can be rearranged to get

$$(1 - u)(u\mu - \lambda) = 0.$$

Thus there are two solutions to the key functional equation

$$u = \rho, \qquad u = 1.$$

This verifies Theorem 8.11. $\qquad \square$

Unlike the above example, it is not always possible to solve the key functional equation of the $G/M/1$ queue analytically. In such cases the solution can be obtained numerically by using the following recursive computation:

$$u_0 = 0, \qquad u_{n+1} = \tilde{G}(\mu(1 - u_n)), \qquad n \geq 0.$$

Then, if $\rho < 1$,

$$\lim_{n \to \infty} u_n = \alpha,$$

where α is the unique solution in $(0, 1)$ to the key functional equation.

Example 8.16 (Degenerate Distribution). Suppose the inter-arrival times are constant equal to $1/\lambda$. Using the results from Example 8.14, the key functional equation becomes

$$u = e^{-\mu(1-u)(1/\lambda)} = e^{-(1-u)/\rho}.$$

This is a nonlinear equation, and no closed form solution is available. It is, however, easy to obtain a numerical solution by the recursive method described above. For example, suppose $\lambda = 5$ and $\mu = 8$. Then $\rho = \frac{5}{8} = .625$ and the above recursion yields $\alpha = .3580$. □

The solution to the key functional equation plays an important part in the computation of the limiting distributions as shown next.

Embedded Distributions. We next study the distribution of the number of customers in a $G/M/1$ queue as seen by an arrival in steady state. Since the $\{X(t), t \geq 0\}$ process jumps by ± 1, it follows from Theorem 8.1 that the arrival time distribution is the same as the departure time distribution, i.e.,

$$\pi_j = \pi_j^*, \qquad j \geq 0.$$

Using the solution to the key functional equation, we give the main results in the next theorem.

Theorem 8.12 (Arrival Time Distribution). *In a stable $G/M/1$ queue, the limiting distribution of the number of customers as seen by an arrival is given by*

$$\pi_j^* = (1 - \alpha)\alpha^j, \qquad j \geq 0,$$

where α is the unique solution in $(0, 1)$ to the key functional equation (8.41).

Idea of Proof. We refer the reader to an advanced text for the complete proof. The main idea is as follows: define X_n^* as the number of customers in the $G/M/1$ queue as seen by the nth arrival (not including the arrival itself). Then we first show that $\{X_n^*, n \geq 0\}$ is an irreducible and aperiodic DTMC on state space $\{0, 1, 2, \ldots\}$ and compute the transition probabilities

$$p_{i,j} = P(X_{n+1}^* = j | X_n^* = i), \qquad i, j \geq 0,$$

in terms of μ and $G(\cdot)$. Then we show that $\pi_j^* = (1 - \alpha)\alpha^j$ satisfies the balance equations

$$\pi_j^* = \sum_{i=0}^{\infty} \pi_i^* p_{i,j}, \qquad j \geq 0.$$

This, along with aperiodicity of the DTMC, shows that

$$\pi_j^* = \lim_{n \to \infty} P(X_n^* = j), \qquad j \geq 0.$$

This proves the theorem. □

Limiting Distribution. Next we study the limiting distribution

$$p_j = \lim_{t \to \infty} P(X(t) = j), \qquad j \geq 0.$$

Since the arrival process in a $G/M/1$ queue is not Poisson, PASTA is not applicable, and $p_j \neq \pi_j^*$. The next theorem gives the main result.

Theorem 8.13 (Limiting Distribution). *In a stable $G/M/1$ queue, the limiting distribution of the number of customers in the system is given by*

$$p_0 = 1 - \rho, \qquad p_j = \rho \pi_{j-1}^*, \qquad j \geq 1,$$

where π_j^* are as in Theorem 8.12.

Idea of Proof. It follows from Example 8.5 that, in steady state, the expected number of busy servers is ρ. Since the system is served by a single server, in the long run we must have,

$$P(\text{server is busy}) = \rho,$$

and hence

$$P(\text{server is idle}) = p_0 = 1 - \rho.$$

Now, fix $j \geq 1$. Consider the following cost structure: whenever the number of customers in the system jumps from $j - 1$ to j, the system earns \$1, and whenever it jumps from j to $j - 1$, it loses \$1. Then R_u, the long-run rate at which it earns dollars, is given by the product of λ, the long-run arrival rate of customers, and π_{j-1}^*, the long-run fraction of customers that see $j - 1$ customers ahead of them in the system. Thus

$$R_u = \lambda \pi_{j-1}^*.$$

On the other hand, R_d, the long-run rate at which it loses dollars, is given by the product of μ, the long-run departure rate of customers (when the system is in state j), and p_j, the long-run fraction of the time that the system is in state j. Thus

$$R_d = \mu p_j.$$

Since the system is stable, we expect these two rates to be the same. Hence we have

$$\lambda \pi_{j-1}^* = \mu p_j,$$

which yields the result in Theorem 8.13.   □

Expected Number in the System. The expected number in the system is given by

$$
\begin{aligned}
L &= \sum_{i=0}^{\infty} i p_i \\
&= \sum_{i=1}^{\infty} i \pi_{i-1}^* \\
&= \sum_{i=1}^{\infty} i \rho (1 - \alpha) \alpha^{i-1} \\
&= \frac{\rho}{1 - \alpha}.
\end{aligned}
\tag{8.42}
$$

Expected Waiting Time. Using Little's Law we get

$$W = \frac{L}{\lambda} = \frac{1}{\mu}\frac{1}{1-\alpha}. \qquad (8.43)$$

Example 8.17 (Manufacturing). A machine manufactures two items per hour in a deterministic fashion, i.e., the manufacturing time is exactly .5 hours per item. The manufactured items are stored in a warehouse with infinite capacity. Demands for the items arise according to a Poisson process with a rate of 2.4 per hour. Any demand that cannot be satisfied instantly is lost.

Let $X(t)$ be the number of items in the warehouse at time t. Then $X(t)$ increases by one every 30 minutes, while it decreases by one whenever a demand occurs. If $X(t)$ is zero when a demand occurs, it stays zero. Thus $\{X(t), t \geq 0\}$ can be thought of as the queue length process of a $G/M/1$ queue with deterministic inter-arrival times with an arrival rate (production rate) of $\lambda = 2$ per hour, and exponential service times with a rate of $\mu = 2.4$ per hour. This makes

$$\rho = \lambda/\mu = 2/2.4 = .8333.$$

Thus the queue is stable. The key functional equation for the $G/M/1$ queue in this case is (see Example 8.16)

$$u = e^{-2.4*.5*(1-u)} = e^{-1.2(1-u)}.$$

Solving the above equation numerically we get

$$\alpha = .6863.$$

Using this parameter we can answer several questions about the system.

(1) What is the average number of items in the warehouse in steady state?
  The answer is given by

$$L = \rho/(1-\alpha) = 5.4545.$$

(2) On average, how long does an item stay in the warehouse?
  The answer is given by

$$W = 1/\mu(1-\alpha) = 2.7273 \text{ hours.}$$

(3) What is the probability that the warehouse has more than five items in it?
  The answer is given by

$$\sum_{j=5}^{\infty} p_j = \rho(1-\alpha)\sum_{j=5}^{\infty}\alpha^{j-1} = \rho\alpha^4 = .4019.$$

□

8.7. Networks of Queues

So far in this chapter we have studied single-station queueing systems. In practice, service systems can consist of multiple service stations. Customers can move from service station to service station many times before leaving the system. Such multiple service-station systems are called *networks of queues*. For example, patients wait at several different queues inside a hospital during a single visit to the hospital. An electronic message may wait at many intermediate nodes in the Internet before it is delivered to its destination. Material in a manufacturing facility may wait at many different work stations until the final product emerges at the end. Paperwork in many organizations has to wait at different offices undergoing scrutiny and collecting signatures until it is finally done. Students registering for a new semester may wait in multiple queues during the registration process.

8.7.1. Jackson Networks

In this subsection we study a special class of queueing networks, called the *Jackson networks*. A network of queues is called a Jackson network if it satisfies all of the following assumptions:

(1) The network has N single-station queues.

(2) The ith station has s_i servers.

(3) There is an unlimited waiting room at each station.

(4) Customers arrive at station i from outside the network according to PP(λ_i). All arrival processes are independent of each other.

(5) Service times of customers at station i are iid Exp(μ_i) random variables.

(6) Customers finishing service at station i join the queue at station j with probability $p_{i,j}$, or leave the network altogether with probability r_i, independently of each other.

The *routing probabilities* $p_{i,j}$ can be put in a matrix form as follows:

$$
P = \begin{bmatrix}
p_{1,1} & p_{1,2} & p_{1,3} & \cdots & p_{1,N} \\
p_{2,1} & p_{2,2} & p_{2,3} & \cdots & p_{2,N} \\
p_{3,1} & p_{3,2} & p_{3,3} & \cdots & p_{3,N} \\
\vdots & \vdots & \vdots & \ddots & \vdots \\
p_{N,1} & p_{N,2} & p_{N,3} & \cdots & p_{N,N}
\end{bmatrix}.
\tag{8.44}
$$

The matrix P is called the *routing matrix*. Since a customer leaving station i either joins some other station in the network, or leaves, we must have

$$
\sum_{j=1}^{N} p_{i,j} + r_i = 1, \qquad 1 \le i \le N.
$$

We illustrate with two examples:

FIGURE 8.3. A tandem queueing network.

Example 8.18 (Tandem Queue). Consider a queueing network with four service stations as shown in Figure 8.3. Such linear networks are called tandem queueing networks.

Customers arrive at station 1 according to a Poisson process at a rate of 10 per hour and get served at stations 1, 2, 3, and 4 in that order. They depart after getting served at station 4. The mean service time is 10 minutes at station 1, 15 minutes at station 2, 5 minutes at station 3, and 20 minutes at station 4. The first station has two servers, the second has three servers, the third has a single server, while the last has four servers.

In order to model this as a Jackson network, we shall assume that the service times are independent exponential random variables with the means given above, and that there is an infinite waiting room at each service station. Thus if the mean service time is 20 minutes, that translates into a service rate of three per hour. Then the parameters of this Jackson network are as follows:

$$N = 4,$$

$$s_1 = 2, \quad s_2 = 3, \quad s_3 = 1, \quad s_4 = 4,$$

$$\lambda_1 = 10, \quad \lambda_2 = 0, \quad \lambda_3 = 0, \quad \lambda_4 = 0,$$

$$\mu_1 = 6, \quad \mu_2 = 4, \quad \mu_3 = 12, \quad \mu_4 = 3,$$

$$P = \begin{bmatrix} 0 & 1 & 0 & 0 \\ 0 & 0 & 1 & 0 \\ 0 & 0 & 0 & 1 \\ 0 & 0 & 0 & 0 \end{bmatrix},$$

$$r_1 = 0, \quad r_2 = 0, \quad r_3 = 0, \quad r_4 = 1. \qquad \square$$

Example 8.19 (Amusement Park). An amusement park has six rides: a Roller Coaster, a Merry-go-Round, a Water-Tube ride, a Fantasy ride, a Ghostmountain ride, and a Journey to the Moon ride. (See Figure 8.4.)

Visitors arrive at the park according to a Poisson process at a rate of 600 per hour and immediately fan out to the six rides. A newly arriving visitor is equally likely to go to any of the six rides first. From then on each customer visits the rides in a random fashion and then departs. We have the following data: the roller coaster ride lasts 2 minutes (including the loading and unloading times). Two roller coaster cars, each carrying 12 riders, are on the track at any one time. The Merry-go-Round can handle 35 riders at one time, and the

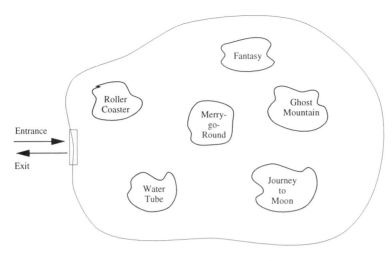

FIGURE 8.4. The amusement park.

ride lasts 3 minutes. The Water-Tube ride lasts 1.5 minutes and there can be 10 tubes on the waterway at any one time, each carrying two riders. The Fantasy Ride is a 5-minute ride that can carry 60 persons at any one time. The Ghostmountain ride can handle 16 persons at a time and lasts for 90 seconds. Finally, the Journey to the Moon ride takes 100 seconds, and 20 riders can be on it at any one time.

We shall model this as a Jackson network with six service stations: 1. Roller Coaster (R); 2. Merry-go-Round (M); 3. Water-Tubes (W); 4. Fantasy (F); 5. Ghostmountain (G); and 6. Journey to the Moon (J). (See Figure 8.5.)

Clearly, we need several assumptions. We shall assume that the ride times are exponentially distributed, although this is typically untrue. Thus if a ride takes 5 minutes, this translates to a service rate of 12 per hour. If a ride can handle s riders simultaneously, we shall assume the corresponding service station has s servers. Although the service times of all riders sharing a ride must be the same, we shall assume the service times to be iid. Routing of riders is especially problematic. We are given that new customers go to any of the six rides with equal probability. Hence the rate of new arrivals at each ride is 100 per hour. We assume that after each ride, the riders choose to join one of the remaining five rides, or leave in a completely random fashion. Note that under this model some riders may not visit all the rides, while others may visit some rides more than once! Also, we assume that riders behave independently of each other, while in practice, they move in small groups. With these assumptions, we get a Jackson network with the following

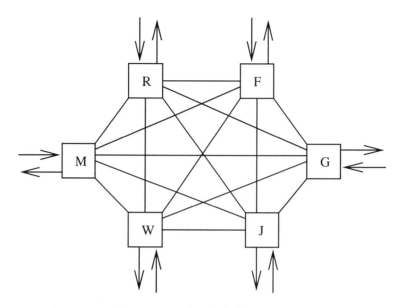

FIGURE 8.5. Jackson network model of the amusement park.

parameters:

$N = 6,$

$s_1 = 24, \qquad s_2 = 35, \qquad s_3 = 20, \qquad s_4 = 60, \qquad s_5 = 16, \qquad s_6 = 20,$

$\lambda_1 = \lambda_2 = \lambda_3 = \lambda_4 = \lambda_5 = \lambda_6 = 100,$

$\mu_1 = 30, \qquad \mu_2 = 20, \qquad \mu_3 = 40, \qquad \mu_4 = 12, \qquad \mu_5 = 40, \qquad \mu_6 = 36,$

$$
P = \begin{bmatrix}
0 & 1/6 & 1/6 & 1/6 & 1/6 & 1/6 \\
1/6 & 0 & 1/6 & 1/6 & 1/6 & 1/6 \\
1/6 & 1/6 & 0 & 1/6 & 1/6 & 1/6 \\
1/6 & 1/6 & 1/6 & 0 & 1/6 & 1/6 \\
1/6 & 1/6 & 1/6 & 1/6 & 0 & 1/6 \\
1/6 & 1/6 & 1/6 & 1/6 & 1/6 & 0
\end{bmatrix},
$$

$r_1 = r_2 = r_3 = r_4 = r_5 = r_6 = 1/6.$ □

8.7.2. Stability

Since the stations in a Jackson network have infinite capacities, we are faced with the possibility of instability as discussed in the context of infinite capacity queues earlier. Hence, in this subsection we establish conditions under which a Jackson network is stable.

Consider the jth station in a Jackson network. Customers arrive at this station from outside at rate λ_j. Customers also arrive at this station from other stations in the network. These are called internal arrivals. Let b_j be the rate of internal arrivals to station j. Then the total arrival rate to station j, denoted by a_j, is given by

$$a_j = \lambda_j + b_j, \quad 1 \le j \le N.$$

Now, if the service stations are all stable, the departure rate of customers from station i will be same as the total arrival rate to station i, namely, a_i. A fraction $p_{i.j}$ of these departing customers go to station j. Hence the arrival rate of internal customers from station i to station j is $a_i p_{i.j}$. Hence the internal arrival rate to station j from all the stations in the network is given by

$$b_j = \sum_{i=1}^{N} a_i p_{i.j}, \quad 1 \le j \le N.$$

Substituting in the previous equation, we get

$$a_j = \lambda_j + \sum_{i=1}^{N} a_i p_{i.j}, \quad 1 \le j \le N. \tag{8.45}$$

These simultaneous equations for a_j are called the *traffic equations*. Using

$$a = [a_1 \quad a_2 \quad \cdots \quad a_N]$$

and

$$\lambda = [\lambda_1 \quad \lambda_2 \quad \cdots \quad \lambda_N]$$

the traffic equations can be written in matrix form as

$$a = \lambda + aP$$

or

$$a(I - P) = \lambda.$$

Suppose the matrix $I - P$ is invertible. Then the above equation has a unique solution given by

$$a = \lambda(I - P)^{-1}. \tag{8.46}$$

The next theorem states the stability condition for Jackson networks in terms of the above solution.

Theorem 8.14 (Stability of Jackson Networks). *A Jackson network with external arrival rate vector λ, and routing matrix P is stable if $I - P$ is invertible, and*

$$a_i < s_i \mu_i$$

for all $i = 1, 2, \ldots, N$, where $a = [a_1, a_2, \ldots, a_N]$ is as given in (8.46).

Intuition. If $I - P$ is invertible, the traffic equations have a unique solution given in (8.46). Now, the ith service station is a single-station queue with s_i servers, arrival rate a_i, and mean service time $1/\mu_i$. Hence from Theorem 8.4, it follows that the queue is stable if

$$a_i/\mu_i < s_i.$$

This equation has to hold for all $i = 1, 2, \ldots, N$ for the traffic equations to be valid. This yields the above theorem. □

We illustrate with two examples.

Example 8.20 (Tandem Queue). Consider the tandem queueing network of Example 8.18. The traffic equations (8.45) for this example are

$$a_1 = 10,$$
$$a_2 = a_1,$$
$$a_3 = a_2,$$
$$a_4 = a_3.$$

These can be solved to get

$$a_1 = a_2 = a_3 = a_4 = 10.$$

Using the parameters derived in Example 8.18, we see that

$$s_1\mu_1 = 12 > a_1,$$
$$s_2\mu_2 = 12 > a_2,$$
$$s_3\mu_3 = 12 > a_3,$$
$$s_4\mu_4 = 12 > a_4.$$

Thus, from Theorem 8.14, the network is stable. □

Example 8.21 (Amusement Park). Consider the Jackson network model of the amusement park of Example 8.19. Using the arrival rates given there we get

$$\lambda = [100 \quad 100 \quad 100 \quad 100 \quad 100 \quad 100].$$

Using the routing matrix given there, we see that the solution to the traffic equations is given by

$$a = [600 \quad 600 \quad 600 \quad 600 \quad 600 \quad 600].$$

The total arrival rate at each ride is the same, due to the symmetry of the problem. The fact that the total arrival rate at each ride is the same as the total external arrival rate at the park can be thought of as an indication that each customer visits each ride once on average! Using the other parameters given there, we see that

$$s_1\mu_1 = 720 > a_1,$$
$$s_2\mu_2 = 700 > a_2,$$

$$s_3 \mu_3 = 800 > a_3,$$
$$s_4 \mu_4 = 720 > a_4,$$
$$s_5 \mu_5 = 640 > a_5,$$
$$s_6 \mu_6 = 720 > a_6.$$

Thus, from Theorem 8.14, we see that this network is stable.   □

8.7.3. Limiting Behavior

Having established the conditions of stability for Jackson networks, we proceed with the study of the limiting behavior of the Jackson networks in this subsection.

First we introduce the relevant notation: Let $X_i(t)$ be the number of customers in the ith service station in a Jackson network at time t. Then the state of the network at time t is given by $X(t) = [X_1(t), X_2(t), \ldots, X_N(t)]$. Now suppose the Jackson network is stable, with a as the unique solution to the traffic equations. Let

$$p(n_1, n_2, \ldots, n_N) = \lim_{t \to \infty} P(X_1(t) = n_1, X_2(t) = n_2, \ldots, X_N(t) = n_N)$$

be the limiting distribution of the state of the network. Let $p_i(n)$ be the limiting probability that there are n customers in an $M/M/s_i$ queue with arrival rate a_i and service rate μ_i. From Theorem 8.9 we get

$$p_i(n) = p_i(0)\rho_i(n), \qquad i \geq 0,$$

where

$$\rho_i(n) = \begin{cases} \dfrac{1}{n!}\left(\dfrac{a_i}{\mu_i}\right)^n & \text{if } 0 \leq n \leq s_i - 1, \\[4mm] \dfrac{s_i^{s_i}}{s_i!}\left(\dfrac{a_i}{s_i \mu_i}\right)^n & \text{if } n \geq s_i, \end{cases}$$

and

$$p_i(0) = \left[\sum_{n=0}^{s_i-1} \frac{1}{n!}\left(\frac{a_i}{\mu_i}\right)^n + \frac{(a_i/\mu_i)^{s_i}}{s_i!} \cdot \frac{1}{1 - a_i/(s_i \mu_i)}\right]^{-1}.$$

The main result is given in the next Theorem.

Theorem 8.15 (Limiting Behavior of Jackson Networks). *The limiting distribution of a stable Jackson network is given by*

$$p(n_1, n_2, \ldots, n_N) = p_1(n_1)p_2(n_2) \cdots p_N(n_N),$$

for $n_i = 0, 1, 2, \ldots$ and $i = 1, 2, \ldots, N$.

Idea of Proof. The idea is to show that $\{X(t), t \geq 0\}$ is an N-dimensional CTMC and to show that the limiting distribution displayed above satisfies the balance equations. The reader is referred to an advanced book for the details. $\square$

The above theorem has many interesting implications. We mention three important ones:

(1) The limiting marginal distribution of $X_i(t)$ is given by $p_i(\cdot)$, i.e.,

$$\lim_{t \to \infty} \mathsf{P}(X_i(t) = n) = p_i(n), \qquad n \geq 0.$$

Thus the limiting distribution of the number of customers in the ith station is the same as that in an $M/M/s_i$ queue with arrival rate a_i and service rate μ_i.

(2) Since the limiting joint distribution of $[X_1(t), X_2(t), \ldots, X_N(t)]$ is a product of the limiting marginal distributions of $X_i(t)$, it follows that, in the limit, the queue lengths at various stations in a Jackson network are independent of each other.

(3) In steady state, we can treat a Jackson network as N independent multiserver birth and death queues. They are coupled via the traffic equations (8.45).

We illustrate with several examples.

Example 8.22 (Jackson Network of Single-Server Queues). Consider a Jackson network with $s_i = 1$ for all $i = 1, 2, \ldots, N$. Thus all service stations are single-server queues. Let a be the solution to the traffic equations. Assume that the network is stable, i.e.,

$$a_i < \mu_i, \qquad i = 1, 2, \ldots, N.$$

We shall answer several questions regarding such a network:

(1) What is the expected number of customers in the network in steady state?
Let L_i be the expected number of customers in station i in steady state. Since station i behaves like an $M/M/1$ queue in steady state L_i can be computed from (8.25) as follows:

$$L_i = \frac{\rho_i}{1 - \rho_i},$$

where

$$\rho_i = a_i/\mu_i.$$

Then the expected number of customers in the network is given by

$$L = \sum_{i=1}^{N} L_i = \sum_{i=1}^{N} \frac{\rho_i}{1 - \rho_i}.$$

(2) What is the probability that the network is empty?
Using Theorem 8.15, the required probability is given by

$$p(0, 0, \ldots, 0) = p_1(0)p_2(0) \cdots p_N(0) = \prod_{i=1}^{N}(1 - \rho_i),$$

where we have used

$$p_i(0) = 1 - \rho_i,$$

from (8.24).

Example 8.23 (Tandem Queue). Consider the tandem queueing network of Example 8.18. We solved the corresponding traffic equations in Example 8.20 and saw that the network is stable. Using the results of Theorem 8.15 we see that, in steady state, station i behaves as an $M/M/s_i$ queue with arrival rate a_i and service rate μ_i, where s_i and μ_i are as given in Example 8.18. Using this, and the results of the $M/M/s$ queue, we compute the expected number in the network here. We get

$$L_1 = 5.4545, \qquad L_2 = 6.0112, \qquad L_3 = 5, \qquad L_4 = 6.6219,$$
$$L = L_1 + L_2 + L_3 + L_4 = 23.0877. \qquad \qquad \square$$

Example 8.24 (Amusement Park). Consider the Jackson network model of the amusement park in Example 8.19. We solved the corresponding traffic equations in Example 8.21 and saw that the network is stable. Using the results of Theorem 8.15 we see that, in steady state, station i behaves as an $M/M/s_i$ queue with arrival rate a_i and service rate μ_i, where s_i and μ_i are as given in Example 8.19.

(1) Compute the expected number of visitors in the park in steady state.

Using (8.31) we compute L_i, the expected number of customers in the ith ride. We get

$$L_1 = 21.4904, \qquad L_2 = 31.7075, \qquad L_3 = 15.4813,$$
$$L_4 = 50.5865, \qquad L_5 = 25.9511, \qquad L_6 = 18.3573.$$

Hence the expected total number of customers in the park is given by

$$L = L_1 + L_2 + L_3 + L_4 + L_5 + L_6 = 163.5741.$$

(2) Which ride has the longest queue?

Here we need to compute L_i^q, the expected number of customers in the queue (not including those in service). This is given by

$$L_i^q = L_i - a_i/\mu_i.$$

Numerical calculations yield:

$$L_1^q = 1.4904, \qquad L_2^q = 1.7075, \qquad L_3^q = 0.4813,$$
$$L_4^q = 0.5865, \qquad L_5^q = 10.9511, \qquad L_6^q = 1.6906.$$

Thus queues are longest in service station 5, the Ghostmountain! Thus if the management wants to invest money to expand the capacities of the rides, it should start with the Ghostmountain ride! $\qquad \square$

8.8.   Problems

CONCEPTUAL PROBLEMS

8.1. Consider a single-server queueing system that is initially empty. A customer arrives at time 0, and successive inter-arrival times are exactly x minutes. The successive service times are exactly y minutes. Verify Theorem 8.1. Do the limiting probabilities p_j exist? Does PASTA (Theorem 8.2) hold? (*Hint:* Consider the cases $y \le x$ and $y > x$ separately.)

8.2. Verify Little's Law for the system in Conceptual Problem 8.1.

8.3. Show that the system in Conceptual Problem 8.1 is stable if $y \le x$ and unstable otherwise.

8.4. Consider a stable single-station queue. Suppose it costs \$$c$ to hold a customer in the system for one unit of time. Show that the long-run holding cost rate is given by cL, where L is the mean number of customers in steady state.

8.5. Show that the blocking probability in an $M/M/1/K$ queue is a monotonically decreasing function of K and that

$$\lim_{K \to \infty} p_K(K) = \begin{cases} 0 & \text{if } \rho \le 1, \\ \dfrac{\rho - 1}{\rho} & \text{if } \rho > 1. \end{cases}$$

What is the intuitive interpretation of this result?

8.6. Show that L_q, the expected number in the queue (not including those in service) in an $M/M/1/K$ queue, is given by

$$L_q = L - \frac{1 - \rho^K}{1 - \rho^{K+1}} \rho,$$

where L is as given in (8.15).

8.7. Show that the blocking probability $p_K(K)$ in an $M/M/K/K$ queue satisfies the following recursive relationship: $p_0(0) = 1$ and

$$p_K(K) = \frac{(\lambda/\mu) p_{K-1}(K - 1)}{K + (\lambda/\mu) p_{K-1}(K - 1)}.$$

8.8. A machine produces items one at a time, the production times being iid exponential with mean $1/\lambda$. The produced items are stored in a warehouse of capacity K. When the warehouse is full the machine is turned off, and it is turned on again when the warehouse has space for at least one item. Demands for the items arise according to a PP(μ). Any demand that cannot be satisfied is lost. Show that the number of items in the warehouse is an $M/M/1/K$ queueing system.

8.9. A bank uses s tellers to handle its walk-in business. All customers form a single line in front of the tellers and get served in a first-come–first-served fashion by the next available teller. Suppose the customer service times are iid Exp(μ) random variables, and the arrival process is a PP(λ). The bank lobby can hold at the most K customers, including those in service. Model the number of customers in the lobby as an $M/M/s/K$ queue.

8.10. A machine shop has K machines and a single repair person. The lifetimes of the machine are iid Exp(λ) random variables. When the machines fail they are repaired by the single repair person in a first-failed–first-repaired fashion. The repair times are iid Exp(μ) random variables. Let $X(t)$ be the number of machines under repair at time t. Model $\{X(t), t \geq 0\}$ as a finite birth and death queue.

8.11. Do Conceptual Problem 8.10 if there are s identical repair persons. Each machine can be repaired by any repair person.

8.12. Telecommunication networks often use the following traffic control mechanism, called the Leaky Bucket: Tokens are generated according to a PP(λ) and stored in a token pool of infinite capacity. Packets arrive from outside according a PP(μ). If an incoming packet finds a token in the token pool, it takes one token and enters the network. If the token pool is empty when a packet arrives, the packet is dropped. Let $X(t)$ be the number of tokens in the token pool at time t. Show that $\{X(t), t \geq 0\}$ is the queue length process of an $M/M/1$ queue.

8.13. In Conceptual Problem 8.12 we described a Leaky Bucket with an infinite capacity token pool, but with no buffer for data packets. Another alternative is to use an infinite capacity buffer for the packets, but a zero capacity token pool. This creates a traffic smoothing control mechanism: Packets arrive according to a PP(λ), and tokens are generated according to a PP(μ). Packets wait in the buffer for the tokens. As soon as a token is generated, a waiting packet grabs it and enters the network. If there are no packets waiting when a token is generated, the token is lost. Let $X(t)$ be the number of packets in the buffer at time t. Show that $\{X(t), t \geq 0\}$ is the queue length process of an $M/M/1$ queue.

8.14. In a stable $M/M/1$ queue with arrival rate λ and service rate μ, show that

$$L_q = \frac{\rho^2}{1 - \rho},$$

and

$$W_q = \frac{1}{\mu} \frac{\rho}{1 - \rho}.$$

8.15. Consider an $M/M/s$ queue with arrival rate λ and service rate μ. Suppose it costs r per hour to hire a server, an additional b per hour if the server is busy, and h per hour to keep a customer in the system. Show that the long-run cost per hour is given by

$$rs + b(\lambda/\mu) + hL,$$

where L is the expected number in the queue.

8.16. In a stable $M/M/s$ queue with arrival rate λ and service rate μ, show that

$$L_q = p_s \frac{\rho}{(1-\rho)^2},$$

and

$$W_q = \frac{p_s}{\mu} \frac{1}{(1-\rho)^2}.$$

8.17. Show that, among all service time distributions with a given mean τ, the constant service time minimizes the expected number of customers in an $M/G/1$ queue.

8.18. Consider the traffic smoothing mechanism described in Conceptual Problem 8.13. Suppose the tokens are generated in a deterministic fashion, with a constant inter-token time. Let $X(t)$ be as defined there. Is $\{X(t), t \geq 0\}$ the queue length process in an $M/G/1$ queueing system with deterministic service times?

8.19. Consider the traffic control mechanism described in Conceptual Problem 8.12. Suppose the tokens are generated in a deterministic fashion, with a constant inter-token time. Let $X(t)$ be as defined there. Is $\{X(t), t \geq 0\}$ the queue length process in a $G/M/1$ queueing system with deterministic inter-arrival times?

8.20. Consider a queueing system with two identical servers. Customers arrive according to a Poisson process with rate λ, and are assigned to the two servers in an alternating fashion (i.e., numbering the arrivals consecutively, all even-numbered customers go to server 1, and odd-numbered customers go to server 2.) Let $X_i(t)$ be the number of customers in front of server i, $i = 1, 2$. (No queue jumping is allowed.) Show that the queue in front of each server is a $G/M/1$ queue.

8.21. Show that the key functional equation for a $G/M/1$ queue with $U(a, b)$ inter-arrival times, and $\text{Exp}(\mu)$ service times is

$$u = \frac{e^{-a\mu(1-u)} - e^{-b\mu(1-u)}}{\mu(1-u)(b-a)}.$$

8.22. Customers arrive at a department store according to a $PP(\lambda)$. They spend iid $\text{Exp}(\mu)$ amount of time picking their purchases and join one of the K checkout queues at random. (This is typically untrue: customers will join nearer and shorter queues more often, but we will ignore this.) The service times at the checkout queues are iid $\text{Exp}(\theta)$. A small fraction p of the customers go to the customer service station for additional service after leaving the checkout queues, while others leave the store. The service times at the customer service counter are iid $\text{Exp}(\alpha)$. Customers leave the store from the customer service counter. Model this as a Jackson network.

8.23. Derive the condition of stability for the network in Conceptual Problem 8.22.

8.24. Assuming stability, compute the expected number of customers in the department store of Conceptual Problem 8.22.

COMPUTATIONAL PROBLEMS

8.1. Customers arrive at a single station queue at a rate of five per hour. Each customer needs 78 minutes of service on the average. What is the minimum number of servers needed to keep the system stable?

8.2. Consider the system in Computational Problem 8.1. What is the expected number of busy servers if the system employs s servers, $1 \leq s \leq 10$.

8.3. Consider the system in Computational Problem 8.1. How many servers are needed if the labor laws stipulate that a server cannot be kept busy more than 80% of the time?

8.4. Consider a single-station queueing system with arrival rate λ. What is the effect on the expected number in the system if the arrival rate is doubled, but the service rate is increased so that the expected time spent in the system is left unchanged?

8.5. Consider a single-station queueing system with arrival rate λ and mean service time τ. Discuss the effect on L, W, L_q, W_q if the arrival rate is doubled, and the service times are simultaneously cut in half.

8.6. Customers arrive at a barber-shop according to a Poisson process at a rate of eight per hour. Each customer requires 15 minutes on average. The barber-shop has four chairs and a single barber. A customer does not wait if all chairs are occupied. Assuming exponential distribution for service times, compute the expected time an entering customer spends in the barber-shop.

8.7. Consider the barber-shop of Computational Problem 8.6. Suppose the barber charges $12 for service. Compute the long-run rate of revenue for the barber. (*Hint:* What fraction of the customers enter?)

8.8. Consider the barber-shop of Computational Problem 8.6. Suppose the barber hires an assistant, so that now there are two barbers. What is the new rate of revenue?

8.9. Consider the barber-shop of Computational Problem 8.6. Suppose the barber installs one more chair for customers to wait in. How much does the revenue increase due to the extra chair?

8.10. Consider the production model of Conceptual Problem 8.8. Suppose the mean manufacturing time is 1 hour, and the demand rate is 20 per day. Suppose the warehouse capacity is 10. Compute the fraction of the time the machine is turned off in the long run.

8.11. Consider the production model of Computational Problem 8.10. Compute the fraction of the demands lost in the long run.

8.12. Consider the production model of Computational Problem 8.10. When an item enters the warehouse, its value is $100. However, it loses value at a rate of $1 per hour as it waits in the warehouse. Thus if an item was in the warehouse for 10 hours when it is sold, it fetches only $90. Compute the long-run revenue per hour.

8.13. Consider the call center of Example 8.7. How many additional holding lines should be installed if the airline desires to lose no more than 5% of the incoming calls? What is the effect of this decision on the queueing time of customers?

8.14. Consider the queueing model of the bank in Conceptual Problem 8.9. Suppose the arrival rate is 18 per hour, and mean service time is 10 minutes and the lobby can handle 15 customers at one time. How many tellers should be used if the aim is to keep the mean queueing time less than 5 minutes?

8.15. Consider the queueing model of the bank in Computational Problem 8.14. How many tellers should the bank use if the aim is to lose no more than 5% of the customers?

8.16. Cars arrive at a parking garage according to a Poisson process at a rate of 60 per hour and park there if there is space, or else they go away. The garage has a capacity to hold 75 cars. A car stays in the garage on average for 45 minutes. Assume that the times spent by the cars in the garage are iid exponential random variables. What fraction of the incoming cars are turned away in the steady state?

8.17. Consider the garage in Computational Problem 8.16. Suppose each car pays at a rate of $3 per hour, based on the exact amount of time spent in the garage. Compute the long-run revenue rate for the garage.

8.18. Consider the taxi stand of Example 8.9. What is the expected number of customers waiting at the stand in steady state? What is the probability that there are at least three customers waiting?

8.19. Consider the Leaky Bucket traffic control mechanism of Conceptual Problem 8.12. Suppose 150,000 tokens are generated per second, while the packet arrival rate is 200,000 per second. Compute the long-run probability that a packet is dropped.

8.20. Consider the Leaky Bucket traffic smoothing mechanism of Conceptual Problem 8.13. Suppose 200,000 tokens are generated per second, while the packet arrival rate is 150,000 per second. Compute the long-run expected time that a packet waits in the buffer.

8.21. Consider the manufacturing operation described in Conceptual Problem 8.8, but with an infinite capacity warehouse. Suppose the demand rate is 12 per hour. What is the largest mean production time so that at most 10% of the demands are lost? What is the expected number of items in the warehouse under this production rate?

8.22. Consider a queueing system where customers arrive according to a Poisson process at a rate of 25 per hour and demand iid exponential service times. We have a choice of using either two servers, each of whom can process a customer in 4 minutes on average, or a single server who can process a customer in 2 minutes on average. What is the optimal decision if the aim is to minimize the expected wait in the system? in the queue?

8.23. Consider the Post Office in Example 8.10. How many server windows should be kept open to keep the probability of waiting for service to begin to be at most .10?

8.24. Consider a service system where customers arrive according to a Poisson process at a rate of 20 per hour, and require exponentially distributed service times with mean

10 minutes. Suppose it costs $15 per hour to hire a server, and an additional $5 per hour when the server is busy. It costs $1 per minute to keep a customer in the system. How many servers should be employed to minimize the long-run expected cost rate? (*Hint:* See Conceptual Problem 8.15.)

8.25. Consider the service system of Computational Problem 8.24 with six servers. How much should each customer be charged so that it is profitable to operate the system?

8.26. Customers arrive at the Drivers License Office according to a Poisson process at a rate of 15 per hour. Each customer takes on average 12 minutes of service. The office is staffed by four employees. Assuming that the service times are iid exponential random variables, what is the expected time that a typical customer spends in the office?

8.27. A computer laboratory has 20 personal computers. Students arrive at this laboratory according to a Poisson process at a rate of 36 per hour. Each student needs on average half an hour at a PC. When all the PCs are occupied, students wait for the next available one. Compute the expected time a student has to wait before getting access to a PC. Assume that the time spent at a PC by students are iid exponential random variables.

8.28. Users arrive at a nature park in cars according to a Poisson process at a rate of 40 cars per hour. They stay in the park for a random amount of time that is exponentially distributed with mean 3 hours and leave. Assuming the parking lot is sufficiently big so that nobody is turned away, compute the expected number of cars in the lot in the long run.

8.29. Students arrive at a bookstore according a Poisson process at a rate of 80 per hour. Each student spends on average 15 minutes in the store, independently of each other. (The time in the checkout counter is negligible, since most of them come to browse!) What is the probability that there are more than 25 students in the store?

8.30. Components arrive at a machine shop according to a Poisson process at a rate of 20 per day. It takes exactly 1 hour to process a component. The components are shipped out after processing. Compute the expected number of components in the machine shop in the long run.

8.31. Consider the machine shop of Computational Problem 8.30. Suppose the processing is not always deterministic, and 10% of the components require an additional 10 minutes of processing. Discuss the effect of this on the expected number of components in the shop.

8.32. Customers arrive at a single-server service station according to a Poisson process at a rate of three per hour. The service time of each customer consists of two stages: each stage lasts for an exponential amount of time with mean 5 minutes, the stages being independent of each other and of other customers. Compute the mean number of customers in the system in steady state.

8.33. Consider Computational Problem 8.32. Now suppose there are two servers, and the two stages of service can be done in parallel, one by each server. However, each server is half as efficient, so that it takes on average 10 minutes to complete one stage. Until both stages are finished, service cannot be started for the next customer. Now compute the expected number of customers in the system in steady state.

8.34. Compute the expected queueing time in an $M/G/1$ queue with arrival rate of 10 per hour, and the following service time distributions, all with mean 5 minutes:

(1) Exponential;

(2) Uniform over $(0, a)$;

(3) Deterministic.

Which distribution produces the smallest W_q, and which one the largest?

8.35. Customers arrive at a post office with a single server. Fifty percent of the customers buy stamps, and take 2 minutes to do so. Twenty percent come to pick up mail and need 3 minutes to do so. The rest take 5 minutes. Assuming all these times are deterministic, and that the arrivals form a Poisson process with a rate of 18 per hour, compute the expected number of customers in the post office in steady state.

8.36. Consider the queueing system described in Conceptual Problem 8.20. Suppose the arrival rate is 10 per hour, and the mean service times are 10 minutes. Compute the long-run expected number of customers in front of each server. Will it reduce congestion if the customers form a single line served by the two servers?

8.37. A machine produces items one at a time, the production times being exactly 1 hour. The produced items are stored in a warehouse of infinite capacity. Demands for the items arise according to a Poisson process at a rate of 30 per day. Any demand that cannot be satisfied is lost. Model the number of items in the warehouse as a $G/M/1$ queue and compute the expected number of items in the warehouse in steady state.

8.38. Compute the fraction of demands lost in the production system of Computational Problem 8.37. Also compute the expected amount of time an item stays in the warehouse.

8.39. Compute the expected queueing time in a $G/M/1$ queue with a service rate of 10 per hour, and the following inter-arrival time distributions, all with mean 8 minutes:

(1) Exponential;

(2) Deterministic.

Which distribution produces the smallest W_q, and which one the largest?

8.40. Customers arrive in a deterministic fashion at a single-server queue. The arrival rate is 12 per hour. The service times are iid exponential with mean 4 minutes. It costs the system $2 per hour to keep a customer in the system, and $40 per hour to keep the server busy. How much should each customer be charged for the system to break even?

8.41. Consider a tandem queue with two stations. Customers arrive at the first station according to a Poisson process at a rate of 24 per hour. The service times at the first station are iid exponential random variables with mean 4 minutes. After service completion at the first station the customers move to station 2, where the service times are iid exponential with mean 3 minutes. The network manager has six servers at her disposal. How many servers should be stationed at each station so that the expected number of customers in the network is minimized in the long run?

8.42. Redo Computational Problem 8.41 if the mean service times are 6 minutes at station 1, and 2 minutes at station 2.

8.43. In the amusement park of Example 8.21, what is the maximum arrival rate the park is capable of handling?

8.44. Consider the Jackson Network 1 as shown in Figure 8.6. The external arrival rate is 25 per hour. The probability of return to the first station from the last is .3. There are two servers at each station. The mean service time at station 1 is 3 minutes, at station 2 it is 2 minutes, and at station 3 it is 2.5 minutes.

(1) Is the network stable?

(2) What is the expected number of customers in the system?

(3) What is the expected wait at station 3?

8.45. Consider the Jackson Network 2 as shown in Figure 8.7. The external arrival rate is 20 per hour. The probability of return to the first station from the second is .3, and from the

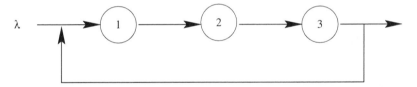

FIGURE 8.6. Jackson Network 1.

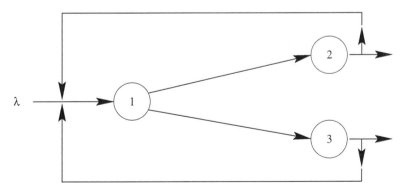

FIGURE 8.7. Jackson Network 2.

third station is .2. There are four servers at station 1, and two each at stations 2 and 3. Forty percent of the customers from station 1 go to station 2, while the rest go to station 3. The mean service times at station 1 is 5 minutes, at station 2 it is 3 minutes, and at station 3 it is 4 minutes.

(1) Is the network stable?

(2) What is the expected number of customers in the system?

(3) What is the expected wait at station 3?

9

Optimal Design

9.1. Introduction

So far in this book we have developed methods of quantifying the performance of stochastic systems. For example, we developed discrete- and continuous-time Markov chains to study manufacturing systems, weather systems, inventory systems, manpower systems, financial systems, telecommunication systems, etc. In Chapter 7 we developed semi-Markov processes to study the above systems under less restrictive assumptions. We also used these new processes to study replacement policies, warranty policies, maintenance policies, etc. In all these chapters we developed methods of computing various relevant performance measures of the systems under study: limiting distributions, occupancy distributions, total costs, long-run cost rates, first passage times, etc.

All these tools help us quantify or predict the performance of a given system. In practice we need the ability not only to predict the future but also to control the future of a system so that its performance meets certain goals. This control is generally classified as *nonadaptive* (*or static*) and *adaptive* (*or dynamic*).

In static control the parameters of the system are set once and for all so that the system achieves the desired behavior. Once set, they are left unchanged, no matter what the actual system performance is observed to be. In the static setting, the parameters under control are called the *design variables*, and the problem of finding the best values of the design variables is called the *optimal design problem*.

In dynamic control, the parameters of the system are changed dynamically over time in response to the actual evolution of the system. In such a case, the parameters under control are called the *decision or control variables* and the problem of determining the best decision variables is called the *optimal decision or control problem*.

Example 9.1 (Multiserver Queue). Customers arriving at a bank form a single queue for service and are served by the available tellers. The bank has space for at most six tellers to be open simultaneously. The bank manager has to decide how many tellers to keep open.

Of course, the answer depends upon what the bank manager wants to achieve. For example, the manager may want to minimize the waiting for the customers, as well as the operating cost of the tellers. But these are conflicting objectives. Using more tellers will increase the operating costs, but decrease the waiting times.

Under static control, the manager will decide to use a fixed number, say k, of tellers at all times. The value of k can be decided as follows: compute the long-run cost rate of a queueing system with k servers. (This includes the cost of waiting as well as the cost of operating k tellers.) Choose a value of k that minimizes this cost rate. This will be the static control problem, with the parameter k as the design variable. Note that once this value of k is set, say $k = 4$, then four tellers will always be open, regardless of the number of customers waiting for service.

Under dynamic control, the manger has a choice of varying the number of tellers depending on the number of customers waiting for service. In this case, the number of tellers can be changed adaptively: the longer the waiting line, the more the number of tellers. This dynamic control problem is much harder: there is no simple set of parameters to determine. Instead, we need to decide when to use how many servers. In effect, we need to find an "optimal policy." For example, the policy may be to use one teller as long as there are three or fewer customers in the queue, two tellers if the number is between four and six, three tellers if the number is between seven and ten, and four tellers thereafter. Note that for a given policy, we may be able to evaluate the long-run cost rate using the tools of previous chapters. However, we need better tools to find an optimal policy, since the number of possible policies is typically very large.   □

The basic method of deriving the optimal static control is the following: we first analyze the descriptive models to derive a functional relationship between the performance of the system (long-run cost rate, for example) and the system parameters. Once the functional relationship is available, we can use numerical or analytical techniques to obtain the values of the parameters that will optimize the performance. We shall study the static control, or the design problem, in this chapter; and the dynamic control problem in the next chapter.

9.2.   Optimal Order Quantity

A department store stocks a seasonal product. Due to the constraints of the procurement process, the entire stock of this product for the season has to be ordered in a single order of size K (called the order quantity). The demand for the product during the season is unknown. However, its distribution is known from experience. The procurement cost is \$$c$ per item, while the selling price is \$$p$. If the demand exceeds K, the unmet demand is lost. If it is less than K, the remaining items have to be disposed off at a post-season sale at a price of $\$s < \$c$. Thus if the order quantity is too small, then there will be too many lost sales thus reducing profits; and if it is too big, there will be too many unsold items, again reducing profits. Thus, the problem is to find the optimal order quantity K.

We first compute $G(K)$, the expected profit if the order quantity is K. Let D be the demand for the product during one season. If $D \geq K$, all items are sold, and the store makes a profit of $\$(p - c)K$. If $D < K$, D items are sold at price p and the remaining $K - D$ items have to be sold at price s. Hence the profit is $(p - c)D + (s - c)(K - D)$. Thus, in general, the profit, as a function of K, is given by $(p - c)\min(D, K) + (s - c)\max(K - D, 0)$. To compute its expected value, we use the following equalities (see Conceptual Problems 9.1 and 9.2):

$$\mathsf{E}(\min(D, K)) = \sum_{d=0}^{K-1} \mathsf{P}(D > d),\tag{9.1}$$

and

$$\mathsf{E}(\max(K - D, 0)) = \sum_{d=0}^{K-1} \mathsf{P}(D \leq d).\tag{9.2}$$

We have

$$G(K) = \mathsf{E}((p - c)\min(D, K) + (s - c)\max(K - D, 0))$$
$$= (p - c)\sum_{d=0}^{K-1} \mathsf{P}(D > d) + (s - c)\sum_{d=0}^{K-1} \mathsf{P}(D \leq d).\tag{9.3}$$

In order to find the optimal value of K, we need to study $G(K)$ as a function of K. From (9.3) we get

$$G(K + 1) - G(K) = (p - c)\sum_{d=0}^{K} \mathsf{P}(D > d) - (p - c)\sum_{d=0}^{K-1} \mathsf{P}(D > d)$$
$$+ (s - c)\sum_{d=0}^{K} \mathsf{P}(D \leq d) - (s - c)\sum_{d=0}^{K-1} \mathsf{P}(D \leq d)$$
$$= (p - c)\mathsf{P}(D > K) + (s - c)\mathsf{P}(D \leq K)$$
$$= (p - c)(1 - \mathsf{P}(D \leq K)) + (s - c)\mathsf{P}(D \leq K)$$
$$= (s - p)\mathsf{P}(D \leq K) + (p - c).$$

Thus

$$\mathsf{P}(D \leq K) < \frac{p - c}{p - s} \quad \Rightarrow \quad G(K + 1) > G(K),$$

and

$$\mathsf{P}(D \leq K) \geq \frac{p - c}{p - s} \quad \Rightarrow \quad G(K + 1) \leq G(K).$$

Since $P(D \leq K)$ increases to 1 as $K \to \infty$, and $0 < (p - c)/(p - s) < 1$ from our assumption $s < c < p$, it follows that there is a K^* such that

$$P(D \leq K) < \frac{p - c}{p - s} \qquad \text{for} \quad K < K^*,$$

and

$$P(D \leq K) \geq \frac{p - c}{p - s} \qquad \text{for} \quad K \geq K^*.$$

Then

$$G(K + 1) > G(K) \qquad \text{for} \quad K < K^*,$$

and

$$G(K + 1) \leq G(K) \qquad \text{for} \quad K \geq K^*.$$

Hence $G(K)$ is maximized at $K = K^*$. Thus K^*, the smallest value of K when $P(D \leq K)$ reaches or exceeds $(p - c)/(p - s)$, is the optimal quantity.

Example 9.2 (Numerical Example). Suppose a store sells plastic Christmas trees. The procurement cost is \$15, the selling price is \$29.99. Any unsold trees have to be sold at an after-Christmas sale for \$9.99. What is the optimal lot size, if the demand is estimated to be a Poisson random variable with mean 300?

We have the following numerical values of the parameters:

$$p = 29.99, \qquad c = 15, \qquad s = 9.99.$$

Hence

$$\frac{p - c}{p - s} = \frac{14.99}{20} = .7495.$$

Using the fact $D \sim P(300)$ we see that

$$P(D \leq 311) = .7484, \qquad P(D \leq 312) = .7662.$$

Hence $K^* = 312$. Thus the store should order 312 Christmas trees. $\square$

9.3. Optimal Leasing of Phone Lines

A small private telephone company leases lines from AT&T and charges its customers a toll to place calls on them. If all the leased lines are busy when a new call arrives, then the new call is lost; otherwise it is accepted. Each accepted call pays c cents per minute. The value of c is fixed due to federal regulations, as well as competition. The cost of leasing is \$$D$ per year per line. The value of D is not under the company's control. Hence the only variable under the company's control is the number of lines leased. How many lines should the company lease?

Suppose calls arrive according to a Poisson process with rate λ, and that an accepted call lasts for an $\text{Exp}(\mu)$ amount of time. All call durations are independent of each other. Suppose the company leases K lines. Let $X(t)$ be the number of calls in progress at time t. Then, we see that $\{X(t), t \geq 0\}$ is the queue-length process of an $M/M/K/K$ queue with arrival rate λ and service rate μ, as analyzed in Section 8.3.

The value of K should be chosen so as to maximize the net revenue per minute. Hence we compute the long-run rate of revenue first. The leasing cost of $\$D$ per year per line is equivalent to

$$d = 100D/(365 \cdot 24 \cdot 60)$$

cents per minute per line. Note that the leasing cost is always Kd cents per minute, regardless of the number of calls in the system. Each call in the system produces a revenue of c cents per minute. Hence, when there are i calls in the system, the net revenue rate is given by

$$r(i) = ic - Kd.$$

We compute the long-run revenue rate by using Theorem 6.11. For this we first need to compute

$$p_i = \lim_{t \to \infty} P(X(t) = i).$$

This is given by Theorem 8.6 as follows:

$$p_i = \frac{\rho_i}{\sum_{j=0}^{K} \rho_j}, \qquad 0 \leq i \leq K,$$

where

$$\rho_i = \frac{(\lambda/\mu)^i}{i!}, \qquad 0 \leq i \leq K.$$

Using Theorem 6.11 we get $g(K)$, the long-run net revenue per minute from using K lines, as follows:

$$g(K) = \sum_{i=0}^{K} r(i)p_i = c \sum_{i=0}^{K} ip_i - Kd.$$

Simplifying the above, we get

$$g(K) = c\rho \left(1 - \frac{\dfrac{\rho^K}{K!}}{\sum_{j=0}^{K} \dfrac{\rho^j}{j!}} \right) - Kd,$$

where

$$\rho = \lambda/\mu.$$

To find the optimal value of K, we study $g(K)$ as a function of K. We can show that $g(K)$ initially increases with K, and then decreases. Hence it is an easy numerical exercise to find the value of K that maximizes it.

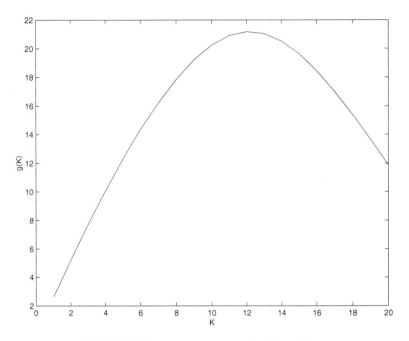

FIGURE 9.1. Net revenue rate as a function of K.

Example 9.3 (Numerical Example). Suppose the calls are charged at a flat rate of 5 cents per minute. The leasing charge is $10,000 per line per year. Suppose the demand for calls is about 120 per hour, and the average call duration is 5 minutes. Compute the number of lines that should be leased so as to maximize the net revenue.

The above data yield the following values for the parameters: $c = 5$ cents per minute, $\lambda = 2$ per minute, $\mu = .2$ per minute, $d = (100)(10,000)/(365)(24)(60) = 1.9026$ cents per minute. Figure 9.1 shows the graph of $g(K)$ versus K. The revenue is maximized at $K = 12$, and the maximum net revenue is 21.18 cents per minute. Thus the company should lease 12 lines at a cost of $120,000 per year. The net revenue is $111,332 per year.

9.4.   Optimal Number of Tellers

Consider the bank manager's problem as described in Example 9.1. Here we build a simple stochastic model to determine the number of tellers to use in the bank.

Suppose the customers arrive according to a Poisson process with parameter λ, and require iid $\text{Exp}(\mu)$ service times. Assume that the manager has decided to use s servers. The bank lobby can hold at most K customers. No customer can enter the bank when the lobby is full. Suppose the tellers cost $\$C_t$ per hour, while the cost of waiting is quantified as $\$C_w$ per customer per hour.

Let $X(t)$ be the number of customers in the lobby at time t. Then $\{X(t), t \geq 0\}$ is a queue-length process in an $M/M/s/K$ queue as analyzed in Section 8.3. The aim is to find the value of s that minimizes the expected cost rate over the long run. We first compute the stationary distribution of $\{X(t), t \geq 0\}$. Let

$$p_i = \lim_{t \to \infty} P(X(t) = i), \qquad 0 \leq i \leq K,$$

and

$$\rho = \frac{\lambda}{s\mu}.$$

Using Theorem 8.5 we get

$$p_i(K) = p_0(K)\rho_i, \qquad i = 0, 1, 2, \ldots, K,$$

where

$$\rho_i = \begin{cases} \dfrac{1}{i!}\left(\dfrac{\lambda}{\mu}\right)^i & \text{if } 0 \leq i \leq s-1, \\[3mm] \dfrac{s^s}{s!}\rho^i & \text{if } s \leq i \leq K, \end{cases}$$

and

$$p_0(K) = \left[\sum_{i=0}^{s-1}\frac{1}{i!}\left(\frac{\lambda}{\mu}\right)^i + \frac{(\lambda/\mu)^s}{s!}\cdot\frac{1-\rho^{K-s+1}}{1-\rho}\right]^{-1}.$$

Next, the cost rate in state i is

$$c(i) = iC_w + sC_t, \qquad 0 \leq i \leq K.$$

Hence the long-run cost rate, from Theorem 6.11, is

$$g(s) = \sum_{i=0}^{K} C(i)p_i = sC_t + C_w \sum_{i=0}^{K} ip_i.$$

To find the optimal value of s we plot $g(s)$ as a function of s and pick the value of s that minimizes the cost rate. We explain with a numerical example below.

Example 9.4 (Numerical Example). Suppose the tellers cost $15 per hour, while the cost of waiting is estimated to be $10 per customer per hour. The arrival rate is 10 customers per hour, and the mean service time is 10 minutes. The capacity of the lobby is 15 customers. Compute the optimal number of tellers to employ so as to minimize the long-run cost rate.

The above data yield $K = 15$, $C_t = \$15$ per hour, $C_w = \$10$ per hour, $\lambda = 10$ customers per hour, and $\mu = 6$ customers per hour. Figure 9.2 shows the graph of $g(s)$ versus s. We see that the cost rate is minimized at $s = 3$, implying that it is optimal to use three tellers. The minimum cost rate is $65.39 per hour. Note that $g(0) < g(1)$, however $g(3)$ is the global minimum.

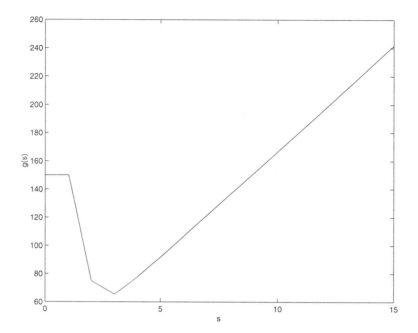

FIGURE 9.2. Net cost rate as a function of s.

Now, the cost of waiting for the customers is hard to estimate, since it involves non-quantitative aspects of customer psychology, etc. In such a case it is important to do a sensitivity analysis with that parameter. Numerical calculations show that it is optimal to use one teller for $0 < C_w \le 1.66$, two tellers for $1.67 \le C_w \le 6.14$, and three tellers for $6.15 \le C_w \le 50.10$. After that, four servers are optimal up to $C_w = 258.53$, and so on. This kind of sensitivity analysis is useful when we are not sure of the values of certain parameters. $\square$

9.5.   Optimal Replacement

Consider the battery replacement problem faced by Mr Smith of Example 7.2. Mr Smith follows the age replacement policy of replacing the battery upon failure or upon reaching the age of T years. Find the optimal value of T that minimizes the maintenance cost per year for Mr Smith.

We first build a general cost model. Suppose the lifetimes of successive batteries are iid random variables with common cdf $G(\cdot)$. Each new battery costs $\$C$, including installation. If the battery fails before it reaches age T, there is an additional cost of $\$D$ that includes cost of towing, lost work time, etc. Using this we first compute $g(T)$, the cost per year as a function of T. Following the notation of Example 7.10, let L_n be the lifetime of the nth battery, and let $T_n = \min\{L_n, T\}$ be the nth inter-replacement time. Then C_n is the cost incurred over the nth interval (the time between the nth and the $(n-1)$st replacement). Now, we have

$$C_n = \begin{cases} C + D & \text{if } L_n < T, \\ C & \text{if } L_n \geq T. \end{cases}$$

Let $C(t)$ be the total battery replacement cost up to time t. We have seen in Example 7.10 that $\{C(t), t \geq 0\}$ is a cumulative process, with the duration of the nth cycle being T_n, and the cost over the nth cycle being C_n. Using Theorem 7.3, we get

$$g(T) = \lim_{t \to \infty} \frac{C(t)}{t} = \frac{\mathsf{E}(C_1)}{T_1}.$$

We have

$$\mathsf{E}(C_1) = C\mathsf{P}(L_1 > T) + (C + D)\mathsf{P}(L_1 \leq T)$$
$$= C + DG(T),$$

and, using Conceptual Problem 2.23,

$$\mathsf{E}(T_1) = \mathsf{E}(\min(L_1, T))$$
$$= \int_0^T (1 - G(t)) \, dt.$$

Substituting, we get

$$g(T) = \frac{C + DG(T)}{\int_0^T (1 - G(t)) \, dt}.$$

We can now plot $g(T)$ as a function of T and find the value of T that minimizes it. We illustrate it with a numerical example.

Example 9.5 (Numerical Example). Suppose the lifetimes (in years) of the successive batteries are Erlang distributed with mean 4 years, and variance 8 (years)^2. The cost of replacement is $25, while the additional cost of failure is $200. Compute the optimal age replacement parameter T.

Suppose the lifetimes are Erl(k, λ). Then the mean lifetime is k/λ and the variance of the lifetime is k/λ^2. Using the data given above, this implies that

$$\lambda = .5, \qquad k = 2.$$

Hence,

$$G(T) = 1 - \exp(-T/2)[1 + T/2],$$

and

$$\int_0^T (1 - G(t)) \, dt = 4 - (4 + T) \exp(-T/2).$$

Also,

$$C = 75, \qquad D = 25.$$

Hence

$$g(T) = \frac{75 + 25 * (1 - \exp(-T/2)[1 + T/2])}{4 - (4 + T)\exp(-T/2)}.$$

Numerical evaluation of $g(T)$ as a function of T shows that the cost rate decreases from infinity at $T = 0$ to \$42.56 per year at about $T = 1.41$, and then increases to \$56.25 per year as t increases to infinity. Hence it is optimal to replace the battery upon reaching an age of 1.41 years. □

9.6. Optimal Server Allocation

Customers arrive according to a Poisson process with rate λ at a service facility consisting of two queues in tandem. They wait at the first service station and, after service completion, join the second service station. After completing the service at the second station, they depart. There is an infinite waiting room at each service station. The service times at station i are exponentially distributed with parameter μ_i, $i = 1, 2$. There are s identical servers available in total, and we have to decide how many of them should be employed at station 1, and how many at station 2, so that the expected total number of customers in the system in steady state is minimized.

The system described above can be seen to be a Jackson network as in Section 8.7.1. More specifically, this is a tandem queue with two nodes. (See Example 8.18.) The arrival rates are

$$\lambda_1 = \lambda, \qquad \lambda_2 = 0.$$

The routing matrix is

$$P = \begin{bmatrix} 0 & 1 \\ 0 & 0 \end{bmatrix}.$$

The traffic equations are

$$a_1 = \lambda, \qquad a_2 = a_1.$$

Hence the effective inputs to the two nodes are

$$a_1 = a_2 = \lambda.$$

Now suppose we assign s_1 servers to station 1, and $s_2 = s - s_1$ servers to station 2. This assignment is feasible (i.e., it leads to a stable network) if

$$a_i < s_i \mu_i, \qquad i = 1, 2.$$

This implies that we must have

$$\lambda < s_1 \mu_1, \qquad \lambda < s_2 \mu_2 = (s - s_1)\mu_2,$$

which can be rewritten as

$$\frac{\lambda}{\mu_1} < s_1 < s - \frac{\lambda}{\mu_2}. \tag{9.4}$$

If there is no integer s_1 satisfying the above constraints, there are not enough servers to make the system stable. So, suppose there is at least one integer s_1 satisfying the above constraints. For such an s_1, the results of Section 8.7.1 imply that, in steady state, station i is an $M/M/s_i$ queue with arrival rate λ and service rate μ_i. Hence $L_i = L_i(s_i)$, the expected number of customers in station i, can be computed by using the analysis of the $M/M/s$ queue in Section 8.4. The problem of finding the optimal s_i then reduces to minimizing $L_1(s_1) + L_2(s - s_1)$ over all values of s_1 satisfying constraints in (9.4). This can be done by direct enumeration. We describe this with a numerical example below.

Example 9.6 (Airport Check-in). The check-in process at an airport consists of the security check, followed by the baggage check-in. The times required for the security checks of passengers are iid exponential random variables with mean 3 minutes, and the baggage check-in takes an exponential amount of time with mean 5 minutes. The passengers arrive according to a Poisson process at a rate of 100 per hour. There is a total of 20 agents available to handle the security check and the baggage check-in. How many of them should be assigned to the baggage check-in so as to minimize the total expected number of customers in the system in steady state?

The check-in process can be thought as a tandem queue of two stations: the security check and the baggage check-in. The arrival rate (in customers per hour) is

$$\lambda = 100.$$

The service rates (in customers per hour per server) are

$$\mu_1 = 20, \qquad \mu_2 = 12.$$

The total number of servers is 20, of which we assign s_1 to the security check, and $s_2 = 20 - s_1$ to the baggage check-in. The stability constraint of (9.4) becomes

$$5 < s_1 < 20 - 8.5.$$

Thus

$$s_1 \in \{6, 7, \ldots, 11\}.$$

Now, in steady state, station 1 behaves like an $M/M/s_1$ queue with arrival rate 100 and service rate 20, and station 2 behaves like an $M/M/s_2$ queue with arrival rate 100 and service rate 12. Using the results on the $M/M/s_i$ queue from Section 8.4, we can compute L_1 and L_2 as shown in Table 9.1.

We see that the expected number of customers in the security check and baggage check-in is minimized if we allocate eight agents to the security area and 12 agents to the baggage

TABLE 9.1. The expected number of customers in the airport check-in example.

s_1	s_2	L_1	L_2	$L_1 + L_2$
6	14	7.9376	8.4107	16.3483
7	13	5.8104	8.5094	14.3198
8	12	5.2788	8.7333	14.0120
9	11	5.1006	9.2696	14.3703
10	10	5.0361	10.7714	15.8075
11	9	5.0126	17.8382	22.8508

check-in area. Under this allocation there will be 5.2788 customers in the security line, and 8.7333 customers in the baggage check-in line, on average.   □

Another possible objective could be to minimize the expected waiting time of a customer in the tandem queue in steady state. Let W_i be the expected waiting time at station i. Then the total expected waiting time of the tandem queue is $W = W_1 + W_2$. Now, Little's Law (see Theorem 8.3), applied to each station separately, yields

$$W_i = L_i/a_i = L_i/\lambda, \quad i = 1, 2.$$

Thus,

$$W = (L_1 + L_2)/\lambda.$$

Hence, minimizing W is the same as minimizing $L_1 + L_2$. In the airport check-in example above, the minimum expected waiting time is

$$W = (14.0120)/100 = .14 \text{ hours} = 8.4 \text{ minutes.}$$

9.7. Problems

CONCEPTUAL PROBLEMS

9.1. Prove (9.1).

9.2. Prove (9.2).

9.3. Newspaper-Vendor Problem. A newspaper-vendor buys newspapers from the distributor for C cents per copy, and sells them for P cents per copy ($P > C$). He buys a fixed number K of newspapers at the beginning of the day and sells them until 5:00 p.m. He quits early if he sells all K copies, otherwise he quits at 5:00 p.m. and recycles the rest at no cost or revenue. Based on experience, he knows the cdf of the random demand D for newspapers on any day. Compute $G(K)$, the net expected profit earned by the newspaper vendor if he buys K papers at the beginning of the day. How many papers should he buy to maximize his profits?

9.4. Consider the following extension of the newspaper-vendor problem of Conceptual Problem 9.3. Suppose the newspaper vendor gets R cents per copy that he recycles at the end of the day, $R < C$. (If he has to pay to dispose of the unsold copies, R is negative.) Now compute the expected net profit $G(K)$ and the optimal ordering quantity K^*.

9.5. Optimal Dispatching. Suppose passengers arrive at a depot according to a renewal process with mean inter-arrival time τ hours. A shuttle is dispatched as soon as there are K passengers waiting, where $K \leq B$, the capacity of the shuttle. Suppose the cost of keeping a passenger waiting is $\$C_w$ per hour per passenger, while the cost of dispatching the shuttle is $\$D$, regardless of the number of passengers carried by the shuttle. Compute $g(K)$, the long-run hourly cost of operating this system as a function of dispatch level K. Let $\tilde{K}$ be the smallest K such that

$$\frac{K(K+1)}{2} \geq \frac{D}{C_w \tau}.$$

Show that the optimal dispatching level is

$$K^* = \min(B, \tilde{K}).$$

9.6. Consider the optimal replacement model of Section 9.5 applied to a machine replacement problem as follows. Suppose the lifetimes (in years) of successive machines are iid random variables with common cdf $G(\cdot)$, and mean τ. A new machine costs $\$C$ (including installation). The machine is replaced upon failure or upon reaching age T. A working machine of age t can be sold in the "used machine" market for $\$Re^{(-t/\tau)}$. The failed machine has no salvage value. Compute $g(T)$, the long-run cost per year of following this age replacement policy. Write a computer program to compute the optimal age replacement parameter T if the lifetimes are iid $U(a, b)$ random variables, with $0 \leq a < b$.

9.7. Consider the optimal server allocation problem of Section 9.6. Suppose the cost of waiting at station i is $\$c_i$ per unit time per customer. Show how to find the optimal server allocation that minimizes the expected total waiting cost.

9.8 (Optimal Machine Operation). Consider the simple manufacturing system described in Example 6.15. Suppose the warehouse capacity K is fixed. The cost of keeping an item in the warehouse for one unit of time is h. It costs nothing to turn the machine off, but costs $\$c$ to turn it on. Each item sold generates a revenue of $\$p$. Show how to compute the optimal k (the turn-on level) so as to minimize the long-term cost per unit time.

9.9 (Optimal Staffing of a Call Center). Consider the call center described in Example 6.12. Suppose the total number of lines K is fixed. Each agent costs $\$C$ per unit time. Each admitted call produces a revenue of $\$D$. Show how to compute the optimal number of agents M to maximize the net revenue per unit time.

9.10 (Optimal (s, S) Inventory Policies). Consider the inventory system of Example 5.6 with the following generalization. Let $0 < s < S$ be two given integers. The store clerk orders enough PCs to bring the stock up to S at 8:00 a.m. next Monday if there are less than s PCs in stock at 5:00 p.m. on Friday. Such an inventory policy is called an (s, S) inventory policy. The aim is find the optimal values of the parameters (s, S) assuming the following cost structure: The weekly storage cost is $\$h$ per PC in stock at the beginning of the week. The purchase price of a PC is $\$c$, while the selling price is $\$p$. Placing an order costs $b, regardless of the size of the order. Suppose the size of the warehouse is K, a given number, so that we must have $S \le K$. Show how to compute the optimal parameters (s, S) to minimize the long-run weekly cost of operating the store.

COMPUTATIONAL PROBLEMS

9.1. Consider the newspaper-vendor problem of Conceptual Problem 9.3. Suppose the newspaper vendor pays 40 cents per copy for the newspapers, and sells them for 50 cents per copy. He estimates the demand to be binomially distributed with parameters $n = 200$ and $p = .8$. Compute the optimum number of papers he should buy in order to maximize his profits.

9.2. Consider Computational Problem 9.1. Suppose the newspaper vendor gets 10 cents per copy that he recycles. Compute the optimal order quantity for this case. (See Conceptual Problem 9.4.)

9.3. Farmer Brown sells real Christmas trees from Nov. 25 to Dec. 24 every year. It costs him $20 per tree to cut and transport them to the selling place. He sells them for $30 each. Any trees that are left over at the end of Dec. 24 have to be disposed off at a cost of $2 each. From experience Farmer Brown estimates that the demand for trees this year is Poisson with mean 150. How many trees should he stock in order to maximize net expected profits?

9.4. A department store sells Christmas cards during the Christmas season for $3 per pack. Any unsold cards after Christmas are sold at half-price. The cost of procurement of the cards is $2.50 per pack. How many card packs should the store stock for the season if the demand D is distributed with the following pmf:

$$P(D = m) = cm(400 - m), \qquad 101 \le m \le 300,$$

where c is an appropriate normalization constant, and the aim is to maximize the expected profits from the sale of cards?

9.5. Consider Example 9.3. For what range of the arrival rate λ does the decision to lease 12 lines remain optimal? Assume all the other parameters remain unchanged.

9.6. Consider Example 9.3. For what range of the call duration μ does the decision to lease 12 lines remain optimal? Assume all the other parameters remain unchanged.

9.7. One-Stop Laundromat offers fancy machines that do the washing, drying, and folding automatically, with no intervention, in exactly 1 hour. The demand is estimated to be 20 per hour. Assume the arrival process is Poisson. A customer who finds all the machines in the Laundromat busy goes away. The cost of operating the Laundromat is $10 per machine per day. The cost to the users is $2.50 for this service. How many machines should the owner of the One-Step Laundromat install so as to maximize the profit per day? (*Hint*: Suppose the number of machines in the Laundromat is K. Let $X(t)$ be the number of machines in use at time t. It is known that the limiting distribution $\{X(t), t \ge 0\}$ depends on the service time (in this case 1 hour) only via its mean, and is independent of the distribution. Thus you may assume that the service times are exponentially distributed.)

9.8. Suppose the One-Stop Laundromat of Computational Problem 9.7 already has 30 machines. For what range of the arrival rate is this number optimal? Assume the rest of the data remains unchanged.

9.9. A small business has to decide how many parking spaces to build. The total cost of building and maintaining one parking space is $600 per year. The customer arrival rate is 20 per hour, and each customer occupies a parking space for 20 minutes on average, and produces a revenue of $20 on average. A customer who cannot find parking goes away. How many parking spaces should the business owner build? Assume the arrival process is Poisson, and the customer visit lengths are exponentially distributed.

9.10. Suppose the business of Computational Problem 9.9 already has 10 parking spaces. For what range of the arrival rate is this number of spaces optimal?

9.11. Consider Example 9.4. For what range of the arrival rate λ does the decision to hire three tellers remain optimal? Assume all the other parameters remain unchanged.

9.12. Consider Example 9.4. For what range of the mean service time $1/\mu$ does the decision to hire three tellers remain optimal? Assume all other parameters remain unchanged.

9.13. A telemarketing company operates a TV channel that advertises products, and customers call in to place orders. Currently the company's equipment can handle a maximum of 16 calls. It employees eight operators, so that up to eight customers can be processed simultaneously, and the others are put on hold. The cost of the operators is $20 per hour, while the cost of the equipment is $150 per hour. The company wants to upgrade its telephone system. It is considering a new system that has a capacity to handle 24 calls, and has 12 lines so 12 of them can be processed simultaneously, while the rest are put on hold. The new system will cost $200 per hour to operate. The per operator cost will remain the same. The company expects to get about 30 calls per hour, each call lasting on average 20 minutes, and producing a net revenue of $40. (The cost of the operators and the system has to be covered from this revenue.) Should the company switch to the new system? If yes, how many operators should the company employ?

9.14. Quick-Change provides a quick oil change service to its customers. It currently has two service bays and a parking lot with 10 spaces. Customers arrive according to a Poisson process at a rate of 12 per hour. A customer who cannot find parking goes away. The oil change takes on average 12 minutes and is approximately exponentially distributed. The customers pay $20 for the oil change. It costs $60 per hour per service bay to operate. The manager is considering adding one more service bay, which will reduce the number of parking spaces by two. Should the manager add the third bay?

9.15. Consider the Quick-Change company of Computational Problem 9.14. For what range of the arrival rates will it be optimal to add the third bay? Assume all other data remain unchanged.

9.16. Consider the optimal replacement problem of Example 9.5. Compute the optimal age replacement parameter T if the lifetime (in years) of the battery is uniformly distributed over $(3, 5)$.

9.17. Consider the optimal replacement problem of Example 9.5. Compute the optimal age replacement parameter T if the pdf of the lifetime (in years) of the battery is given by

$$f(t) = kt^2, \quad 2 \le t \le 5,$$

where k is the appropriate normalization constant.

9.18. Consider the optimal dispatching model of Conceptual Problem 9.5. Suppose the arrival rate is 20 customers per hour, the waiting cost is $5 per hour per customer, and the dispatching cost is $100. What is the optimal dispatching level K^*? Assume the shuttle capacity is 25.

9.19. Consider the shuttle of Computational Problem 9.18. For what range of the arrival rates is it optimal to dispatch the shuttle as soon as it is full? Assume that the rest of the data remain unchanged.

9.20. The libraries at UNC Chapel Hill and Duke University participate in an inter-library loan program. Requests accumulate at the UNC library for books from the Duke library, and vice versa, and are communicated to the other library by email almost instantly. As soon as 30 books are collected at UNC to be transported to Duke, a van is dispatched. Dispatching the van costs $150. The requests accumulate at Duke for books from UNC according to a Poisson process at a rate of 5 per day. What is the range of the assumed cost of waiting per book per day for which this dispatching policy is optimal?

9.21. Consider the machine replacement problem of Conceptual Problem 9.6. Compute the optimal age replacement parameter if the new machine costs $10,000, and the lifetimes are uniformly distributed over $(2, 5)$ years. Assume the resale parameter $R = 5,000$.

9.22. Consider the optimal machine operation problem described in Conceptual Problem 9.8. Suppose the warehouse capacity is 12, the storage cost is $1 per hour, each item sells for $25, and the cost of turning the machine on is $10. Compute the optimal value of k, if the production rate is two per hour and the demand rate is also two per hour.

9.23. Consider the optimal machine operation problem described in Computational Problem 9.22. Suppose the system uses $k = 6$. For what range of the selling price will this policy minimize the long-run cost per unit time? Assume the rest of the data are as in Computational Problem 9.22.

10

Optimal Control

10.1. Introduction

The optimal control problem was defined in the last chapter as the problem of dynamically changing the system parameters in response to the system evolution so as to optimize its performance. It is more common to use the phrase "choose an action" in place of "set the parameters" in the context of control problems. The rule that specifies what action to choose as a function of the system evolution is called a *policy*. In this chapter we develop methods of determining optimal policies to optimize the performance of systems, and illustrate this methodology with several examples. We begin with two examples.

Example 10.1 (Optimal Group Maintenance). A machine shop consists of K independent and identical machines. The machines are subject to failures and repairs. We check the number of working machines at the beginning of each day. We have an option of replacing any number of machines based upon the observed state of the system. Replacing a machine costs $\$C_r$. The repair person charges $\$C_v$ per visit and takes exactly 1 day to complete the replacements regardless of how many machines are replaced. (Thus, it costs nothing and takes no time to replace zero machines, while it costs $\$C_v + kC_r$ and takes 1 day to replace k machines, $1 \leq k \leq K$.) During the replacement operation all machines are turned off. Each machine earns a revenue of $\$R$ per day that it is functioning, and earns no revenue when it is in a failed condition, or is under maintenance, or is turned off. If a machine is turned off, it does not fail. If a machine is working (and not turned off) at the beginning of the nth day, it is in working condition at the beginning of the $(n + 1)$st day with probability p (or in a failed condition with probability $q = 1 - p$), regardless of its age or state of the other machines. A maintenance policy specifies how many machines to replace on the nth day based on how many are working on that day. The aim is to determine an optimal maintenance policy which minimizes the expected cost per day in the long run. □

Example 10.2 (Optimal Processor Scheduling). Consider a multiprocessor multiprogramming computer system with K identical processors and multiprogramming level M (i.e., it can handle at most M jobs at a time). Suppose jobs (computer programs) arrive according a PP(λ) to this computer system and form a single queue. The processing times are iid Exp(μ) random variables. Each program needs exactly one processor, and each processor can execute at most one program at a time. If an arriving job finds M jobs in the system, it is lost. At each arrival and departure epoch we have to decide how many of the K processors should be assigned to execute the waiting programs. Switching a processor on or off does not cost anything. Keeping a job in the system costs \$$C_h$ per hour, while keeping a processor on costs \$$C_p$ per hour. Each lost job costs \$$C_l$. The scheduling policy specifies how many processors should continue to execute the programs after observing the number of jobs in the system after an arrival or departure epoch. The aim is to find an optimal scheduling policy that minimizes waiting costs plus the processor operating costs plus the cost of lost jobs per hour in the long run.   □

In the next section we describe a general model of dynamic control.

10.2. Discrete-Time Markov Decision Processes: DTMDPs

Let X_n be the state of the system at time n ($n = 0, 1, 2, \ldots$). Let $S = \{1, 2, \ldots, M\}$ be the state space of $\{X_n, n \geq 0\}$. Let A_n be the action chosen at time n. The set of allowable actions is given by

$$\mathcal{A} = \{1, 2, \ldots, K\}.$$

$\mathcal{A}$ is called the *action space*. If the system is in state i at time n, and action $a \in \mathcal{A}$ is chosen, the system moves to state j at time $n + 1$ with probability $p_{i,j}(a)$ (independent of n or of the history up to time n). Then, mathematically,

$$\mathsf{P}(X_{n+1} = j | X_n = i, A_n = a, X_{n-1}, A_{n-1}, \ldots, X_0, A_0) = p_{i,j}(a), \quad i, j \in S, \quad a \in \mathcal{A}.$$

The process $\{(X_n, A_n), n \geq 0\}$ is called a *Discrete-Time Markov Decision Process* or DTMDP for short. We need to specify how the actions are chosen in order to describe the evolution of the DTMDP. A rule that describes how actions are to be chosen is called a *policy*. In general, the action chosen at time n may be based upon the history of the DTMDP up to time n. However, we shall restrict ourselves to policies that choose the action at time n based only on the state of the system at time n, possibly in a randomized fashion. To be specific, let $f(i, a)$ be the probability that action $a \in \mathcal{A}$ is chosen at time n if the state of the system at time n is $i \in S$. Mathematically

$$f(i, a) = \mathsf{P}(A_n = a | X_n = i, X_{n-1}, A_{n-1}, \ldots, X_0, A_0), \quad i \in S, \quad a \in \mathcal{A}.$$

Since $f(i, a)$ is a probability, and we can choose one and only one action, we must have

$$f(i, a) \geq 0, \quad i \in S, \quad a \in \mathcal{A}, \tag{10.1}$$

and

$$\sum_{a=1}^{K} f(i, a) = 1, \qquad i \in S. \tag{10.2}$$

A policy is thus described by the function $f = \{f(i, a) : i \in S, a \in \mathcal{A}\}$.

Now we can describe the evolution of the DTMDP under a policy f. Suppose the system starts in state i at time 0. Then the action a is chosen with probability $f(i, a)$ at time 0. If action a is chosen, the system moves to state j at time 1 with probability $p_{i,j}(a)$. If the state is j at time 1, a new action a' is chosen with probability $f(j, a')$, and the system proceeds accordingly. Thus a DTMDP is described by a policy f, and the K transition probability matrices of size $M \times M$ as defined below:

$$P(a) = \begin{bmatrix} p_{1,1}(a) & p_{1,2}(a) & p_{1,3}(a) & \cdots & p_{1,M}(a) \\ p_{2,1}(a) & p_{2,2}(a) & p_{2,3}(a) & \cdots & p_{2,M}(a) \\ \vdots & \vdots & \vdots & \ddots & \vdots \\ p_{M-1,1}(a) & p_{M-1,2}(a) & p_{M-1,3}(a) & \cdots & p_{M-1,M}(a) \\ p_{M,1}(a) & p_{M,2}(a) & p_{M,3}(a) & \cdots & p_{M,M}(a) \end{bmatrix}, \quad a \in \mathcal{A}.$$

Next we describe the cost structure. If the system is in state i at time n, and action a is chosen, it incurs an expected cost $c(i, a)$. We define an $M \times K$ matrix of costs as follows:

$$C = \begin{bmatrix} c(1, 1) & c(1, 2) & c(1, 3) & \cdots & c(1, K) \\ c(2, 1) & c(2, 2) & c(2, 3) & \cdots & c(2, K) \\ \vdots & \vdots & \vdots & \ddots & \vdots \\ c(M-1, 1) & c(M-1, 2) & c(M-1, 3) & \cdots & c(M-1, K) \\ c(M, 1) & c(M, 2) & c(M, 3) & \cdots & c(M, K) \end{bmatrix}.$$

We illustrate with an example.

Example 10.3 (Optimal Group Maintenance). Consider the optimal group maintenance problem of Example 10.1. Let X_n be the number of working machines at the beginning of day n. The state space is

$$S = \{0, 1, 2, \ldots, K\}.$$

The decision to replace machines is made at the beginning of each day after observing the number of working machines. The action is the number of machines to replace. Thus the action space is

$$\mathcal{A} = \{0, 1, 2, \ldots, K\}.$$

Next we compute the probability transition matrices. We assume that no failures take place while the replacement is being done. Suppose $X_n = i$, and action $a \in \mathcal{A}$ is chosen. If $a = 0$, no replacement is done. The failed machines remain failed, while the working machines fail independently with probability q. Hence

$$p_{i,j}(0) = \binom{i}{j} p^j q^{i-j}, \qquad 0 \le j \le i \le K.$$

If $1 \le a \le K - i$, all the working machines are shut down, and a of the failed machines are replaced. The working machines remain working. Hence $X_{n+1} = X_n + a$ with probability 1, i.e.,

$$p_{i,i+a}(a) = 1, \qquad 1 \le a \le K - i.$$

Similarly, if $K - i < a \le K$, $X_{n+1} = K$ with probability 1, and hence

$$p_{i,K}(a) = 1, \qquad K - i < a \le K.$$

For example, if the machine shop consists of two machines, we have

$$P(0) = \begin{bmatrix} 1 & 0 & 0 \\ q & p & 0 \\ q^2 & 2pq & p^2 \end{bmatrix},$$

$$P(1) = \begin{bmatrix} 0 & 1 & 0 \\ 0 & 0 & 1 \\ 0 & 0 & 1 \end{bmatrix},$$

and

$$P(2) = \begin{bmatrix} 0 & 0 & 1 \\ 0 & 0 & 1 \\ 0 & 0 & 1 \end{bmatrix}.$$

Next consider the costs. Suppose $X_n = i$ and action a is chosen. If $a = 0$, the working machines continue to operate, producing a revenue of iR. There are no other costs. Hence

$$c(i, 0) = -iR, \qquad 0 \le i \le K.$$

If $a \ge 1$, the working machines are shut down for 1 day. The repair person's visit costs C_v, and replacing the a machines (we always replace nonworking machines first) costs aC_r. Hence

$$c(i, a) = C_v + aC_r, \qquad 0 \le i \le K, \quad 1 \le a \le K.$$

For the case of $K = 2$ machines, we get the following cost matrix:

$$C = \begin{bmatrix} 0 & C_r + C_v & 2C_r + C_v \\ -R & C_r + C_v & 2C_r + C_v \\ -2R & C_r + C_v & 2C_r + C_v \end{bmatrix}. \qquad \square$$

A policy is said to be *optimal* if it achieves the lowest long-run cost per unit time among all policies. We shall develop methods of obtaining such optimal policies in the next section.

10.3. Optimal Policies for DTMDPs

Let $\{(X_n, A_n), n \geq 0\}$ be a DTMDP with state space $S = \{1, 2, \ldots, M\}$, action space $\mathcal{A} = \{1, 2, \ldots, K\}$, transition probability matrices $\{P(a), a \in \mathcal{A}\}$, and cost matrix C. Let $f = \{f(i, a) : i \in S, a \in \mathcal{A}\}$ be a policy that chooses action a with probability $f(i, a)$ whenever the system state is i. The next theorem describes the structure of the $\{X_n, n \geq 0\}$ process under policy f.

Theorem 10.1. *Under policy f, $\{X_n, n \geq 0\}$ is a DTMC with transition probability matrix $P^f = [p_{i,j}^f]$, where*

$$p_{i,j}^f = \mathsf{P}(X_{n+1} = j | X_n = i) = \sum_{a \in \mathcal{A}} f(i, a) p_{i,j}(a). \tag{10.3}$$

Proof. We have

$$\mathsf{P}(X_{n+1} = j | X_n = i) = \sum_{a \in \mathcal{A}} \mathsf{P}(X_{n+1} = j | X_n = i, A_n = a) \mathsf{P}(A_n = a | X_n = i)$$

$$= \sum_{a \in \mathcal{A}} p_{i,j}(a) f(i, a). \qquad \square$$

Next we describe how to evaluate the "performance" of such a policy. We define the performance of a policy f as the long-run expected cost per unit time. We have studied this quantity in Section 5.7. We saw in Theorem 5.12 that the long-run cost rate is independent of the starting state if the DTMC is irreducible. Assuming that f yields an irreducible DTMC $\{X_n, n \geq 0\}$, we define the long-run cost rate of following policy f, denoted by G^f, as follows:

$$G^f = \lim_{N \to \infty} \mathsf{E}\left(\frac{1}{N+1} \sum_{n=0}^{N} c(X_n, A_n)\right). \tag{10.4}$$

We describe how to compute G^f in the next theorem.

Theorem 10.2. *Let $\{(X_n, A_n), n \geq 0\}$ be a DTMDP with policy f, and transition probability matrices $\{P(a), a \in \mathcal{A}\}$. Suppose the DTMC $\{X_n, n \geq 0\}$ is irreducible, with occupancy distribution $\hat{\pi} = [\hat{\pi}_1, \hat{\pi}_2, \ldots, \hat{\pi}_M]$. Then, the long-run cost rate G^f, starting from any state i, is given by*

$$G^f = \sum_{i \in S} \hat{\pi}_i \sum_{a \in \mathcal{A}} f(i, a) c(i, a). \tag{10.5}$$

Proof. Theorem 10.1 says that $\{X_n, n \geq 0\}$ is a DTMC with transition probability matrix P^f. When this DTMC visits state i, action a is chosen with probability $f(i, a)$ and that results in a cost of $c(i, a)$. Hence the expected cost incurred upon each visit to state i is

$$c(i) = \sum_{a \in \mathcal{A}} f(i, a) c(i, a).$$

Thus the long-run cost G^f of following policy f can be found by using the results of Theorem 5.12. This yields (10.5).   □

We illustrate the above material with the following example.

Example 10.4 (Optimal Group Maintenance). Consider the optimal group maintenance problem described in Examples 10.1 and 10.3. Now consider the policy of replacing only the failed machines at the beginning of each day. This policy can be described by the following function f:

$$f(i, K - i) = 1; \quad 0 \le i \le K.$$

Consider the case $K = 2$. From Theorem 10.1 we see that under this policy $\{X_n, n \ge 0\}$ is a DTMC with transition probability matrix

$$P^f = \begin{bmatrix} 0 & 0 & 1 \\ 0 & 0 & 1 \\ q^2 & 2pq & p^2 \end{bmatrix}.$$

The DTMC is irreducible, with occupancy distribution

$$\hat{\pi}_0 = \frac{q^2}{2 - p^2}, \qquad \hat{\pi}_1 = \frac{2pq}{2 - p^2}, \qquad \hat{\pi}_2 = \frac{1}{2 - p^2}.$$

The costs of visiting state i under this policy are

$$c(0) = 2C_r + C_v, \qquad c(1) = C_r + C_v, \qquad (2) = -2R.$$

The long-run expected cost per day of following this policy is given by (see Theorems 5.12 or 10.2)

$$\frac{q^2(2C_r + C_v) + 2pq(C_r + C_v) - 2R}{2 - p^2}.$$ □

Now that we know how to evaluate the performance of a given policy f, we address the question of finding the best one. In effect we need to solve the following optimization problem:

OPT: **Minimize** G^f.

Subject to: f is a policy that chooses action a in state i with probability $f(i, a)$.

We shall formulate the above optimization problem using the following notation:

$$x_{i,a} = \hat{\pi}_i f(i, a), \qquad i \in S, \qquad a \in \mathcal{A}. \tag{10.6}$$

Note that $x_{i,a}$ is the long-run fraction of the time that the system is in state i *and* action a is chosen. We can recover the quantities $\hat{\pi}_i$ and $f(i, a)$ from the quantities $x_{i,a}$ as follows.

Using

$$\sum_{a \in \mathcal{A}} f(i, a) = 1$$

we get

$$\hat{\pi}_i = \sum_{a \in \mathcal{A}} \hat{\pi}_i f(i, a) = \sum_{a \in \mathcal{A}} x_{i,a}, \qquad i \in S. \tag{10.7}$$

Using the above equality in (10.6), we get

$$f(i, a) = \frac{x_{i,a}}{\sum_{b \in \mathcal{A}} x_{i,b}}. \tag{10.8}$$

We now express the objective function G^f in terms of $x_{i,a}$. Using (10.5) we get

$$\begin{aligned} G^f &= \sum_{i \in S} \hat{\pi}_i \sum_{a \in \mathcal{A}} f(i, a) c(i, a) \\ &= \sum_{i \in S} \sum_{a \in \mathcal{A}} \hat{\pi}_i f(i, a) c(i, a) \\ &= \sum_{i \in S} \sum_{a \in \mathcal{A}} x_{i,a} c(i, a). \end{aligned} \tag{10.9}$$

Thus the objective function is a linear function of the variables $\{x_{i,a}, i \in S, a \in \mathcal{A}\}$.

Next, we know that the occupancy distribution must satisfy (5.47) and (5.48), viz.,

$$\hat{\pi}_j = \sum_{i \in S} \hat{\pi}_i p_{i,j}^f, \qquad j \in S, \tag{10.10}$$

and

$$\sum_{j \in S} \hat{\pi}_j = 1. \tag{10.11}$$

Using (10.7) we see that the right-hand side of (10.10) reduces to

$$\begin{aligned} \sum_{i \in S} \hat{\pi}_i p_{i,j}^f &= \sum_{i \in S} \hat{\pi}_i \sum_{a \in \mathcal{A}} f(i, a) p_{i,j}(a) \\ &= \sum_{i \in S} \sum_{a \in \mathcal{A}} \hat{\pi}_i f(i, a) p_{i,j}(a) \\ &= \sum_{i \in S} \sum_{a \in \mathcal{A}} x_{i,a} p_{i,j}(a), \end{aligned}$$

and hence (10.10) can be written as

$$\sum_{a \in \mathcal{A}} x_{j,a} = \sum_{i \in S} \sum_{a \in \mathcal{A}} x_{i,a} p_{i,j}(a). \tag{10.12}$$

Similarly, (10.11) reduces to

$$\sum_{j \in S} \sum_{a \in \mathcal{A}} x_{j,a} = 1. \tag{10.13}$$

Using (10.9), (10.12), and (10.13), the optimization problem can be rewritten as

$$\textbf{LP:} \quad \textbf{Minimize} \quad \sum_{i \in S} \sum_{a \in \mathcal{A}} x_{i,a} c(i, a).$$

$$\textbf{Subject to:} \quad \sum_{a \in \mathcal{A}} x_{j,a} = \sum_{i \in S} \sum_{a \in \mathcal{A}} x_{i,a} p_{i,j}(a), \quad j \in S,$$

$$\sum_{i \in S} \sum_{a \in \mathcal{A}} x_{i,a} = 1,$$

$$x_{i,a} \geq 0, \quad i \in S, \quad a \in \mathcal{A}.$$

The above problem is a *Linear Programming* problem (hence the title LP for the problem formulation), since the objective function as well as the constraints are linear in the unknown variables $\{x_{i,a}, i \in S, a \in \mathcal{A}\}$. The total number of unknowns are MK, the number of states times the number of actions. Note that we have explicitly added the nonnegativity constraints $x_{i,a} \geq 0$. Excluding these nonnegativity constraints, there are $K + 1$ constraints. However, one of the first K constraints is redundant. There are very efficient numerical packages available on computers that can solve linear programs involving several thousand unknown variables and several thousand constraints. Such packages can be used to solve the above linear program.

A linear programming package yields the optimal values of the variables $x_{i,a}^*$ along with the optimal value of the objective function. How do we obtain the optimal policy from this information? First let

$$T = \left\{ i \in S : \sum_{a \in \mathcal{A}} x_{i,a}^* > 0 \right\}.$$

Consider two cases:

Case 1: $T = S$. In this case the optimal policy is given by

$$f^*(i, a) = \frac{x_{i,a}^*}{\sum_{u \in \mathcal{A}} x^*(i, u)}, \quad i \in S, \quad a \in \mathcal{A}.$$

The theory of linear programming can be used to show that, for a given $i \in S$, there is exactly one $a^*(i) \in \mathcal{A}$ such that

$$x_{i,a}^* \begin{cases} > 0 & \text{if } a = a^*(i), \\ = 0 & \text{if } a \neq a^*(i). \end{cases}$$

Thus,

$$f^*(i, a) = \begin{cases} 1 & \text{if } a = a^*(i), \\ 0 & \text{if } a \neq a^*(i). \end{cases}$$

Thus the optimal policy chooses action $a^*(i)$ whenever the system visits state i. The optimal value of the objective function is the long-run cost rate of following the optimal policy f^* starting from any initial state i.

Case 2: $T \neq S$. In this case $\sum_{a \in A} x^*_{i.a} = 0$ for $i \notin T$. In such a case we compute $f^*(i, a)$ as in Case 1 for $i \in T$. The $f^*(i, a)$ for $i \notin T$ are not defined. If the initial state is in T, then f^* is optimal. Otherwise, the analysis is more complicated, and the reader is referred to more advanced texts for further information.

We illustrate the methodology with a few examples.

Example 10.5 (Optimal Group Maintenance). Consider the problem described in Example 10.1. It is modeled as a DTMDP and the corresponding parameters are given in Example 10.3. Here we formulate the linear program to compute the optimal maintenance policy for the case $K = 2$.

The decision variables are $\{x_{i.a}, i = 0, 1, 2; a = 0, 1, 2\}$. The linear program is as follows:

$$\textbf{Minimize} \qquad (C_r + C_v)x_{0.1} + (2C_r + C_v)x_{0.2} - Rx_{1.0} + (C_r + C_v)x_{1.1}$$
$$+ (2C_r + C_v)x_{1.2} - 2Rx_{2.0} + (C_r + C_v)x_{2.1} + (2C_r + C_v)x_{2.2}.$$

$$\textbf{Subject to:} \qquad x_{0.0} + x_{0.1} + x_{0.2} = x_{0.0} + qx_{1.0} + q^2x_{2.0}$$
$$x_{1.0} + x_{1.1} + x_{1.2} = x_{1.1} + px_{1.0} + 2pqx_{2.0}$$
$$x_{2.0} + x_{2.1} + x_{2.2} = x_{0.2} + x_{1.1} + x_{1.2} + p^2x_{2.0} + x_{2.1} + x_{2.2}$$
$$x_{0.0} + x_{0.1} + x_{0.2} + x_{1.0} + x_{1.1} + x_{1.2} + x_{2.0} + x_{2.1} + x_{2.2} = 1$$
$$x_{i.a} \geq 0, \quad i, a = 0, 1, 2.$$

As a numerical example, suppose the probability of failure of a machine is .2, and the revenue per machine is \$100 per day. It costs \$200 to replace a machine, and a visit by the repair person costs \$150. This implies

$$p = .8, \qquad R = 100, \qquad C_r = 200, \qquad C_v = 150.$$

Using a linear programming package, we get the following optimal solution:

$$X = [x_{i.a}] = \begin{bmatrix} 0 & 0 & 0.1216 \\ 0.5405 & 0 & 0 \\ 0.3378 & 0 & 0 \end{bmatrix}.$$

Hence the optimal policy is

$$f^* = [f^*_{i.a}] = \begin{bmatrix} 0 & 0 & 1 \\ 1 & 0 & 0 \\ 1 & 0 & 0 \end{bmatrix}.$$

Thus it is optimal to replace only when both machines are down. The DTMC $\{X_n, n \geq 0\}$ under this policy has the following transition probability matrix:

$$P* = \begin{bmatrix} 0 & 0 & 1 \\ .2 & .8 & 0 \\ .04 & .32 & .64 \end{bmatrix}.$$

This is an irreducible DTMC, hence the optimal policy f^* minimizes the long-run cost per day staring from any initial state. This cost per day is $-\$54.73$ per day, i.e., the maximum revenue is $\$54.73$ per day.   □

10.4.   Optimal Inventory Control

Inventory is a stock of goods. Inventories are kept at various places in a system. A manufacturer keeps an inventory of manufactured items in the factory warehouse, from which they are shipped to the retailers. The retailers keep inventory at their store warehouse, from which they satisfy customer demands. Customers themselves keep an inventory at their homes. There is a variety of reasons for maintaining an inventory: to handle the uncertainty of demand, the uncertainty of price, to reduce the cost of too-frequent ordering, etc. There are also reasons for not keeping too much inventory: holding inventory costs money through storage costs, tied-up capital, spoilage, insurance, taxes, etc. Hence, there is a problem of deciding optimal inventory levels.

In this section we shall illustrate the inventory control problem within the context of the Computers-R-Us company of Example 5.6. The store operates from Monday to Friday every week, and sells Pentium PCs. The demand for the PCs is random, and any demand that cannot be satisfied immediately is lost, i.e., there is no back-logging or rain checks. After the store closes on Friday, the store manager checks the inventory of the Pentium PCs and decides how many to order. The order is delivered over the weekend so that the new PCs are available on Monday morning when the store opens. Here we develop a DTMDP to find the optimum ordering policy.

Let X_n be the number of PCs in the store at 5:00 p.m. on the nth Friday. (Note that this is different than the DTMC in Example 5.6.) Let A_n be the number of PCs ordered after the store closes on the nth Friday. We shall assume that the store can have at most B PCs on the shelf (due to capacity limitation). If there is no space for the PCs when they are delivered to the store, the extra ones will have to be thrown away. Of course it wouldn't make sense to order PCs for which there is no space, but the cost structure of the problem will automatically ensure that the optimal policy will not do such a dumb thing. Thus it makes sense to set the action space as $\mathcal{A} = \{0, 1, \ldots, B\}$. Let D_n be the demand for the PCs during the nth week. With this notation we see that

$$X_{n+1} = \max\{\min(X_n + A_n, B) - D_{n+1}, 0\}, \qquad n \geq 0.$$

Now assume that $\{D_n, n \geq 1\}$ are iid random variables with pmf

$$p_k = \mathsf{P}(D_n = k), \qquad k = 0, 1, 2, 3, \ldots .$$

This implies that $\{(X_n, A_n), n \geq 0\}$ is a DTMDP. We compute the transition probabilities below. For $1 \leq j \leq \min(i + a, B)$, we have

$$
\begin{aligned}
p_{i,j}(a) &= \mathsf{P}(X_{n+1} = j | X_n = i, A_n = a) \\
&= \mathsf{P}(\max\{\min(X_n + A_n, B) - D_{n+1}, 0\} = j | X_n = i, A_n = a) \\
&= \mathsf{P}(D_{n+1} = \min(i + a, B) - j) \\
&= p_{\min(i+a.B)-j}.
\end{aligned}
$$

Similarly, for $j = 0$, we get

$$p_{i,0}(a) = \mathsf{P}(D_{n+1} \geq \min(i + a, B)) = \sum_{k=\min(i+a.B)}^{\infty} p_k.$$

Next we describe the cost structure. Suppose the store buys the PCs for $\$c$ and sells them for $\$p > c$. The weekly storage cost is $\$h$ per Pentium PC that is in the store at the end of the week. There is also a fixed delivery charge of $\$d$ (regardless of the size of the delivery). Using these cost components we compute the one-step costs. Suppose $X_n = i$ and $A_n = a$. Then the holding cost is hi, the cost of the order is 0 if $a = 0$, and $d + ca$ if $a > 0$. The inventory at the beginning of the following Monday is $\min(i + a, B)$, and the demand during the following week is D_{n+1}. Hence the expected revenue during the next week is $p\mathsf{E}(\min(D_{n+1}, \min(i + a, B)))$. (Note that it is irrelevant that the costs are incurred in week n while the revenues are incurred in week $n + 1$, since we are interested in long-run costs per week.) Combining these costs and revenues, and using (9.1), we get

$$
\begin{aligned}
c(i, 0) &= hi - p\mathsf{E}(\min(D_{n+1}, i)) \\
&= hi - p \sum_{k=0}^{i-1} \mathsf{P}(D_{n+1} > k),
\end{aligned}
\tag{10.14}
$$

and, for $a = 1, 2, \ldots, B$,

$$
\begin{aligned}
c(i, a) &= hi + d + ca - p\mathsf{E}(\min(D_{n+1}, \min(i + a, B))) \\
&= hi + d + ca - p \sum_{k=0}^{\min(i+a.B)-1} \mathsf{P}(D_{n+1} > k).
\end{aligned}
\tag{10.15}
$$

In Example 5.6 the store followed the policy of ordering up to five (i.e., ordering enough PCs so that the store had five PCs at 8:00 a.m. on Monday) if the number of PCs in stock at 5:00 p.m. on Friday was 0 or 1, and ordering nothing otherwise. This policy is described by the following f:

$$
\begin{aligned}
f(i, 5 - i) &= 1, & i &= 0, 1, \\
f(i, 0) &= 1, & i &= 2, 3, 4, 5.
\end{aligned}
$$

This policy is analyzed in Examples 5.6 and 5.29. (The state of the system X_n defined in those examples is actually $X_n + A_n$ defined here.) The optimal policy can be computed by solving the linear programming problem using a computer package. We illustrate with an example.

Example 10.6 (Numerical Example). Consider the data given in Examples 5.6 and 5.29. The weekly demands are iid P(3) random variables. The store buys computers for $1500 and sells them for $1750. The weekly storage charge is $50 per computer. Let us further assume that the delivery charge is $500. Suppose the store has space for at most five computers on its shelves. Compute the ordering policy that minimizes the long-run cost per week.

The above data produces the following values for the parameters of our model:

$$p = 1750, \qquad c = 1500, \qquad h = 50, \qquad d = 500, \qquad B = 5,$$

and

$$p_k = e^{-3}\frac{3^k}{k!}, \qquad k = 0, 1, 2, \ldots.$$

Using this we get the following cost matrix:

$$C = \begin{bmatrix} 0 & 337.13 & 435.64 & 926.22 & 1808.88 & 2985.59 \\ -1612.87 & -1014.36 & -523.78 & 358.88 & 1535.59 & 3035.59 \\ -2964.36 & -1973.78 & -1091.12 & 85.59 & 1585.59 & 3085.59 \\ -3923.78 & -2541.12 & -1364.41 & 135.59 & 1635.59 & 3135.59 \\ -4491.12 & -2814.41 & -1314.41 & 185.59 & 1685.59 & 3185.59 \\ -4764.41 & -2764.41 & -1264.41 & 235.59 & 1735.59 & 3235.59 \end{bmatrix},$$

and the following transition probability matrices:

$$P(0) = \begin{bmatrix} 1.0000 & 0 & 0 & 0 & 0 & 0 \\ .9502 & .0498 & 0 & 0 & 0 & 0 \\ .8009 & .1494 & .0498 & 0 & 0 & 0 \\ .5768 & .2240 & .1494 & .0498 & 0 & 0 \\ .3528 & .2240 & .2240 & .1494 & .0498 & 0 \\ .1847 & .1680 & .2240 & .2240 & .1494 & .0498 \end{bmatrix},$$

$$P(1) = \begin{bmatrix} .9502 & .0498 & 0 & 0 & 0 & 0 \\ .8009 & .1494 & .0498 & 0 & 0 & 0 \\ .5768 & .2240 & .1494 & .0498 & 0 & 0 \\ .3528 & .2240 & .2240 & .1494 & .0498 & 0 \\ .1847 & .1680 & .2240 & .2240 & .1494 & .0498 \\ .1847 & .1680 & .2240 & .2240 & .1494 & .0498 \end{bmatrix},$$

and so on until

$$P(5) = \begin{bmatrix} .1847 & .1680 & .2240 & .2240 & .1494 & .0498 \\ .1847 & .1680 & .2240 & .2240 & .1494 & .0498 \\ .1847 & .1680 & .2240 & .2240 & .1494 & .0498 \\ .1847 & .1680 & .2240 & .2240 & .1494 & .0498 \\ .1847 & .1680 & .2240 & .2240 & .1494 & .0498 \\ .1847 & .1680 & .2240 & .2240 & .1494 & .0498 \end{bmatrix}.$$

The solution to the linear programming model yields

$$X = [x_{i.a}] = \begin{bmatrix} 0 & 0 & 0 & 0 & 0 & .3706 \\ 0 & 0 & 0 & 0 & .1782 & 0 \\ .1812 & 0 & 0 & 0 & 0 & 0 \\ .1504 & 0 & 0 & 0 & 0 & 0 \\ .0908 & 0 & 0 & 0 & 0 & 0 \\ .0288 & 0 & 0 & 0 & 0 & 0 \end{bmatrix}.$$

Hence the policy f is given by

$$f = [f_{i.a}] = \begin{bmatrix} 0 & 0 & 0 & 0 & 0 & 1 \\ 0 & 0 & 0 & 0 & 1 & 0 \\ 1 & 0 & 0 & 0 & 0 & 0 \\ 1 & 0 & 0 & 0 & 0 & 0 \\ 1 & 0 & 0 & 0 & 0 & 0 \\ 1 & 0 & 0 & 0 & 0 & 0 \end{bmatrix}.$$

Thus it is optimal to order nothing if there are two, three, four, or five computers in stock at 5:00 p.m. Friday, otherwise order five computers if there are none in stock, and four if only one is in stock. That is, if the stock is down to zero or one, order enough to bring the inventory up to five at 8:00 a.m. the following Monday morning. This was precisely the policy studied in Examples 5.6 and 5.29. (Of course there was no delivery cost in those examples.) The DTMC $\{X_n, n \geq 0\}$ under this policy can be seen to be irreducible, hence the long-run average cost under this policy is the same from all initial states. The linear programming package gives the optimal cost to be $-\$292.11$, i.e., the store makes a profit of $\$291.11$ per week under the optimal policy.   □

10.5. Semi-Markov Decision Processes: SMDPs

In this section we study continuous-time MDPs. Let $X(t)$ be the state of the system at time t $(t \geq 0)$. Let $S = \{1, 2, \ldots, M\}$ be the state space of $\{X(t), t \geq 0\}$. Let S_n be the time of the nth transition in the state of the system. Assume that a control action can be taken at times S_n $(n = 0, 1, 2, \dots)$, with $S_0 = 0$. At time S_n, an action can be chosen from the set of allowable actions

$$\mathcal{A} = \{1, 2, \dots, K\},$$

called the *action space*. We shall assume that the rule for choosing an action is described by a function $f = \{f(i, a) : i \in S, a \in \mathcal{A}\}$ as follows: if the system is in state i at time S_n, action $a \in \mathcal{A}$ is chosen with probability $f(i, a)$, regardless of the history of the system up to time S_n. Clearly, f satisfies (10.1) and (10.2). Let X_n be the state, and A_n the action chosen at time S_n. Also assume that if $X_n = i$ and $A_n = a$, the future of the system from time S_n onward depends only upon i and a, and not on the history of the system up to time S_n. The next decision epoch occurs at time $S_{n+1} > S_n$. Again, the distribution of S_{n+1} depends only upon X_n and A_n, and not upon the history. We introduce the following notation:

$$w_i(a) = \mathsf{E}(S_{n+1} - S_n | X_n = i, A_n = a), \qquad i \in S, \quad a \in \mathcal{A},$$

and

$$p_{i,j}(a) = \mathsf{P}(X_{n+1} = j | X_n = i, A_n = a), \qquad i, j \in S, \quad a \in \mathcal{A}.$$

Thus, if $X_n = i$ and $A_n = a$, the next decision epoch occurs after a random amount of time with mean $w_i(a)$ and at the next decision epoch the system is in state j with probability $p_{i,j}(a)$ (independent of the history up to time S_n). Let $A(t)$ be the most recent action chosen at or before t. The process $\{(X(t), A(t)), t \geq 0\}$ is called a *Semi-Markov Decision Process* or SMDP for short.

For our analysis, an SMDP is described by a policy f, and the K transition probability matrices of size $M \times M$

$$P(a) = \begin{bmatrix} p_{1,1}(a) & p_{1,2}(a) & p_{1,3}(a) & \cdots & p_{1,M}(a) \\ p_{2,1}(a) & p_{2,2}(a) & p_{2,3}(a) & \cdots & p_{2,M}(a) \\ \vdots & \vdots & \vdots & \ddots & \vdots \\ p_{M-1,1}(a) & p_{M-1,2}(a) & p_{M-1,3}(a) & \cdots & p_{M-1,M}(a) \\ p_{M,1}(a) & p_{M,2}(a) & p_{M,3}(a) & \cdots & p_{M,M}(a) \end{bmatrix}, \quad a \in \mathcal{A},$$

and the $M \times K$ matrix of sojourn times given below:

$$W = \begin{bmatrix} w_1(1) & w_1(2) & w_1(3) & \cdots & w_1(K) \\ w_2(1) & w_2(2) & w_2(3) & \cdots & w_2(K) \\ \vdots & \vdots & \vdots & \ddots & \vdots \\ w_{M-1}(1) & w_{M-1}(2) & w_{M-1}(3) & \cdots & w_{M-1}(K) \\ w_M(1) & w_M(2) & w_M(3) & \cdots & w_M(K) \end{bmatrix}.$$

Finally, suppose the following cost structure holds: if the system is in state i at time S_n, and action a is chosen, it incurs an expected total cost of $c(i, a)$ over the interval $[S_n, S_{n+1})$. We define an $M \times K$ matrix of costs as follows:

$$
C = \begin{bmatrix}
c(1, 1) & c(1, 2) & c(1, 3) & \cdots & c(1, K) \\
\vdots & \vdots & \vdots & \ddots & \vdots \\
c(M-1, 1) & c(M-1, 2) & c(M-1, 3) & \cdots & c(M-1, K) \\
c(M, 1) & c(M, 2) & c(M, 3) & \cdots & c(M, K)
\end{bmatrix}.
$$

Let $C(T)$ be the total cost incurred by the system up to time T. The long-run expected cost per unit time, if policy f is followed, is defined to be

$$
G^f = \lim_{T \to \infty} \frac{\mathsf{E}(C(T))}{T}.
$$

Thus, an SMDP is similar to a DTMDP except that it stays in the current state for a random amount of time depending upon the state and the action chosen. We illustrate the above formulation with an example.

Example 10.7 (Optimal Processor Scheduling). Consider the optimal processor scheduling problem of Example 10.2. Let $X(t)$ be the number of jobs in the computer system at time t. The state space is then $S = \{0, 1, \ldots, M\}$. The nth decision epoch S_n is the nth event (arrival or departure) in the $\{X(t), t \geq 0\}$ process (some arrivals may not enter the system if it is full). The action at time S_n is the number of processors to be kept "on" during the interval $[S_n, S_{n+1})$. The maximum number of processors we can use is K. Hence the action space is $\mathcal{A} = \{0, 1, \ldots, K\}$. Since it does not make sense to use more processors than the number of jobs in the system (since each processor can serve only one job at a time), we can assume that $K \leq M$.

Next we compute the transition probabilities. Suppose $X_n = i$ $(0 \leq i \leq K)$ and $A_n = a$ $(0 \leq a \leq M)$. That is, there are i jobs in the system after the nth event, and we have decided to use a processors to execute them. The number of jobs in service then is $\min(i, a)$. Hence the next event will take place after an $\text{Exp}(\lambda + \min(i, a)\mu)$ amount of time. It will be an arrival with probability $\lambda/(\lambda + \min(i, a)\mu)$, and a departure with probability $\min(i, a)\mu/(\lambda + \min(i, a)\mu)$. The expected time until the next event is $1/(\lambda + \min(i, a)\mu)$, and hence the expected cost is $(aC_p + iC_h)/(\lambda + \min(i, a)\mu)$. Calculations like these yield the following transition probabilities:

$$
p_{i,i+1}(a) = \frac{\lambda}{\lambda + \min(i, a)\mu}, \qquad 0 \leq i < M, \quad 0 \leq a \leq K,
$$

$$
p_{M,M}(a) = \frac{\lambda}{\lambda + a\mu}, \qquad 0 \leq a \leq K,
$$

$$
p_{i,i-1}(a) = \frac{\min(i, a)\mu}{\lambda + \min(i, a)\mu}, \qquad 0 < i \leq M, \quad 0 \leq a \leq K,
$$

and the following costs:

$$c(i, a) = \frac{aC_p + iC_h}{\lambda + \min(i, a)\mu}, \qquad 0 \le i < M, \quad 0 \le a \le K,$$

$$c(M, a) = \frac{aC_p + MC_h + \lambda * C_l}{\lambda + a\mu}, \qquad 0 < a \le K.$$

The sojourn times are given by

$$w_i(a) = \frac{1}{\lambda + \min(i, a)\mu}, \qquad 0 \le i \le M, \quad 0 \le a \le K.$$

Note that it does not make sense to use more processors than there are jobs in the system, although the above model allows it. However, an optimal policy will not choose action a in state i if $i < a$!   □

Our aim is to develop methods of computing an optimal policy f^* that minimizes this long-run cost rate. This is done in the next section.

10.6.   Optimal Policies for SMDPs

Let $\{(X(t), A(t)), t \ge 0\}$ be an SMDP with state space $S = \{1, 2, \ldots, M\}$, action space $\mathcal{A} = \{1, 2, \ldots, K\}$, transition probability matrices $\{P(a), a \in \mathcal{A}\}$, sojourn time matrix W, and cost matrix C. Let $f = \{f(i, a) : i \in S, a \in \mathcal{A}\}$ be a policy that chooses action a with probability $f(i, a)$ whenever the system enters state i. The next theorem describes the structure of the $\{X(t), t \ge 0\}$ process under policy f.

Theorem 10.3. *Under policy f, $\{X(t), t \ge 0\}$ is an SMP. The embedded DTMC $\{X_n, n \ge 0\}$ has transition probability matrix $P^f = [p_{i,j}^f]$, where*

$$p_{i,j}^f = \mathsf{P}(X(S_{n+1}) = j | X(S_n) = i) = \sum_{a \in \mathcal{A}} f(i, a) p_{i,j}(a), \qquad i, j \in S, \quad a \in \mathcal{A}, \quad (10.16)$$

The sojourn time vector $w^f = [w_i^f]$ is given by

$$w_i^f = \mathsf{E}(S_1 | X_0 = i) = \sum_{a \in \mathcal{A}} f(i, a) w_i(a). \qquad (10.17)$$

Proof. The $\{X(t), t \ge 0\}$ process has the Markov property at the transition epochs $0 = S_0 < S_1 < S_2 < \cdots$. Hence it is an SMP. (See Section 7.4.) The rest of the proof follows along the same lines as that of Theorem 10.1.   □

Next we describe how to evaluate the "performance" of a policy f. We define the performance as the long-run expected cost rate per unit time, denoted by G^f, as given below:

$$G^f = \lim_{T \to \infty} \mathsf{E}\left(\frac{C(T)}{T}\right), \qquad (10.18)$$

where $C(T)$ is the total cost incurred up to time T. We have studied this quantity in Section 7.5. We saw in Theorem 7.7 that the long-run cost rate is independent of the starting state of an SMP if the embedded DTMC is irreducible. From Theorem 10.3 we see that the embedded discrete-time process $\{X_n, n \geq 0\}$ is a DTMC with transition probability matrix P^f. We describe how to compute G^f in the next theorem.

Theorem 10.4. *Let* $\{(X(t), A(t)), t \geq 0\}$ *be an SMDP with policy* f, *transition probability matrices* $\{P(a), a \in \mathcal{A}\}$, *sojourn time matrix* W, *and cost matrix* C. *Suppose the embedded DTMC* $\{X_n, n \geq 0\}$ *is irreducible, with occupancy distribution* $\hat{\pi} = [\hat{\pi}_1, \hat{\pi}_2, \ldots, \hat{\pi}_M]$. *Then, the long-run cost rate* G^f, *starting from any state* i, *is given by*

$$G^f = \frac{\sum_{i \in S} \sum_{a \in \mathcal{A}} \hat{\pi}_i f(i, a) c(i, a)}{\sum_{i \in S} \sum_{a \in \mathcal{A}} \hat{\pi}_i f(i, a) w_i(a)}. \tag{10.19}$$

Proof. Theorem 10.3 implies that $\{X_n, n \geq 0\}$ is a DTMC with transition matrix P^f. When this DTMC visits state i, action a is chosen with probability $f(i, a)$ and that results in a cost of $c(i, a)$. Hence the expected cost incurred upon each visit to state i is

$$c^f(i) = \sum_{a \in \mathcal{A}} f(i, a) c(i, a).$$

The sojourn time in state i is given by

$$w_i^f = \sum_{a \in \mathcal{A}} f(i, a) w_i(a).$$

Thus the long-run cost G^f of following policy f can be found by using (7.22). This yields (10.19). $\qquad\square$

We illustrate the above material with an example.

Example 10.8 (Optimal Processor Scheduling). Consider the optimal processor problem described in Example 10.7. Consider the policy: "never keep a job waiting, if possible." This means we activate as many processors as there are jobs. Mathematically, this policy is described by the following f

$$f(i, a) = \begin{cases} 1 & \text{if } a = \min(i, K), \\ 0 & \text{otherwise.} \end{cases}$$

Using Theorem 10.3, the $\{X(t), t \geq 0\}$ under this policy is an SMP (actually it is a CTMC!) with transition probability $P^f = [p_{i,j}^f]$ given by

$$p_{i,i-1}^f = \frac{\min(i, K)\mu}{\lambda + \min(i, K)\mu}, \qquad 1 \leq i \leq M,$$

$$p_{i,i+1}^f = \frac{\lambda}{\lambda + \min(i, K)\mu}, \qquad 0 \leq i \leq M - 1,$$

and

$$p^f_{M,M} = \frac{\lambda}{\lambda + K\mu}.$$

The sojourn times are given by

$$w^f_i = \frac{1}{\lambda + \min(i, K)\mu}, \qquad 0 \le i \le M.$$

The cost incurred upon a visit to state i is given by

$$c^f(i) = (\min(i, K)C_p + iC_h)w^f_i, \qquad 0 \le i < M,$$

and

$$c^f(M) = (KC_p + MC_h + \lambda C_l)w^f_M.$$

We could compute the occupancy distribution $\hat{\pi}$ for the embedded DTMC and use (7.22) to compute the long-run cost rate of this policy.

Alternately, we see that, under this policy, the system reduces to an $M/M/K/M$ queueing system with arrival rate λ and service rate μ. The occupancy distribution $[p_i, 0 \le i \le M]$ for this system can be computed using the results of Theorem 8.5. The cost rate can be computed from Theorem 6.11 as

$$G^f = \sum_{i=0}^{M}(\min(i, K)C_p + iC_h)p_i + \lambda C_l p_M. \qquad \Box$$

As in the case of DTMDP we next address the question of finding the optimal policy f^* that minimizes the long-run cost rate per unit time. We face the same optimization problem **OPT** displayed in Section 10.3. We use the notation $x_{i,a}$ as defined in (10.6).

We now express the objective function G^f in terms of $x_{i,a}$. Using (10.19) we get

$$G^f = \frac{\sum_{i \in S}\sum_{a \in \mathcal{A}}\hat{\pi}_i f(i, a)c(i, a)}{\sum_{i \in S}\sum_{a \in \mathcal{A}}\hat{\pi}_i f(i, a)w_i(a)}$$

$$= \frac{\sum_{i \in S}\sum_{a \in \mathcal{A}}x_{i,a}c(i, a)}{\sum_{i \in S}\sum_{a \in \mathcal{A}}x_{i,a}w_i(a)}. \qquad (10.20)$$

Thus the objective function is a ratio of two linear functions of the variables $\{x_{i,a}, i \in S, a \in \mathcal{A}\}$.

Next, following the same analysis as in the DTMDP case, we see that (10.12) and (10.13) continue to hold in the SMDP case as well. Using (10.20), (10.12), and (10.13), the optimization problem can be rewritten as

FRACLP: Minimize $\dfrac{\sum_{i\in S}\sum_{a\in\mathcal{A}}x_{i.a}c(i,a)}{\sum_{i\in S}\sum_{a\in\mathcal{A}}x_{i.a}w_i(a)}.$

Subject to: $\sum_{a\in\mathcal{A}}x_{j.a}=\sum_{i\in S}\sum_{a\in\mathcal{A}}x_{i.a}p_{i.j}(a),\qquad j\in S,$

$$\sum_{i\in S}\sum_{a\in\mathcal{A}}x_{i.a}=1,$$

$$x_{i.a}\geq 0,\quad i\in S,\quad a\in\mathcal{A}.$$

The above problem is a *Fractional Linear Programming* problem (hence the title FRACLP for the problem formulation), since the objective function is a ratio of two linear functions, and the constraints are linear in the unknown variables $\{x_{i.a}, i\in S, a\in\mathcal{A}\}$. Fortunately, fractional linear programming problems can be reduced to linear programming problems by properly scaling the decision variables. We describe this transformation below.

Define

$$y_{i.a}=\frac{x_{i.a}}{\alpha},$$

where α is an unknown nonnegative scaling variable. Using this transformation FRACLP can be written as

FRACLP: Minimize $\dfrac{\sum_{i\in S}\sum_{a\in\mathcal{A}}y_{i.a}c(i,a)}{\sum_{i\in S}\sum_{a\in\mathcal{A}}y_{i.a}w_i(a)}.$

Subject to: $\sum_{a\in\mathcal{A}}y_{j.a}=\sum_{i\in S}\sum_{a\in\mathcal{A}}y_{i.a}p_{i.j}(a),\qquad j\in S,$

$$\sum_{i\in S}\sum_{a\in\mathcal{A}}y_{i.a}=\alpha,$$

$$\alpha\geq 0,\quad y_{i.a}\geq 0,\quad i\in S,\quad a\in\mathcal{A}.$$

Since the scaling variable α is arbitrary, we can choose it so that

$$\sum_{i\in S}\sum_{a\in\mathcal{A}}y_{i.a}w_i(a)=1.$$

With this added constraint the denominator in the objective function can be ignored. Also the two constraints

$$\sum_{i\in S}\sum_{a\in\mathcal{A}}y_{i.a}=\alpha,\qquad \alpha\geq 0,$$

can be replaced by a single constraint

$$\sum_{i\in S}\sum_{a\in\mathcal{A}}y_{i.a}\geq 0.$$

However, this constraint is automatically satisfied, since the variables $y_{i.a}$ are constrained to be nonnegative. Thus we can eliminate this constraint. With these modifications, the **FRACLP** reduces to the following linear program:

LPSMDP: **Minimize** $\sum_{i \in S} \sum_{a \in A} y_{i.a} c(i, a).$

Subject to: $\sum_{a \in A} y_{j.a} = \sum_{i \in S} \sum_{a \in A} y_{i.a} p_{i.j}(a),$ $j \in S,$

$$\sum_{i \in S} \sum_{a \in A} y_{i.a} w_i(a) = 1,$$

$$y_{i.a} \geq 0, \quad i \in S, \quad a \in A.$$

The above problem has $K * M$ variables, and $M + 1$ constraints. However, one of the first M constraints is redundant, and hence can be deleted. Thus there are M constraints. Using a linear programming package to solve the above problem we obtain the optimal values of the variables $y_{i.a}^*$ along with the optimal value of the objective function. We obtain the optimal policy from this information in the same way as we did in the DTMDP case, except that we use $y_{i.a}^*$ instead of $x_{i.a}^*$. We illustrate with an example.

Example 10.9 (Optimal Processor Scheduling). Consider the optimal processor scheduling problem of Example 10.8. Suppose the computer system has four processors, and a multi-programming level of 8. The jobs arrive according to a Poisson process at a rate of 150 per hour. Mean processing time is 2 minutes. Suppose the waiting cost for a job is $20 per minute. Cost of lost jobs is $200 per job. Compute the optimal scheduling policy as the operating cost of the processors varies from $0 per minute to $160 per minute.

The above data yield

$$K = 4, \quad M = 8, \quad \lambda = 2.5 \text{ per min}, \quad \mu = .5 \text{ per min},$$

$$C_h = \$20 \text{ per job per min}, \quad C_l = \$200 \text{ per job}.$$

The state space is $\{0, 1, \ldots, 8\}$ and the action space is $\{0, 1, \ldots, 4\}$. The linear programming formulation has $9 \cdot 5 = 45$ variables, and nine constraints. The above linear program can be solved easily using a standard the linear programming package. Solving the problem for various values of $0 \leq C_p \leq 160$, we see that the optimal policy is to use no processors at all if $C_p \geq 111.75$. If $C_p < 111.75$, the optimal policy is to use $\min(i, 4)$ processors when there are i jobs in the system. This makes intuitive sense: if the processor operating cost is too high ($111.75 per minute or more in this case), it is better to pay the cost of holding the jobs and losing the jobs, rather than the cost of operating a processor. If the processor operating cost is low (less than $111.75 per minute in this case), it is better to use all the processors as needed. It is interesting to note that it is never optimal to keep a processor idle if there are waiting jobs, in the region $C_p < 111.75$! Table 10.1 shows the cost g (dollars

TABLE 10.1. Optimal cost as a function of C_p.

$C_p \rightarrow$	0	20	40	60	80	100	120	140	160
g	241.66	316.53	391.41	466.29	541.16	616.04	660.00	660.00	660.00

per minute) of the optimal policy as a function of C_p. Note that the optimal cost increases linearly from 241.66 to 660.00 as C_p increases from 0 to 111.75, and then stays at 660.   □

10.7.   Optimal Machine Operation

In this section we consider the manufacturing system consisting of a single machine as described in Example 6.15. Recall that the machine takes an exponential amount of time to produce one item, with mean production time being $1/\lambda$ hours. The manufactured items are stored in a warehouse of capacity K. If an item is manufactured and there is no space to store it, it has to be thrown away. The demands for the items arrive according to a Poisson process with rate μ per hour. Any demand that cannot be satisfied immediately is lost. Each satisfied demand produces a revenue of $\$r$. Keeping an item in the warehouse costs $\$h$ per hour. The operating cost of the machine is $\$c$ per hour when it is on, and $\$0$ when it is off. The machine can be switched on and off whenever needed. Turning the machine off costs nothing, but it costs $\$s$ to switch it on.

The aim is to minimize the long-run cost per hour by turning the machine on and off intermittently. For example, it makes sense to turn the machine off if the warehouse is full, and turn it on again if the number of items in the warehouse falls below a certain level. Such a policy was analyzed in Examples 6.15, 6.23, and 6.29 by using CTMCs. Here we use the SMDP to decide when to turn the machine on and off in an optimal fashion. We assume that the decision to turn the machine on or off can be taken whenever a new item is produced or sold.

Let $Z(t)$ be the number of items in the warehouse at time t. Let $Y(t)$ be 1 if the machine is on, and 0 if it is off at time t. Thus the state of the system is described by the bivariate process $\{(Z(t), Y(t)), t \geq 0\}$. The action space is

$$\mathcal{A} = \{0, 1\},$$

where action 0 means "turn the machine off," and action 1 means "turn the machine on." Let (Z_n, Y_n) be the state at the nth transition epoch.

First we compute the sojourn time matrix $W = [w_i(a)]$. Suppose at the nth decision epoch, the machine is on, and there are i items in storage. Thus the state of the system is $(Z_n, Y_n) = (i, 1)$. Consider two cases:

Case 1: Action chosen $= 0$, i.e., turn the machine off. In this case the next decision epoch occurs when a new demand occurs, i.e., after an $\text{Exp}(\mu)$ amount of time. Hence

$$w_{(i,1)}(0) = 1/\mu, \qquad 0 \leq i \leq K.$$

Case 2: Action chosen $= 1$, i.e., leave the machine on. In this case the next decision epoch occurs when a new demand occurs or a new item is produced, i.e., after an $\text{Exp}(\lambda + \mu)$ amount of time. Hence

$$w_{(i,1)}(1) = 1/(\lambda + \mu), \qquad 0 \le i \le K.$$

Similar analysis shows that

$$w_{(i,0)}(0) = 1/\mu, \qquad 0 \le i \le K,$$
$$w_{(i,0)}(1) = 1/(\lambda + \mu), \qquad 0 \le i \le K.$$

Next we compute the transition probabilities $p_{i,j}(a)$. Suppose at the nth decision epoch, the machine is on, and there are i items in storage, i.e., the state of the system is $(Z_n, Y_n) = (i, 1)$. Consider two cases:

Case 1: Action chosen $= 0$, i.e., turn the machine off. In this case the next decision epoch occurs when a new demand occurs. At that time there will be $\max(i - 1, 0)$ items in the warehouse, and the machine will be off. Hence $(Z_{n+1}, Y_{n+1}) = (\max(i - 1, 0), 0)$.

Case 2: Action chosen $= 1$, i.e., keep the machine on. In this case the next decision occurs when the next demand occurs or a new item is produced. If the new production occurs before the new demand, the new state becomes $(Z_{n+1}, Y_{n+1}) = (\min(K, i + 1), 1)$. If the new demand occurs before the new production, the new state becomes $(Z_{n+1}, Y_{n+1}) = (\max(i - 1, 0), 1)$. This analysis yields the following transition probabilities:

$$p_{(i,1),(i-1,0)}(0) = 1, \qquad 1 \le i \le K,$$
$$p_{(0,1),(0,0)}(0) = 1,$$
$$p_{(K,1),(K,1)}(1) = \lambda/(\lambda + \mu),$$
$$p_{(i,1),(i+1,1)}(1) = \lambda/(\lambda + \mu), \qquad 0 \le i \le K - 1,$$
$$p_{(i,1),(i-1,1)}(1) = \mu/(\lambda + \mu), \qquad 1 \le i \le K,$$
$$p_{(0,1),(0,1)}(1) = \mu/(\lambda + \mu).$$

Similar analysis can be done when $(Z_n, Y_n) = (i, 0)$ to obtain

$$p_{(i,0),(i-1,0)}(0) = 1, \qquad 1 \le i \le K,$$
$$p_{(0,0),(0,0)}(0) = 1,$$
$$p_{(K,0),(K,1)}(1) = \lambda/(\lambda + \mu),$$
$$p_{(i,0),(i+1,1)}(1) = \lambda/(\lambda + \mu), \qquad 0 \le i \le K - 1,$$
$$p_{(i,0),(i-1,1)}(1) = \mu/(\lambda + \mu), \qquad 1 \le i \le K,$$
$$p_{(0,0),(0,1)}(1) = \mu/(\lambda + \mu).$$

Next we compute the costs. Suppose $(Z_n, Y_n) = (i, 1)$, i.e., there are i items in the warehouse and the machine is on. Consider two cases:

Case 1: Action chosen $= 0$, i.e., turn the machine off. The expected holding cost until the next decision epoch (an arrival of a demand) is hi/μ. The demand itself will generate a revenue of r if it can be satisfied. Hence

$$c_{(i,1)}(0) = -r + hi/\mu, \qquad 1 \le i \le K,$$

$$c_{(0,1)}(0) = 0.$$

Case 2: Action chosen $= 1$, i.e., keep the machine on. The expected holding cost until the next decision epoch (production of an item or an arrival of a demand) is $hi/(\lambda + \mu)$. The expected cost of operating the machine is $c/(\lambda + \mu)$. The expected revenue is $r\mu/(\lambda + \mu)$ if $i > 0$, and 0 if $i = 0$. Thus we get

$$c_{(i,1)}(1) = (c + hi - r\mu)/(\lambda + \mu), \qquad 1 \le i \le K,$$

$$c_{(0,1)}(1) = c/(\lambda + \mu).$$

Similar analysis can be done for the case $(Z_n, Y_n) = (i, 0)$. Here we have to include the cost of switching the machine on. The final result is given as

$$
\begin{aligned}
c_{(i,0)}(0) &= -r + hi/\mu, \qquad 1 \le i \le K, \\
c_{(0,0)}(0) &= 0 \\
c_{(i,0)}(1) &= s + (c + hi - r\mu)/(\lambda + \mu), \qquad 1 \le i \le K, \\
c_{(0,0)}(1) &= s + c/(\lambda + \mu).
\end{aligned}
\tag{10.21}
$$

This completes the description of the parameters of the problem. We can now use the linear programming formulation to find the optimal policy.

Example 10.10 (Numerical Example). Suppose the storage capacity of the warehouse is 4. It takes 1 hour to manufacture one item, and it sells for \$200. The demands arise according to a Poisson process with a rate of one per hour. It takes \$50 to switch the machine from off-mode to on-mode and \$100 per hour to keep it on. The holding cost is \$2 per hour per item. Compute the optimal operating policy for the machine.

The above data yield the following parameters:

$$K = 4, \qquad \lambda = 1, \qquad \mu = 1, \qquad h = 2,$$

$$s = 50, \qquad c = 100, \qquad r = 200.$$

The state space is $S = \{1 = (0, 0), 2 = (1, 0), 3 = (2, 0), 4 = (3, 0), 5 = (4, 0), 6 = (0, 1), 7 = (1, 1), 8 = (2, 1), 9 = (3, 1), 10 = (4, 1)\}$ and the action space is $\mathcal{A} = \{0, 1\}$.

The transition probability matrices are

$$
P(0) = \begin{bmatrix}
1 & 0 & 0 & 0 & 0 & 0 & 0 & 0 & 0 & 0 \\
1 & 0 & 0 & 0 & 0 & 0 & 0 & 0 & 0 & 0 \\
0 & 1 & 0 & 0 & 0 & 0 & 0 & 0 & 0 & 0 \\
0 & 0 & 1 & 0 & 0 & 0 & 0 & 0 & 0 & 0 \\
0 & 0 & 0 & 1 & 0 & 0 & 0 & 0 & 0 & 0 \\
1 & 0 & 0 & 0 & 0 & 0 & 0 & 0 & 0 & 0 \\
1 & 0 & 0 & 0 & 0 & 0 & 0 & 0 & 0 & 0 \\
0 & 1 & 0 & 0 & 0 & 0 & 0 & 0 & 0 & 0 \\
0 & 0 & 1 & 0 & 0 & 0 & 0 & 0 & 0 & 0 \\
0 & 0 & 0 & 1 & 0 & 0 & 0 & 0 & 0 & 0
\end{bmatrix},
$$

$$
P(1) = \begin{bmatrix}
0 & 0 & 0 & 0 & 0 & .5 & .5 & 0 & 0 & 0 \\
0 & 0 & 0 & 0 & 0 & .5 & 0 & .5 & 0 & 0 \\
0 & 0 & 0 & 0 & 0 & 0 & .5 & 0 & .5 & 0 \\
0 & 0 & 0 & 0 & 0 & 0 & 0 & .5 & 0 & .5 \\
0 & 0 & 0 & 0 & 0 & 0 & 0 & 0 & .5 & .5 \\
0 & 0 & 0 & 0 & 0 & .5 & .5 & 0 & 0 & 0 \\
0 & 0 & 0 & 0 & 0 & .5 & 0 & .5 & 0 & 0 \\
0 & 0 & 0 & 0 & 0 & 0 & .5 & 0 & .5 & 0 \\
0 & 0 & 0 & 0 & 0 & 0 & 0 & .5 & 0 & .5 \\
0 & 0 & 0 & 0 & 0 & 0 & 0 & 0 & .5 & .5
\end{bmatrix}.
$$

The cost matrix is

$$
C = \begin{bmatrix}
0 & 100 \\
-198 & 1 \\
-196 & 2 \\
-194 & 3 \\
-192 & 4 \\
0 & 50 \\
-198 & -49 \\
-196 & -48 \\
-194 & -47 \\
-192 & -46
\end{bmatrix},
$$

and the sojourn time matrix is

$$W = \begin{bmatrix} 1 & .5 \\ 1 & .5 \\ 1 & .5 \\ 1 & .5 \\ 1 & .5 \\ 1 & .5 \\ 1 & .5 \\ 1 & .5 \\ 1 & .5 \\ 1 & .5 \end{bmatrix}.$$

The linear program LPSMDP can be solved to obtain the optimal policy. The optimal solution $Y^* = [y^*_{i,a}]$ is given by

$$Y^* = \begin{bmatrix} 0 & 0 \\ 0 & 0 \\ 0 & .1111 \\ .1111 & 0 \\ 0 & 0 \\ 0 & .4444 \\ 0 & .4444 \\ 0 & .3333 \\ 0 & .2222 \\ .1111 & 0 \end{bmatrix}.$$

Thus the optimal policy f^* is as follows:

$$f^* = \begin{bmatrix} ? & ? \\ ? & ? \\ 0 & 1 \\ 1 & 0 \\ ? & ? \\ 0 & 1 \\ 0 & 1 \\ 0 & 1 \\ 0 & 1 \\ 1 & 0 \end{bmatrix}.$$

The "?" marks denote that the policy is undefined in states 1, 2, and 5 (i.e., when the machine is off and there are zero, one, or four items in the warehouse.) In state 3 (the machine is off and there are two items in the warehouse) it is optimal to turn the machine on; while in state 4 (the machine is off and there are three items in the warehouse), it is optimal to keep the machine off (action 0). In states 6, 7, 8, and 9 (i.e., the machine is on and the warehouse is not full) it is optimal to keep the machine on (action 1). In state 10 (the machine is on and the warehouse is full) it is optimal to turn the machine off (action 0).

Let us see how this policy will work. Suppose the machine has just produced a product and the warehouse is now full. Thus the state is $(4, 1)$ (i.e., 10) and hence it is optimal to turn the machine off. Then we wait for the next demand. The inventory reduces to state 3, and the state is $(3, 0)$, i.e., 4. The optimal policy is to keep the machine off. When the next demand occurs, the state is $(2, 0)$ or 3. The optimal policy now is to turn the machine on. At the next decision epoch, the state could be $(1, 1)$ or $(3, 1)$, and it is optimal to keep the machine on. The next state could be $(0, 1)$ or $(2, 1)$ or $(4, 1)$. It is optimal to keep the machine on in the first two states, while it is optimal to turn the machine off in the state $(4, 1)$.

Thus the optimal policy can be described as follows: Once the machine is on, turn it off when the warehouse is full and turn it back on again when the inventory drops to 2. Under this policy we never see states $(0, 0)$, $(1, 0)$, or $(4, 0)$. Hence the optimal policy in those states is not defined. But what if we start in those states? It seems intuitive that we should follow a policy that moves the system into a state where the optimal policy is defined. Hence it is reasonable turn the machine on so that we get into the operating states soon and then follow the optimal policy.

The optimal value of the objective function is -68.67, i.e., under the optimal policy the machine produces a revenue of $68.67 per hour.   □

10.8. Problems

CONCEPTUAL PROBLEMS

10.1. Consider the optimal group maintenance problem of Examples 10.1 and 10.3 with the following modification: when the replacement is going on, the working machines not under replacement continue to produce a revenue of $R per day per machine; and are subject to failures as before. Formulate the problem as a DTMDP for this modified case.

10.2. Consider the optimal group maintenance problem of Examples 10.1 and 10.3 with the following modification: suppose it takes one repair person 1 day to replace one failed machine. We need to decide how many repair persons to hire on the nth day based on the number of failed machines at the beginning of the nth day. Each repair person charges $C per day. Formulate this problem as a DTMDP.

10.3. Let X_n be the state of a machine at the beginning of the nth day. Left to itself, the state of the machine changes according to a DTMC with state space $\{1, 2, \ldots, N\}$ and transition probability matrix $P = [p_{i,j}]$. If the machine is in state i at the beginning of a day, it produces a revenue of $\$r(i)$ during that day. It is possible to replace the machine with a new one at a cost of $\$C_r$. The replacement takes 1 day. The new machine always starts in state 1. We need to decide whether or not to replace the machine at the beginning of each day after observing its state. Formulate this system as a DTMDP.

10.4. Consider Conceptual Problem 10.3. Suppose we also have an option of replacing the machine with a used one at the cost of $\$C_u$. The used machine starts in state i with probability a_i, $1 \leq i \leq N$. Formulate a DTMDP to include this option.

10.5. Consider Conceptual Problem 10.3. Suppose the machine in state i has a salvage value (i.e., it can be sold for) $\$s(i)$. Reformulate the DTMDP to take into account this information.

10.6. This is a simple model of price hedging. The price of an item fluctuates between two price levels, c_1 and c_2, with $c_1 < c_2$. Let Y_n be the price of an item at the beginning of day n. Assume that $\{Y_n, n \geq 0\}$ is a DTMC on state space $\{c_1, c_2\}$. Let X_n be the number of items in the warehouse at the beginning of the nth day. After observing X_n and Y_n, we have to decide A_n, the number of items to purchase at the beginning of the nth day. One item is consumed during each day, and hence we must have $A_n \geq 1$ when $X_n = 0$. The warehouse capacity is B. The aim is to minimize the long-run cost per day of purchasing the items. Model this system as a DTMDP. Intuitively, what policy would you expect to be optimal?

10.7. Consider the model of Conceptual Problem 10.6. Suppose there is an ordering cost of s for buying a nonzero number of items, over and above the cost of the items. Also, each item in the warehouse at the beginning of a day costs $\$h$ in storage costs for that day. Formulate the DTMDP with this cost structure.

10.8. This is a simple model of optimal storage to take advantage of price fluctuations. A machine manufactures one item per day (at a cost of p) and it is stored in a warehouse of capacity B. Let X_n be the number of items in the warehouse at the beginning of the nth day. The selling price of this item fluctuates between two price levels c_1 and c_2, with $c_1 < p < c_2$. Let Y_n be the price of an item at the beginning of day n. Assume that $\{Y_n, n \geq 0\}$ is a DTMC on state space $\{c_1, c_2\}$. After observing X_n and Y_n, we have to decide A_n, the number of items to sell at the beginning of the nth day. The aim is to maximize the long-run net revenue per day from selling the items. Model this system as a DTMDP. Intuitively, what policy would you expect to be optimal?

10.9. Consider the model of Conceptual Problem 10.8. Suppose each item in the warehouse at the beginning of a day incurs a storage cost of h per day. Reformulate the DTMDP with this cost structure.

10.10. The central computer of a bank is always busy processing transactions. Each transaction takes exactly p hours to process. After processing a transaction the computer (instantaneously) adds it to a "processed" file, which can store at most K transactions. The transactions in the "processed" file need to be backed up intermittently to guard against losing them due to system failures. Backing up the "processed" file takes b hours per transaction stored in it, plus the setup time of s hours. Suppose the lifetime of the central computer is exponentially distributed with mean $1/\lambda$ hours, and the mean repair time is

r hours. If the failure takes place during the processing of a transaction, that transaction has to be reprocessed. If the failure takes place during a backup operation, the entire "processed" file has to be backed up again. At the end of the processing of a transaction, or at the end of a backup operation, or at the end of a repair, the computer has an option of processing the next transaction or backing up the entire "processed" file. If there is no space for the processed transaction in the "processed" file, that transaction needs to be reprocessed. The aim is to devise a backing-up policy that will maximize the expected number of backed-up transactions per unit time. Formulate this as an SMDP, i.e., give the cost matrix, the sojourn time matrix, and the transition probability matrices.

10.11. Customers arrive according to PP(λ) at a single-server queue with finite capacity B. Customers arriving at a full system are lost. The customer work loads are iid exponential random variables with parameter μ. The server can process the work load at K different speeds $s_1, s_2, \ldots, s_K$. Thus, if the worker is using speed s_i, it can finish the workload of a single customer in an exponential amount of time with parameter $\mu_i = s_i \mu$. The worker speed can be reset whenever a customer enters or departs, with no switch-over cost. It costs $\$c_i$ to operate the server at speed s_i, and it costs $\$r$ if a customer is lost due to capacity limitation. The aim is to find an admission policy that minimizes the total cost per unit time in the long run. Formulate an SMDP to solve this problem.

10.12. Recompute the cost vector in the optimal machine operation problem of Section 10.7 if it costs $\$s'$ to turn the machine off. The rest of the problem is the same as before.

10.13. Tasks arrive at a processing station according to a renewal process with mean inter-arrival time T. At each arrival instant we need to decide whether to clear all the accumulated tasks instantaneously at a fixed cost of $\$C$ or to wait for a new arrival. A maximum of K tasks can be stored at the station. It costs $\$h$ per unit time to keep a task at the station. Formulate an SMDP to find an optimal clearing policy.

10.14. Customers arrive at a single-server queueing system according to a PP(λ) and request iid Exp(μ) service times. The system controller has to decide whether to admit a customer or reject him. An admitted customer pays a fee of $\$R$ to the system. It costs $\$h$ per unit time to keep a customer in the system. If there are K customers in the system, no more customers can be admitted (i.e., the capacity is K). Formulate an SMDP to compute the optimal admission control policy to minimize the long-run cost rate.

COMPUTATIONAL PROBLEMS

10.1. Consider a DTMDP with state space $\{1, 2, 3\}$ and action space $\{1, 2\}$ with the following transition probability matrices:

$$P(1) = \begin{bmatrix} 1 & 0 & 0 \\ .3 & .7 & 0 \\ .1 & .8 & .1 \end{bmatrix},$$

$$P(2) = \begin{bmatrix} .5 & .5 & 0 \\ 0 & .5 & .5 \\ 0 & .5 & .5 \end{bmatrix},$$

and the following cost matrix:

$$C = \begin{bmatrix} 30 & 10 \\ 30 & 10 \\ 20 & 50 \end{bmatrix}.$$

Consider the policy of choosing action 2 in states 1 and 2, and action 1 in state 3.

(1) State the f function corresponding to this policy.

(2) Compute the transition probability matrix P^f corresponding to this policy.

(3) Compute the cost per unit time in the long run for this policy.

10.2. Compute the optimal policy for the DTMDP of Computational Problem 10.1. Compute the cost per unit time for this policy.

10.3. Do Computational Problem 10.1 for a DTMDP with state space $\{1, 2, 3\}$ and action space $\{1, 2\}$ with the following transition probability matrices:

$$P(1) = \begin{bmatrix} .2 & 0 & .8 \\ .3 & .2 & .5 \\ .1 & .8 & .1 \end{bmatrix},$$

$$P(2) = \begin{bmatrix} .5 & .5 & 0 \\ .8 & 0 & .2 \\ 0 & .3 & .7 \end{bmatrix},$$

and the following cost matrix:

$$C = \begin{bmatrix} 30 & 5 \\ 20 & 10 \\ 20 & 50 \end{bmatrix}.$$

10.4. Compute the optimal policy for the DTMDP of Computational Problem 10.3. Compute the cost per unit time for this policy.

10.5. Consider a DTMDP with state space $\{1, 2, 3\}$ and action space $\{1, 2, 3\}$ with the following transition probability matrices:

$$P(1) = \begin{bmatrix} 1 & 0 & 0 \\ .3 & .7 & 0 \\ .1 & .8 & .1 \end{bmatrix},$$

$$P(2) = \begin{bmatrix} .5 & .5 & 0 \\ 0 & .5 & .5 \\ 0 & .5 & .5 \end{bmatrix},$$

$$P(3) = \begin{bmatrix} .5 & .5 & 0 \\ .8 & 0 & .2 \\ 0 & .3 & .7 \end{bmatrix},$$

and the following cost matrix:

$$C = \begin{bmatrix} 30 & 10 & 40 \\ 30 & 10 & 5 \\ 20 & 50 & 10 \end{bmatrix}.$$

Consider the policy of choosing action 3 in state 1, action 2 in state 2, and action 1 in state 3.

(1) State the f function corresponding to this policy.

(2) Compute the transition probability matrix P^f corresponding to this policy.

(3) Compute the cost per unit time in the long run for this policy.

10.6. Compute the optimal policy for the DTMDP of Computational Problem 10.5. Compute the cost per unit time for this policy.

10.7. Compute the optimal policy for the DTMDP with state space $\{1, 2\}$ and action space $\{1, 2, 3\}$ with the following transition probability matrices:

$$P(1) = \begin{bmatrix} .25 & .75 \\ .3 & .7 \end{bmatrix},$$

$$P(2) = \begin{bmatrix} .5 & .5 \\ .8 & .2 \end{bmatrix},$$

$$P(3) = \begin{bmatrix} .8 & .2 \\ .4 & .6 \end{bmatrix},$$

and the following cost matrix:

$$C = \begin{bmatrix} 10 & 30 & 40 \\ 20 & 70 & 5 \end{bmatrix}.$$

10.8. Compute the optimal policy for the DTMDP with state space $\{1, 2, 3, 4\}$ and action space $\{1, 2\}$ with the following transition probability matrices:

$$P(1) = \begin{bmatrix} .25 & .75 & 0 & 0 \\ 0 & .25 & .75 & 0 \\ 0 & 0 & .25 & .75 \\ .75 & 0 & 0 & .25 \end{bmatrix},$$

$$P(2) = \begin{bmatrix} .25 & 0 & 0 & .75 \\ .75 & .25 & 0 & 0 \\ 0 & .75 & .25 & 0 \\ 0 & 0 & .75 & .25 \end{bmatrix},$$

and the following cost matrix:

$$C = \begin{bmatrix} 10 & 40 \\ 20 & 30 \\ 30 & 20 \\ 40 & 10 \end{bmatrix}.$$

10.9. For what range of $C(1, 1)$ does the policy obtained in Computational Problem 10.2 remain optimal?

10.10. For what range of $C(2, 1)$ does the policy obtained in Computational Problem 10.4 remain optimal?

10.11. For what range of $C(1, 2)$ does the policy obtained in Computational Problem 10.6 remain optimal?

10.12. For what range of $C(1, 3)$ does the policy obtained in Computational Problem 10.7 remain optimal?

10.13. For what range of $C(4, 1)$ does the policy obtained in Computational Problem 10.8 remain optimal?

10.14. Consider the modified group maintenance model of Conceptual Problem 10.1. Using the data from Example 10.5, compute the optimal policy for this model. Compute the cost per unit time for this policy.

10.15. Consider the modified group maintenance model of Conceptual Problem 10.2. Suppose the repair-person charges are $75 per person per day. There is no visit charge. The rest of the data are as in Example 10.5. Compute the optimal policy for this model. Compute the cost per unit time for this policy.

10.16. For what range of R, the revenue per machine per day, does the policy obtained in Computational Problem 10.14 remain optimal?

10.17. For what range of C, the charge per repair person per day, does the policy obtained in Computational Problem 10.15 remain optimal?

10.18. Consider the machine of Conceptual Problem 10.3 with the following parameters:

$$P = \begin{bmatrix} 0.75 & 0.25 & 0 \\ 0 & 0.3 & 0.7 \\ 0 & 0 & 1 \end{bmatrix}.$$

The revenues are

$$r(1) = 800, \qquad r(2) = 500, \qquad r(3) = 0.$$

The cost of replacement is \$500. Compute the optimal policy. What is the cost per unit time in the long run of following the optimal policy?

10.19. Consider the machine of Conceptual Problem 10.4. Suppose a used machine costs \$300, and that it is in state 1 with probability .6 and state 2 with probability .4. All other data are as in Computational Problem 10.18 above. What is the cost per unit time in the long run of following the optimal policy?

10.20. What is the largest price of the used machine for which the optimal policy obtained in Computational Problem 10.19 remains optimal?

10.21. For what range of the revenue earned by a new machine in 1 day, does the policy obtained in Computational Problem 10.18 remain optimal?

10.22. For what range of the replacement cost of the machine, does the policy obtained in Computational Problem 10.18 remain optimal?

10.23. The state of a machine can be classified as: new (or as good as new), good, bad, and down. The state of this machine is observed at the beginning of each day. In the absence of any intervention, the state of the machine evolves according to a DTMC with transition probability matrix:

$$\begin{bmatrix} .75 & .25 & 0 & 0 \\ 0 & .30 & .70 & 0 \\ 0 & 0 & .40 & .60 \\ 0 & 0 & 0 & 1 \end{bmatrix}.$$

After observing the state of the machine we have an option of doing nothing, or replacing the machine with a new one that is statistically identical and independent. The replacement takes 1 day and costs \$2500. The daily revenue from the machine is \$800 per day if it is new or as good as new, \$500 if it is good, \$300 if it is bad, and 0 if it is down. Formulate a DTMDP to model this system. Compute the optimal policy and its cost per day. (See Conceptual Problem 10.3.)

10.24. Consider Computational Problem 10.23. Suppose we also have an option of replacing the machine with a used one at a cost of \$1500. The used machine is as good as new with probability .7, good with probability .2, and bad with probability .1. Formulate a DTMDP to include this option. (See Conceptual Problem 10.4.)

10.25. Compute the optimal policy and its cost per day for Computational Problem 10.24.

10.26. Consider Computational Problem 10.23. Suppose the cost of replacement is $2500. However, there is a salvage value to the used machine as follows: a machine in an "as good as new" state fetches $2000, a good machine fetches $500, a bad machine fetches $250, and a down machine fetches nothing. Recompute the cost matrix of the DTMDP, and find the optimal policy. (See Conceptual Problem 10.5.)

10.27. Consider the price hedging problem of Conceptual Problem 10.6. Suppose the price fluctuates between $10 and $12 per item according to a DTMC with transition probability matrix:

$$\begin{bmatrix} .8 & .2 \\ .4 & .6 \end{bmatrix}.$$

The warehouse can store at most three items. One item is consumed each day. There is no holding cost. Compute the optimal purchase policy that minimizes the cost per day, assuming the stock-out cost to be $15 (i.e., if there is no item in the warehouse, the item needed for consumption has to be purchased at a cost of $15).

10.28. Consider Computational Problem 10.27, but with a stock-out cost of $40. Suppose there is a holding cost as follows: it costs $4 to hold an item overnight in the warehouse. Also there is an ordering cost of $10 every time a nonzero number of items are purchased (over and above the cost of the items purchased). (Ordering cost does not apply to the emergency purchases made to satisfy demand when there are no items in stock.) Recompute the cost matrix of the DTMDP and compute the optimal policy and its cost per day.

10.29. Consider an SMDP with state space $\{1, 2, 3\}$ and action space $\{1, 2\}$ with the following transition probability matrices:

$$P(1) = \begin{bmatrix} 1 & 0 & 0 \\ .3 & .7 & 0 \\ .1 & .8 & .1 \end{bmatrix},$$

$$P(2) = \begin{bmatrix} .5 & .5 & 0 \\ 0 & .5 & .5 \\ 0 & .5 & .5 \end{bmatrix}.$$

The cost matrix is as follows:

$$C = \begin{bmatrix} 30 & 10 \\ 30 & 10 \\ 20 & 50 \end{bmatrix},$$

while the sojourn time matrix is as follows:

$$W = \begin{bmatrix} 1 & 2 \\ 2 & 3 \\ 3 & 1 \end{bmatrix}.$$

Consider the policy of choosing action 2 in states 1 and 3, and action 1 in state 2.

(1) State the f function corresponding to this policy.

(2) Compute the transition probability matrix P^f corresponding to this policy.

(3) Compute the cost per unit time in the long run for this policy.

10.30. Compute the optimal policy for the SMDP of Computational Problem 10.29. Compute the cost per unit time for this policy.

10.31. Consider an SMDP with state space $\{1, 2, 3\}$ and action space $\{1, 2, 3\}$. The C, $P(1)$, $P(2)$, and $P(3)$ matrices are as in Computational Problem 10.5. The sojourn time matrix is

$$W = \begin{bmatrix} 1 & 2 & 5 \\ 2 & 3 & 1 \\ 3 & 1 & 2 \end{bmatrix}.$$

Compute the optimal policy and its cost per unit time.

10.32. Consider an SMDP with state space $\{1, 2\}$ and action space $\{1, 2, 3\}$. The C, $P(1)$, $P(2)$, and $P(3)$ matrices are as in Computational Problem 10.7. The sojourn time matrix is

$$W = \begin{bmatrix} 10 & 12 & 25 \\ 30 & 15 & 20 \end{bmatrix}.$$

Compute the optimal policy and its cost per unit time.

10.33. Consider an SMDP with state space $\{1, 2, 3, 4\}$ and action space $\{1, 2\}$. The C, $P(1)$, and $P(2)$ matrices are as in Computational Problem 10.8. The sojourn time matrix is

$$W = \begin{bmatrix} 5 & 20 \\ 3 & 5 \\ 12 & 8 \\ 7 & 9 \end{bmatrix}.$$

Compute the optimal policy and its cost per unit time.

10.34. Compute the optimal backup policy for Conceptual Problem 10.10 with the following data. Each transaction takes 1 hour, backing up a transaction takes half an hour. The backing up overhead is 1 hour. The backup file capacity is three transactions. The mean uptime is 10 hours, while the average repair time is 3 hours. What is the number of transactions backed up per hour under the optimal policy?

10.35. Consider Conceptual Problem 10.11. Find the optimal policy if the capacity of the system is five, the arrival rate is one per hour, and there are two speeds: two customers per hour (costing $20 per hour) and five customers per hour (costing $35 per hour). The cost of losing a customer is $100.

10.36. For what range of the cost of losing a customer is it optimal to use only the lowest speed in Computational Problem 10.35?

10.37. Consider Conceptual Problem 10.13. Suppose tasks arrive according to a renewal process at a rate of one per minute. We can store at the most ten tasks. It takes $20 to clear the tasks, while it costs $1 per minute to hold a task in the system. Compute the optimal policy that minimizes the long-run cost per minute.

10.38. Consider Computational Problem 10.37. For what range of the clearing cost is it optimal to clear the system when it is full?

Answers to Selected Problems

Chapter 1

CONCEPTUAL PROBLEMS

1.1. (1) $E \cap F \cap G^c$, (2) $E F^c \cup F E^c$, (3) $E^c F^c G^c$, (4) $E F G^c \cup E F^c G \cup E^c F G$.

1.3. Let $\omega \in \left(\bigcup_{i=1}^n E_i \right)^c$, then ω belongs to none of the E_i's, and thus ω belongs to all of the E_i^c's. Therefore, $\omega \in \bigcap_{i=1}^n E_i^c$. Hence $\left(\bigcup_{i=1}^n E_i \right)^c \subseteq \bigcap_{i=1}^n E_i^c$. Conversely, let $\omega \in \bigcap_{i=1}^n E_i^c$. Then ω belongs to none of the E_i's, and thus $\omega \in \left(\bigcup_{i=1}^n E_i \right)^c$. Thus $\bigcap_{i=1}^n E_i^c \subseteq \left(\bigcup_{i=1}^n E_i \right)^c$. This completes the proof of the first statement. The second statement follows similarly.

1.5. If $E \subseteq F$, then $F = E \cup F E^c$. Since E and $F E^c$ are disjoint: $P(F) = P(E) + P(F E^c) \geq P(E)$.

1.7. $P(\bigcup_{n=1}^\infty E_n | F) = \dfrac{P(\bigcup_{n=1}^\infty E_n F)}{P(F)} = \dfrac{\sum_{n=1}^\infty P(E_n F)}{P(F)} = \sum_{n=1}^\infty P(E_n | F).$

1.9. $P(E | F G) P(F | G) = \dfrac{P(E F G)}{P(F G)} \cdot \dfrac{P(F G)}{P(G)} = \dfrac{P(E F G)}{P(G)} = P(E F | G).$

1.11. E is independent of itself implies $P(E \cap E) = P(E) \cdot P(E)$. But $P(E \cap E) = P(E)$. Hence $P(E) = (P(E))^2$, and thus $P(E) = 0$ or 1. Hence E must be either the universal event or the null event.

1.15. $P(\bigcup_{i=1}^{n} E_i) = 1 - P((\bigcup_{i=1}^{n} E_i)^c) = 1 - P(\bigcap_{i=1}^{n} E_i^c) = 1 - \prod_{i=1}^{n} P(E_i^c) = 1 - \prod_{i=1}^{n}(1 - P(E_i))$.

COMPUTATIONAL PROBLEMS

1.1. (1) $\Omega = \{HT, TH, TT, HH\}$, (2) $\Omega = [0, \infty)$.

1.3. (1) $\Omega = \{NN, NG, NB, GN, GG, GB, BN, BG, BB\}$, (2) $\Omega = \{0, 1, 2, \ldots\}$, (3) $\Omega = [0, \infty)$.

1.5. (1) $E = \{13\}$, (2) $E = \{n : n > 10\}$, (3) $E = \{n : 0 \le n \le 5\}$, (4) $E = \{n : n \ge 20\}$.

1.7. $P(E_0) = P(E_3) = \frac{1}{8}$, $P(E_1) = P(E_2) = \frac{3}{8}$.

1.9. .3538.   **1.11.** (1) .75, (2) .25, (3) .375.

1.13. (1) $\dfrac{r}{r+b} + \dfrac{\binom{r}{1}\binom{b}{2} + \binom{r}{3}\binom{b}{0}}{\binom{r+b}{3}} - \dfrac{\binom{r}{3} + \frac{1}{3}\binom{r}{1}\binom{b}{2}}{\binom{r+b}{3}}$, (2) $\dfrac{\binom{r}{3}\binom{b}{0} + \frac{1}{3}\binom{r}{1}\binom{b}{2}}{\binom{r+b}{3}}$,

(3) $2\dfrac{\binom{r}{3}\binom{b}{0} + \frac{1}{3}\binom{r}{1}\binom{b}{2}}{\binom{r+b}{3}} - \dfrac{\binom{r}{3}\binom{b}{0}}{\binom{r+b}{3}}$.

1.15. $1 - 1 + \dfrac{1}{2!} - \dfrac{1}{3!} + \cdots + (-1)^N \dfrac{1}{N!}$.   **1.17.** (1) $\dfrac{r-1}{r+b-1}$, (2) $\dfrac{r-1}{r+b-1}$.

1.19. $\dfrac{N-2}{N-1}$.   **1.21.** $\dfrac{1}{2}$.   **1.23.** $\dfrac{1}{1+p}$.

1.25. $\dfrac{1}{2}$.   **1.27.** $\dfrac{p^2(2-p)}{1-p+p^2}$.   **1.29.** .9596.

Chapter 2

CONCEPTUAL PROBLEMS

2.1. (1) $X \sim \text{Bin}(1, p)$ implies

$$P(X = k) = \binom{1}{k} p^k (1-p)^{1-k}, \qquad k = 0, 1.$$

This reduces to $P(X = 0) = 1 - p$ and $P(X = 1) = p$, hence $X \sim B(p)$.
(2) $X \sim NB(1, p)$ implies

$$P(X = k) = \binom{k-1}{0} p(1-p)^{k-1} = p(1-p)^{k-1} = (1-p)^{k-1}p, \qquad k \geq 1.$$

Hence $X \sim G(p)$.

2.5. The mode of $\text{Bin}(n, p)$ is the largest integer less than or equal to $(n+1)p$. If $(n+1)p$ is an integer, there are two modes: $(n + 1)p - 1$ and $(n + 1)p$.

2.7. $\dfrac{k-1}{\lambda}$.   **2.9.** $\dfrac{\ln 2}{\lambda}$.   **2.11.** μ.

2.13. $h(x) = \dfrac{F'(x)}{1 - F(x)} = -\dfrac{d}{dx}\ln(1 - F(x))$. Hence

$$1 - F(x) = C \exp\left(-\int_{u=0}^{x} h(u)\, du\right).$$

The boundary condition $F(0) = 0$ implies $C = 1$.

2.15. $E(X) = \int_{u=0}^{\infty} u f(u)\, du = \int_{u=0}^{\infty} \int_{x=0}^{u} f(u)\, dx\, du = \int_{x=0}^{\infty} \int_{u=x}^{\infty} f(u)\, du\, dx = \int_{x=0}^{\infty} (1 - F(x))\, dx$.

2.17. $n(n - 1)p^2$.

2.19. $\text{Var}(aX+b) = E((aX+b-E(aX + b))^2) = E((aX - aEX)^2) = a^2 E((X - EX)^2) = a^2 \text{Var}(X)$.

2.21. $Y \sim N(0, 1)$ and $X = \mu + \sigma Y$, implies $X \sim N(\mu, \sigma^2)$. $\text{Var}(X) = \text{Var}(\mu + \sigma Y) = \sigma^2 \text{Var}(Y) = \sigma^2$.

2.23. The cdf of $Y = \min(X, T)$ is given by $G_Y(t) = G(t)$ if $t < T$, and 1 if $t \geq T$. Substituting in Conceptual Problem 2.15 we get $E(Y) = \int_0^{\infty}(1 - G_Y(t))\, dt = \int_0^T (1 - G(t))\, dt$.

COMPUTATIONAL PROBLEMS

2.1. (2) $\Omega = \{(i, j) : 1 \leq i, j \leq 6\}$, $X((i, j)) = |i - j|$.

2.3. $P(X = 0) = .125$, $P(X = 1) = P(X = 2) = .375$, $P(X = 3) = .125$.

2.5. (1) $P(X = -6) = P(X = 9) = \frac{1}{8}$, $P(X = -1) = P(X = 4) = \frac{3}{8}$.

2.7. $P(X = 1) = \frac{1}{5}$, $P(X = -\frac{1}{4}) = \frac{4}{5}$.

2.9. $P(X = 1) = .027$, $P(X = 0) = .973$.

2.11. $P(X = 0) = p^2 + (1 - p)p^2 + (1 - p)p^2 = 1 - P(X = 1)$.

2.13. .7334.   **2.15.** Bin(6, $\frac{1}{2}$).   **2.17.** .8732.

2.19. NB(2, $\frac{1}{2}$).   **2.21.** .0417.   **2.23.** .1088.

2.25. (1) 1.1447. (2) $F(x) = \begin{cases} 0 & \text{if } x < 0, \\ \dfrac{(x-a)^3}{3} + \dfrac{1}{2} & \text{if } 0 \le x \le 2a, \\ 1 & \text{if } x > 2a. \end{cases}$

2.27. (1) $\sqrt[3]{3}$. (2) $F(x) = \begin{cases} 0 & \text{if } x < 0, \\ \dfrac{x^3}{3} & \text{if } 0 \le x \le \sqrt[3]{3}, \\ 1 & \text{if } x > \sqrt[3]{3}. \end{cases}$

2.29. (1) 3. (2) $F(x) = \begin{cases} 0 & \text{if } x < 0, \\ x^3 & \text{if } 0 \le x \le 1, \\ 1 & \text{if } x > 1. \end{cases}$

2.31. 1481.94 years.   **2.33.** .3420.

2.35. $f_Y(y) = \begin{cases} 0 & \text{if } y < 3, \\ \left(\dfrac{y-3}{2} - a\right)^2 & \text{if } 3 \le y \le 4a + 3, \\ 0 & \text{if } y > 4a + 3. \end{cases}$

2.37. $F_X(2 + \sqrt{y}) - F_X(2 - \sqrt{y})$, $y \ge 0$.   **2.39.** U(a, b).

2.41. $P(Y = k) = .8^{k-1}.2$, $1 \le k \le 9$, $P(Y = 10) = .8^9$.

2.43. (1) 0, (2) 1.9444. **2.45.** 0.   **2.47.** $(\frac{3}{2})^{1/3}$.

2.49. .75.   **2.51.** .0780.   **2.53.** 2.7534.

Chapter 3

CONCEPTUAL PROBLEMS

3.1. $P(a_1 < X_1 \le b_1, a_2 < X_2 \le b_2)$

$\qquad = P(a_1 < X_1 \le b_1, X_2 \le b_2) - P(a_1 < X_1 \le b_1, X_2 \le a_2)$

$\qquad = P(X_1 \le b_1, X_2 \le b_2) - P(X_1 \le a_1, X_2 \le b_2)$

$\qquad\quad - [P(X_1 \le b_1, X_2 \le a_2) - P(X_1 \le a_1, X_2 \le a_2)]$

$\qquad = F(b_1, b_2) - F(a_1, b_2) - [F(b_1, a_2) - F(a_1, a_2)].$

3.3. $P(T \le t) = 1 - \sum_{j=k}^{n} \binom{n}{j}(1 - F(t))^j F(t)^{n-j}$.

3.14. $\sum_{i=1}^{r}(1 - p_i)^n$.   **3.15.** $\dfrac{n}{\lambda^2}$.

COMPUTATIONAL PROBLEMS

3.3. $P(J = j, S = s, G = g) = \dfrac{\binom{10}{j}\binom{10}{s}\binom{4}{g}}{\binom{14}{6}}$ for $j + g + s = 6, g \le 4, j, g, s \ge 0$.

3.5. $P(H = h, D = d) = \dfrac{\binom{13}{h}\binom{13}{d}\binom{26}{6-h-d}}{\binom{52}{6}}$ for $h + d \le 6, h, d \ge 0$.

3.7. 180.   **3.9.** 24.

3.11. $F(x_1, x_2)$

$$
= \begin{cases}
20x_1^3 x_2^3 & \text{if } x_1, x_2 \ge 0, x_1 + x_2 \le 1, \\
20x_1^3 - 45x_1^4 + 36x_1^5 - 10x_1^6 & \text{if } 0 \le x_1 \le 1, x_2 > 1, \\
20x_2^3 - 45x_2^4 + 36x_2^5 - 10x_2^6 & \text{if } 0 \le x_2 \le 1, x_1 > 1, \\
20x_2^3(1 - x_2)^3 + 20(x_1^3 - (1 - x_2)^3) - 45(x_1^4 - \\
(1 - x_2)^4) + 36(x_1^5 - (1 - x_2)^5) - 10(x_1^6 - \\
(1 - x_2)^6) & \text{if } 0 \le x_1, x_2 \le 1, x_1 + x_2 > 1, \\
1 & \text{if } x_1, x_2 > 1, \\
0 & \text{otherwise.}
\end{cases}
$$

3.13. $F(x_1, x_2) =$

$$
\begin{cases}
6x_1^2 x_2^2 & \text{if } x_1, x_2 \ge 0, x_1 + x_2 \le 1, \\
6x_1^2 - 8x_1^3 + 3x_1^4 & \text{if } 0 \le x_1 \le 1, x_2 > 1, \\
6x_2^2 - 8x_2^3 + 3x_2^4 & \text{if } 0 \le x_2 \le 1, x_1 > 1, \\
6x_2^2(1 - x_2)^2 + 6(x_1^2 - (1 - x_2)^2) \\
-8(x_1^3 - (1 - x_2)^3) + 3(x_1^4 - \\
(1 - x_2)^4) & \text{if } x_1, x_2 \ge 0, x_1 + x_2 > 1, \\
1 & \text{if } x_1, x_2 > 1, \\
0 & \text{otherwise.}
\end{cases}
$$

3.15. $P(H = h) = \dfrac{\binom{13}{h}\binom{39}{6-h}}{\binom{52}{6}}$.

3.17. $f_{X_1}(x_1) = 60x_1^2(1 - x_1)^3, \ 0 \le x_1 \le 1$.

3.19. $f_{X_1}(x_1) = 12x_1(1 - x_1)^2, \ 0 \le x_1 \le 1$.

3.21. 10/3.

3.23. $f_{U_1+U_2}(x) = \begin{cases} x & \text{if } 0 \le x \le 1, \\ 2-x & \text{if } 1 < x < 2, \\ 0 & \text{otherwise.} \end{cases}$

3.25. $6\left(\frac{5}{6}\right)^{10}$.

3.27. 8.333.

3.29. (1) $E(X_1) = E(X_2) = \frac{3}{4}$. (2) $Var(X_1) = Var(X_2) = \frac{3}{80}$.

3.31. (1) $E(X_1) = E(X_2) = 55(3)^{4/3}/12$. (2) $Var(X_1) = Var(X_2) = 29(3)^{5/3}/4 - (55(3)^{4/3}/12)^2$.

Chapter 4

CONCEPTUAL PROBLEMS

4.1. $P(X = i, Y = j) = \frac{1}{6}\binom{i+j}{i}\left(\frac{1}{2}\right)^{i+j}$ for $1 \le i + j \le 6$.

4.3. $P(T_1 = j | T_2 = k) = \frac{1}{k-1}$, $1 \le j \le k - 1$, $k \ge 2$.

4.5. $P(X_1 = i_1, \ldots, X_k = i_k | X_{k+1} + \cdots + X_r = m) = \frac{(n-m)!}{i_1! \ldots i_k!} \frac{p_1^{i_1} \ldots p_k^{i_k}}{(p_1 + \cdots + p_k)^{n-m}}$.

4.7. $N\left(\frac{z}{2}, \frac{\sigma^2}{2}\right)$.

4.11. $P(X = k) = \frac{1}{2^{k+1}}$, $k = 0, 1, 2, \ldots$.

4.13. 6.

4.15. k/p.

COMPUTATIONAL PROBLEMS

4.1. $P(Y = 0 | X = 0) = .2$; $P(Y = 1 | X = 0) = .4$; $P(Y = 2 | X = 0) = .4$; $P(Y = 0 | X = 1) = .4$; $P(Y = 1 | X = 1) = .2$; $P(Y = 2 | X = 1) = .4$.

4.3. The conditional pmf $P(Y = y | X = k)$ is given as

$k \downarrow y \rightarrow$	0	2	4
1	.250	.625	.125
3	.625	.250	.125
5	.250	.250	.500

4.5. $P(X = 3, Y = 3 | Z = 3) = .5$; $P(X = 1, Y = 2 | Z = 3) = .5$. $P(X = 1 | Y = 1) = 1$; $P(Z = 1 | X = 2) = .5$; $P(Z = 2 | X = 2) = .5$.

4.7. $P(j \text{ hearts} | 2 \text{ diamonds}) = \dfrac{\binom{13}{j}\binom{37}{4-j}}{\binom{50}{4}}$, $0 \le j \le 4$.

4.9. $\binom{6}{j}\left(\frac{6}{14}\right)^j\left(\frac{8}{14}\right)^{6-j}$, $0 \le j \le 6$.

4.11. $f_{X_2|X_1}(x_2|x_1) = \dfrac{3x_2^2}{(1-x_1)^3}$ for $0 \le x_2 \le 1 - x_1, 0 \le x_1 \le 1$.

4.13. $f_{X_2|X_1}(x_2|x_1) = \dfrac{2x_2}{(1-x_1)^2}$ for $0 \le x_2 \le 1 - x_1, 0 \le x_1 \le 1$.

4.15. $f_{X_1|X_2}(x_1|x_2) = \dfrac{3x_1^2}{(1-x_2)^3}$, $0 \le x_1 \le 1 - x_1, \ 0 \le x_1 \le 1$.

4.17. $f_{X_1|X_2}(x_1|x_2) = \dfrac{2x_1}{(1-x_2)^2}$, $0 \le x_2 \le 1 - x_1, \ 0 \le x_1 \le 1$.

4.19. $\frac{5}{11}$.   **4.21.** $\frac{1}{2}$.

4.23. $E(Y|X=0) = 1.2$, $\ E(Y|X=1) = 1$.   **4.25.** $1.75, 1, 2.5$.

4.27. $\frac{3}{4}(1-x_1)$.   **4.29.** $\frac{2}{3}(1-x_1)$.   **4.31.** 4.

Chapter 5

CONCEPTUAL PROBLEMS

5.1. Proof by induction. The statement is true for $k = 1$ by definition of transition probabilities. Suppose it is true for k. We have
$P(X_{k+1} = i_{k+1}, \ldots, X_1 = i_1 | X_0 = i_0)$

$$
\begin{aligned}
&= P(X_{k+1} = i_{k+1} | X_k = i_k, \ldots, X_1 = i_1, X_0 = i_0) \\
&\quad \times P(X_k = i_k, \ldots, X_1 = i_1 | X_0 = i_0) \\
&= P(X_{k+1} = i_{k+1} | X_k = i_k) \\
&\quad \times P(X_k = i_k, \ldots, X_1 = i_1 | X_0 = i_0) \\
&= p_{i_k.i_{k+1}} \, p_{i_{k-1}.i_k} \, p_{i_{k-2}.i_{k-1}} \cdots p_{i_0.i_1}.
\end{aligned}
$$

Here the second equality follows due to the Markov property, and the third one from the induction hypothesis.

5.3. Let $\bar{p} = 1 - p$ and $\bar{q} = 1 - q$. Then

$$
P = \begin{bmatrix}
q^3 & 3q^2\bar{q} & 3q\bar{q}^2 & \bar{q}^3 \\
\bar{p}q^2 & pq^2 + 2\bar{p}q\bar{q} & 2pq\bar{q} + \bar{p}\bar{q}^2 & \bar{q}^2 p \\
q\bar{p}^2 & 2pq\bar{p} + \bar{q}\bar{p}^2 & qp^2 + 2\bar{q}p\bar{p} & \bar{q}p^2 \\
\bar{p}^3 & 3\bar{p}^2 p & 3\bar{p}p^2 & p^3
\end{bmatrix}.
$$

5.5. (1) $a_j^n = P(X_n = j) = \sum_{i=1}^{N} P(X_n = j | X_0 = i) P(X_0 = i) = \sum_{j=1}^{N} a_i [P^n]_{i,j}$. Thus, $a^n = a * P^n$.

(2) Proof is by induction on m. The statement holds for $m = 1$ due to the definition of a DTMC. Suppose it holds for $m \geq 1$. Then

$$P(X_{n+m+1} = j | X_n = i, X_{n-1}, \ldots, X_0)$$

$$= \sum_{k=1}^{N} P(X_{n+m+1} = j | X_{n+m} = k, X_n = i, \ldots, X_0) P(X_{n+m} = k | X_n = i, \ldots, X_0)$$

$$= \sum_{k=1}^{N} p_{kj} p_{ik}^{(m)} = p_{ij}^{(m+1)}$$

Thus the statement follows from induction.

5.7. $P(T_i = k) = P(X_0 = i, X_1 = i, \ldots, X_{k-1} = i, X_k \neq i)$
$$= P(X_k \neq i | X_{k-1} = i) P(X_{k-1} = i | X_{k-2} = i) \cdots P(X_1 = i | X_0 = i) P(X_0 = i)$$
$$= p_{i,i}^{k-1} (1 - p_{i,i}).$$

5.9. $P = \begin{bmatrix} 1 - p & p \\ 1 - p & p \end{bmatrix}$.

5.11. Let

$$X_n = \begin{cases} 1 & \text{if the machine is idle at the beginning of the } n\text{th minute} \\ & \text{and there are no items in the bin,} \\ 2 & \text{if the machine has just started production at the beginning} \\ & \text{of the } n\text{th minute and there are no items in the bin,} \\ 3 & \text{if the machine has been busy for 1 minute at the beginning} \\ & \text{of the } n\text{th minute and there are no items in the bin,} \\ 4 & \text{if the machine has been busy for 1 minute at the beginning} \\ & \text{of the } n\text{th minute and there is one item in the bin.} \end{cases}$$

Then $\{X_n, n \geq 0\}$ is a DTMC on $S = \{1, 2, 3, 4\}$ with the transition probability matrix

$$P = \begin{bmatrix} 1 - p & p & 0 & 0 \\ 0 & 0 & 1 - p & p \\ 1 - p & p & 0 & 0 \\ 0 & 1 & 0 & 0 \end{bmatrix}.$$

5.13. Let

$$X_n = \begin{cases} 1 & \text{if day } n \text{ is sunny,} \\ 2 & \text{if day } n \text{ is cloudy and day } n - 1 \text{ is not,} \\ 3 & \text{if day } n \text{ is rainy,} \\ 4 & \text{if day } n \text{ and } n - 1 \text{ are both cloudy.} \end{cases}$$

$\frac{1}{\lambda}+\frac{1}{\mu}$ $\frac{\mu}{\lambda+\mu}+\frac{\lambda}{\lambda+\mu}$

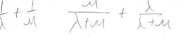

Then $\{X_n, n \geq 0\}$ is a DTMC on state space $S = \{1, 2, 3, 4\}$ with the transition probability matrix

$$P = \begin{bmatrix} .5 & .3 & .2 & 0 \\ .5 & 0 & .3 & .2 \\ .4 & .5 & .1 & 0 \\ 0 & 0 & .2 & .8 \end{bmatrix}.$$

$rate = \frac{1}{\lambda}$ $\frac{\mu}{\lambda+\mu}$ $\frac{\lambda}{\lambda+\mu}$

5.15. Let $A = \{X_n \text{ visits } 1 \text{ before } N\}$. Then $u_1 = 1$, $u_N = 0$, and for $2 \leq i \leq N - 1$,

$$u_i = P(A|X_0 = i) = \sum_{j=1}^{N} P(A|X_1 = j, X_0 = i)P(X_1 = j|X_0 = i)$$

$$= \sum_{j=1}^{N} P(A|X_0 = j)P(X_1 = j|X_0 = i) = \sum_{j=1}^{N} u_j p_{i,j}.$$

5.17. $p_{i,i+1} = \mu/(\lambda + \mu) = 1 - p_{i,i-1}$ for $-4 \leq i \leq 4$ and $p_{5,5} = p_{-5,-5} = 1$.

COMPUTATIONAL PROBLEMS

5.1. 6.3613.   **5.3.** \$4.

5.5. 9.6723, 9.4307, 8.8991, 8.2678.   **5.7.** [47.40   28.01   14.67   9.92]

5.9. .0358.   **5.11.** 1.11 days.   **5.13.** .3528.

5.15. (a)

$$M(10) = \begin{bmatrix} 2.6439 & 2.6224 & 2.8520 & 2.8817 \\ 1.6507 & 3.7317 & 2.9361 & 2.6815 \\ 1.8069 & 2.7105 & 3.6890 & 2.7936 \\ 1.6947 & 2.6745 & 2.8920 & 3.7388 \end{bmatrix}.$$

(c)

$$M(10) = \begin{bmatrix} 3.3750 & 0 & 7.6250 & 0 \\ 0 & 3.6704 & 0 & 7.3296 \\ 2.5417 & 0 & 8.4583 & 0 \\ 0 & 2.6177 & 0 & 8.3823 \end{bmatrix}.$$

5.17. .0336 hours.   **5.19.** (a) irreducible, (c) reducible.

5.21. $\pi = \pi^* = \hat{\pi} = [.1703 \quad .2297 \quad .2687 \quad .3313]$.

5.23. π does not exist. $\pi^* = \hat{\pi} = [.50 \quad .10 \quad .15 \quad .25]$.

5.25. (1) 0.0979. (2) 5.3686.   **5.27.** 0.1428.

5.29. .1977.   **5.31.** 0.7232.   **5.33.** 7.3152.

5.35. Both produce .95 items per minute.

5.37. $454.91 per day. **5.39.** .0003 per day.   **5.41.** $11.02 per week.

5.43. (b) 3, (d) 3.   **5.45.** 60.7056.   **5.47.** 241.2858.

Chapter 6

CONCEPTUAL PROBLEMS

6.5. $S = \{0, 1, 2, \ldots, K\}, r_{i,i-1} = \lambda L, \ 1 \leq i \leq K$ all other $r_{i,j} = 0$.

6.7. $S = \{0, 1, 2, 3, 4, 5\}, r_{i,i-1} = ic\mu, \ 2 \leq i \leq 5, r_{i,0} = i(1-c)\mu, \ 2 \leq i \leq 5 \ r_{1,0} = \mu$, all other $r_{i,j} = 0$.

6.9. $\mu_i = \mu, \ 1 \leq i \leq K, \lambda_i = \lambda_1 + \lambda_2$ for $0 \leq i \leq M, \lambda_1$ for $M + 1 \leq i < K$ and 0 for $i = K$.

6.11. Let $A(t) = (b, l)$ be the number of operational borers and lathes at time t. Then $\{A(t), t \geq 0\}$ is a CTMC with state space

$$S = \{(0, 0), (0, 1), (0, 2), (1, 0), (2, 0), (1, 1), (1, 2), (2, 1), (2, 2)\}$$

and rate matrix

$$R = \begin{bmatrix} 0 & \lambda_l & 0 & \lambda_b & 0 & 0 & 0 & 0 & 0 \\ \mu_l & 0 & \lambda_l & 0 & 0 & \lambda_b & 0 & 0 & 0 \\ 0 & 2\mu_l & 0 & 0 & 0 & 0 & \lambda_b & 0 & 0 \\ \mu_b & 0 & 0 & 0 & \lambda_b & \lambda_l & 0 & 0 & 0 \\ 0 & 0 & 0 & 2\mu_b & 0 & 0 & 0 & 2\lambda_l & 0 \\ 0 & \mu_b & 0 & \mu_l & 0 & 0 & \lambda_l & \lambda_b & 0 \\ 0 & 0 & \mu_b & 0 & 0 & 2\mu_l & 0 & 0 & \lambda_b \\ 0 & 0 & 0 & 0 & \mu_l & 2\mu_b & 0 & 0 & \lambda_l \\ 0 & 0 & 0 & 0 & 0 & 0 & 2\mu_b & 2\mu_l & 0 \end{bmatrix}.$$

6.13. Let $X(t)$ be the number of customers in the system at time t if the system is up, and let $X(t) = d$ if the system is down at time t. $\{X(t), t \geq 0\}$ is a CTMC with transition rates: $r_{i,d} = \theta$ for $0 \leq i \leq K, r_{d,0} = \alpha, r_{i,i-1} = \mu$ for $1 \leq i \leq K, r_{i,i+1} = \lambda$ for $0 \leq i \leq K-1$.

6.15. $S = \{0, 1, 2, 3, 4\}$ where the state is 0 if both components are down and one component is being repaired, state 1 if component 1 is up, component 2 is down, state 2 if component 2 is up, component 1 is down, state 3 if both components are up, and state 4 if the system is down, one component has completed repairs, and the other component is being repaired. The nonzero transition rates are: $r_{0.4} = \alpha$, $r_{1.0} = \lambda p$, $r_{2.0} = \lambda q$, $r_{3.0} = \lambda r$, $r_{3.1} = \lambda q$, $r_{3.2} = \lambda p$, $r_{4.3} = \alpha$.

COMPUTATIONAL PROBLEMS

6.1. $P(\text{Lifetime} \le x) = (1 - e^{-0.2x})^2$. $E(\text{Lifetime}) = 7.5$ years.

6.3. .3050.   **6.5.** 1.667 minutes.

6.7. $P(0.20) = \begin{bmatrix} .3747 & .2067 & .1964 & .1319 & .0903 \\ .2615 & .2269 & .2045 & .1841 & .1230 \\ .2424 & .1964 & .2153 & .1858 & .1601 \\ .1804 & .1947 & .2088 & .2327 & .1835 \\ .1450 & .1544 & .2103 & .2191 & .2711 \end{bmatrix}.$

6.9. $P(0.10) = \begin{bmatrix} .1353 & .1876 & .3765 & .3006 & 0 & 0 \\ 0 & .3229 & .3765 & .3006 & 0 & 0 \\ 0 & .0640 & .6732 & .2627 & 0 & 0 \\ 0 & .1775 & .4143 & .4082 & 0 & 0 \\ .2001 & .0801 & .1307 & .1197 & .2956 & .1738 \\ .1874 & .0696 & .1121 & .1036 & .1738 & .3535 \end{bmatrix}.$

6.11. 18.   **6.13.** .0713.   **6.15.** .0308.

6.17. $M(.20) = \begin{bmatrix} .1167 & .0318 & .0299 & .0134 & .0083 \\ .0400 & .0834 & .0328 & .0307 & .0131 \\ .0371 & .0318 & .0760 & .0291 & .0260 \\ .0184 & .0319 & .0336 & .0864 & .0297 \\ .0132 & .0164 & .0338 & .0358 & .1007 \end{bmatrix}.$

6.19. $M(.10) = \begin{bmatrix} .0432 & .0140 & .0221 & .0207 & 0 & 0 \\ 0 & .0572 & .0221 & .0207 & 0 & 0 \\ 0 & .0028 & .0801 & .0171 & 0 & 0 \\ 0 & .0134 & .0256 & .0609 & 0 & 0 \\ .0165 & .0035 & .0051 & .0051 & .0561 & .0136 \\ .0143 & .0030 & .0042 & .0043 & .0136 & .0607 \end{bmatrix}.$

6.21. 3.95 minutes.   **6.23.** 9.36 hours.   **6.25.** .5289 hours.

6.27. $p = [.2528, .1981, .2064, .1858, .1569]$.

6.29. $p = [.3777, .1863, .0916, .0790, .0939, .0466, .0517, .0732]$.

6.31. .5232.   **6.33.** 4.882.   **6.35.** .4068.

6.37. 65.7519.   **6.39.** 1233.6.   **6.41.** 187.96.

6.43. $25.72.   **6.45.** $28.83.   **6.47.** Yes.

6.49. $223.73 per day. **6.51.** .7425.   **6.53.** .2944.

6.55. 8.   **6.57.** 3.85 years.

Chapter 7

CONCEPTUAL PROBLEMS

7.7. $\lim_{t \to \infty}(Y(t)/t) = \lim_{t \to \infty}(Z(t)/t) = 1/\tau$.

7.9. No. The batch sizes depend upon the inter-arrival times.

7.11. $\dfrac{\tau_i v_i}{1 + \sum_{j=1}^{N} \tau_j v_j}$.   **7.13.** $\dfrac{1 + \sum_{j=1}^{N} \tau_j v_j}{v_i}$.   **7.15.** $\dfrac{\tau}{\tau + \sum_{i=1}^{N} \frac{1}{iv}}$.

7.17. $P = \begin{bmatrix} 0 & 1 \\ 1 & 0 \end{bmatrix}$, $w_1 = \sum_{i=1}^{N}(1/iv)$, $w_0 = \tau$.

7.19. $P = \begin{bmatrix} 1 - A_1(T) & A_1(T) \\ A_2(T) & 1 - A_2(T) \end{bmatrix}$, w_i = the expected lifetime of component i, $i = 1, 2$.

7.21. $P = \begin{bmatrix} .366 & .573 & .061 \\ .3025 & .1512 & .5463 \\ .6 & .3 & .1 \end{bmatrix}$, $w = [5.5, 4.4795, 6]$.

COMPUTATIONAL PROBLEMS

7.1. $\frac{1}{3}$.   **7.3.** .1818.   **7.5.** .019596.

7.7. 0.2597.   **7.9.** Plan: $50.80 per year; Unplan: $50.00 per year.

7.11. $\geq$ \$24 per day.   **7.13.** Contr: \$25.66 per year. No Contr: \$26.35 per year.

7.15. Current: \$376.67 per year. New: \$410.91 per year.   **7.17.** \$5.43 per day.

7.19. .5355.   **7.21.** \$16,800.   **7.23.** 255.15.

7.25. .8392.   **7.27.** .0629.   **7.29.** .16.

7.31. \$31.2 per day.   **7.33.** 0.8125.   **7.35.** 0.5185.

7.37. \$533.33 per day. **7.39.** \$53.37 per day.   **7.41.** 1.4274 fac/year.

7.43. 0.9136 fac/year.

Chapter 8

CONCEPTUAL PROBLEMS

8.1. *Case* 1. $y \leq x$: $\pi_0 = \pi_0^* = \hat{\pi}_0 = 1$, $\pi_j = \pi_j^* = \hat{\pi}_j = 0, j \geq 1$. p_j's do not exist. Thus PASTA does not hold. *Case* 2. $y > x$: The queue is unstable.

8.11. $\lambda_i = (K - i)\lambda$, $0 \leq i \leq K$, $\mu_i = \min(i, s)\mu$, $0 \leq i \leq K$.

8.18. No.   **8.19.** Yes.

8.23. $\lambda < \min\left(K\theta, \dfrac{\alpha}{p}\right)$.

COMPUTATIONAL PROBLEMS

8.1. 7.   **8.3.** 9.

8.5. L and L_q remain unchanged. W and W_q are halved.

8.7. \$46.4544 per hour.   **8.9.** \$.7837 per hour.

8.11. .0311.   **8.13.** 13 additional lines. Old: 1.48 mins, New: 7.74 mins.

8.15. 4.   **8.17.** \$209.86 per hour.   **8.19.** .25.

8.21. 5.5556 mins. 9.   **8.23.** 6.   **8.25.** $\geq$ \$15.89.

8.27. 8.26 mins.   **8.29.** .1122.   **8.31.** 5 to 5.5511.

8.33. 3.6250.   **8.35.** 14.3721.   **8.37.** 2.1540.

8.39. (1) .3 hr, (2) .1201 hr.   **8.41.** 3 per station.

8.43. < 640 per hour.   **8.45.** (1) Yes. (2) 4.4872. (3) 5.62 mins.

Chapter 9

CONCEPTUAL PROBLEMS

9.1. $E(\min(D, K)) = \sum_{i=1}^{\infty} \min(i, K) P(D = i) = \sum_{i=1}^{\infty} \sum_{j=1}^{\min(i,K)} P(D = i) = \sum_{j=1}^{K} \sum_{i=j}^{\infty} P(D = i) = \sum_{j=1}^{K} P(D \geq j) = \sum_{j=0}^{K-1} P(D > j).$

9.3. The smallest K such that $P(D \leq K) \geq \frac{P-C}{P}$.

9.5. $g(K) = \frac{D}{\tau K} + C_w \frac{K-1}{2}$.

9.7. Minimize $c_1 L_1(s_1) + c_2 L_2(s - s_2)$ subject to (9.4).

COMPUTATIONAL PROBLEMS

9.1. 154.   **9.3.** 143.

9.5. $1.8836 \leq \lambda \leq 2.0568$.   **9.7.** 27.

9.9. 18.   **9.11.** $9.00 \leq \lambda \leq 14.16$.

9.13. For the first system, the optimal number of operators is eight, yielding a net revenue of $632.33 per hour. For the second system, the optimal number of operators is 11, yielding a net revenue of $753.73 per hour. Hence the second system is better.

9.15. $\lambda \geq 14.32$.   **9.17.** $T^* = 4.15$.   **9.19.** $\lambda \geq 16.25$.

9.21. $T^* = 4.40$.   **9.23.** $23.2 \leq p \leq 28.6$.

Chapter 10

CONCEPTUAL PROBLEMS

10.1. Same as in Example 10.3 with the following changes:

$$P(1) = \begin{bmatrix} 0 & 1 & 0 \\ 0 & q & p \\ 0 & q & p \end{bmatrix}$$

and

$$
C = \begin{bmatrix} 0 & C_r + C_v & 2C_r + C_v \\ -R & -R + C_r + C_v & 2C_r + C_v \\ -2R & -R + C_r + C_v & 2C_r + C_v \end{bmatrix}.
$$

10.3. $S = \{1, 2, \ldots, N\}$. $\mathcal{A} = \{0, 1\}$, where $a = 1$ (replace the machine) and $a = 0$ (do nothing). $P(0) = P$, $p_{i.1}(1) = 1$, $1 \le i \le N$, $p_{i.j}(1) = 0$, $1 \le i \le N$, $2 \le j \le N$, and $c(i, 0) = -r(i)$, $c(i, 1) = C_r$, $1 \le i \le N$.

10.5. Same as the solution to Conceptual Problem 10.3 with $c(i, 1) = C_r - s(i)$, $1 \le i \le N$.

10.7. The state of the system at time n is given by (X_n, Y_n). Let $p_{i.j} = P(Y_{n+1} = c_j | Y_n = c_i)$, $i, j = 1, 2$. Then

$$P(X_{n+1} = \max(0, \min(m + a, B) - 1), Y_{n+1} = c_j | X_n = m, Y_n = c_i, A_n = a) = p_{ij},$$

$$C(X_n = m, Y_n = c_i, A_n = a) = s + hm + ac_i, \qquad 0 \le m \le B - 1, \quad 1 \le a \le B,$$

$$C(X_n = m, Y_n = c_i, A_n = 0) = hm, \qquad 1 \le m \le B - 1,$$

$$C(X_n = 0, Y_n = c_i, A_n = 0) = s + c_3$$

for some $c_3 > c_2$. This last equation ensures that at least one item will be ordered when the warehouse is empty.

10.11. $X(t) =$ the number of customers in the system at time t. $S = \{0, 1, \ldots, B\}$. The action space is $\mathcal{A} = \{1, 2, \ldots, K\}$.

$$p_{0.1}(a) = 1, \qquad p_{i.i+1}(a) = \lambda/(\lambda + \mu_a), \qquad 1 \le i \le B - 1,$$

$$p_{i.i-1}(a) = \mu_a/(\mu_a + \lambda), \qquad 1 \le i \le B - 1, \qquad p_{B.B-1}(a) = 1,$$

$$w_0(a) = 1/\lambda, \qquad w_i(a) = 1/(\lambda + \mu_a), \qquad 1 \le i \le B - 1, \qquad w_B(a) = 1/\mu_a,$$

$$c(0, a) = \mu_a/\lambda, \qquad c(i, a) = c_a/(\lambda + \mu_a), \qquad 1 \le i \le B - 1,$$

$$c(B, a) = (\lambda r + c_a)/\mu_a.$$

10.13. $X(t) =$ the number of tasks in the system at time t. Let action $1 =$ wait for the next arrival, and action $2 =$ clear all the accumulated tasks.

$$p_{i.\max(i+1.K)}(1) = 1, \qquad w_i(1) = T, \qquad c(i, 1) = i * h * T, \qquad 0 \le i \le K,$$

$$p_{i.0}(2) = 1, \qquad w_i(2) = 0, \qquad c(i, 2) = C.$$

COMPUTATIONAL PROBLEMS

10.1. $f(1, 2) = f(2, 2) = f(3, 1) = 1, P^f = \begin{bmatrix} .5 & .5 & 0 \\ 0 & .5 & .5 \\ .1 & .8 & .1 \end{bmatrix}, G^f = 13.333.$

10.3. $f(1, 2) = f(2, 2) = f(3, 1) = 1, P^f = \begin{bmatrix} .5 & .5 & 0 \\ .8 & 0 & .2 \\ .1 & .8 & .1 \end{bmatrix}, G^f = 7.907.$

10.5. $f(1, 3) = f(2, 2) = f(3, 1) = 1, P^f = \begin{bmatrix} .5 & .5 & 0 \\ 0 & .5 & .5 \\ .1 & .8 & .1 \end{bmatrix}, G^f = 15.333.$

10.7. $a^*(1) = 1, a^*(2) = 3, G^* = 6.7391.$

10.9. $C(1, 1) \geq 13.34.$ **10.11.** $3.31 \leq C(1, 2) \leq 13.12.$

10.13. $C(4, 1) \geq 20.01.$

10.15. $a^*(0) = 2,\ a^*(1) = 1,\ a^*(2) = 0,\ G^* = \$75$ per day.

10.17. Always optimal. **10.19.** $-540.$ **10.21.** $r(1) \geq 750.$

10.23. Replace when down. $G^* = -236.47.$

10.25. Replace by a used one when down. $G^* = -278.84.$

10.27. When the price is $\$10$ and there are i items at the beginning of a day, purchase $3 - i$ items. When the price is $\$12$ then buy a single item only if there are no items in hand, otherwise do not buy any items. The long-run cost rate is 10.24.

10.29. $f(1, 2) = f(3, 2) = f(2, 1) = 1, P^f = \begin{bmatrix} .5 & .5 & 0 \\ .3 & .7 & 0 \\ 0 & .5 & .5 \end{bmatrix}, G^f = 11.25.$

10.31. $a^*(2) = 2, a^*(3) = 3, G^* = 4.2105.$

10.33. $a^*(i) = 2, 1 \leq i \leq 4.\ G^* = 1.9048.$

10.35. $a^*(i) =$ low speed, $i \leq 3,$ high speed, $i \geq 4.\ G^* = \$20.50$ per hour.

10.37. Clear the tasks as soon as there are five tasks waiting. $G^* = \$8$ per minute.

Bibliography

Probability

[1]   W. Feller, *An Introduction to Probability Theory and Its Applications*, Vol. 1, 2nd ed., Wiley, New York, 1959.

[2]   M. F. Neuts, *Probability*, Allyn and Bacon, Boston, 1973.

[3]   E. Parzen, *Modern Probability Theory and Its Applications*, Wiley, New York, 1960.

[4]   S. M. Ross, *A first Course in Probability*, 3rd ed., Macmillan, New York, 1988.

Stochastic Processes

[5]   E. P. C. Kao, *An Introduction to stochastic Processes*, Duxbury Press, 1997.

[6]   S. M. Ross, *Introduction to Probability Models*, Academic Press, 1993.

[7]   H. M. Taylor and S. Karlin, *An Introduction to Stochastic Modeling*, Academic Press, 1984.

[8]   K. S. Trivedi, *Probability and Statistics with Reliability, Queueing and Computer Science Applications*, Prentice Hall, 1982.

Advanced Textbooks

[9]   E. Cinlar, *Introduction to Stochastic Processes*, Prentice Hall, 1975.

[10] D. P. Heyman and M. J. Sobel, *Stochastic Models in Operations Research*, Vol.1, McGraw Hill, 1982.

[11] S. Karlin and H. M. Taylor, *A first Course in Stochastic processes*, Academic Press, 1975.

[12] V. G. Kulkarni, *Modeling and Analysis of Stochastic systems*, Chapman Hall, 1995.

[13] S. M. Ross, *Stochastic Processes*, Wiley, 1983.

[14] S. I. Resnick, *Adventures in Stochastic Processes*, Birkhauser, 1992.

Books on Special Topics

[15] D. Bertsekas and R. Gallager, *Data Networks*, Prentice Hall, 1992.

[16] J. A. Buzzacott and J. G. Shanthikumar, *Stochastic Models of Manufacturing Systems*, Prentice Hall, 1993.

[17] D. Gross and C. M. Harris, *Fundamentals of Queueing Theory*, Wiley, NY 1985.

Index

Springer Texts in Statistics *(continued from page ii)*